Francis Galton

Francis Galton

Pioneer of Heredity and Biometry

Michael Bulmer

THE JOHNS HOPKINS UNIVERSITY PRESS
Baltimore and London

This book has been brought to publication with the generous assistance of the Science Publication Fund.

Printed in the United States of America on acid-free paper
9 8 7 6 5 4 3 2 1

The Johns Hopkins University Press
2715 North Charles Street
Baltimore, Maryland 21218-4363
www.press.jhu.edu

Library of Congress Cataloging-in-Publication Data

Bulmer, M. G.
Francis Galton : pioneer of heredity and biometry / Michael Bulmer.
p. cm.
Includes bibliographical references and index.
ISBN 0-8018-7403-3 (hardcover : alk. paper)
1. Galton, Francis, Sir, 1822–1911. 2. Geneticists—England—Biography.
3. Genetics—History. 4. Biometry—History. I. Title.

OH429.2.G35B856 2003
576.5′092—dc21
[B] 2003047528

A catalog record for this book is available from the British Library.

Contents

Acknowledgments xi

Chronology xiii

Introduction xv

1 A Victorian Life 1

Family Background and Education 1

Travels 6

Eastern Europe, 1840 6

The Near East, 1845–46 7

South West Africa, 1850–52 11

Vacation Tours 18

Scientific Career 21

The Royal Geographical Society 22

Exploration in Central Africa 23

The British Association 27

Inventions 28

Meteorology 30

Heredity and Evolution 32

Psychology 32

Photography 34

Fingerprints 35

Characterization 36

2 Hereditary Ability **42**
"Hereditary Talent and Character" (1865) 44
Hereditary Genius (1869) 46
English Judges 48
Comparison of Results for All Professions 50
Transmission through Male and Female Lines 54
The Reception of *Hereditary Genius* 57
Nature and Nurture 60
English Men of Science: Their Nature and Nurture (1874) 60
"The History of Twins" (1875) 64
Galton's Hereditarianism 67
Epilogue 71
Appendix: Number of Kinsfolk 74

3 Eugenics **79**
Galtonian Eugenics 79
Later History of Eugenics 84
Britain 84
America 87
Germany 92
The Rationale of Eugenics 98

4 The Mechanism of Heredity **102**
Galton's Knowledge of Heredity in 1865 103
Biparental Inheritance 103
The Non-Inheritance of Acquired Characters 105
The Law of Reversion 107
Darwin's Provisional Hypothesis of Pangenesis 108
Reversion 110
The Inheritance of Acquired Characters 112
Xenia and Telegony 113
Galton's Reaction to Pangenesis 114
Galton's Political Metaphor of Pangenesis 114
An Experimental Test of Pangenesis 116

Galton's Theory of Heredity in the 1870s 119
Similarities between Relatives 123
Galton's Ideas on Heredity in 1889 127
Discussion 131
Weismann and the Continuity of the Germ-Plasm 132
De Vries's Theory of Intracellular Pangenesis 133
Segregation 136
Blending Inheritance 138
Fleeming Jenkin and the Problem of Swamping 141

5 **Four Evolutionary Problems** **147**
The Domestication of Animals 147
The Evolution of Gregariousness 150
The Fertility of Heiresses 153
The Extinction of Surnames 156
The Evolution of Sex 160
"A Theory of Heredity" (1875) 161
Three Unpublished Essays 163

6 **The Charms of Statistics** **168**
Quetelet and the Average Man 169
Galton and the Normal Distribution 173
Hereditary Genius (1869) 173
Natural Inheritance (1889) 175
The Importance of the Normal Distribution to Galton 180
Galton's Quincunx 182
Regression and the Bivariate Normal Distribution 184
Correlation 191
Two Concepts of Probability 196
The Development of Statistics 202
Appendix: Regression Theory 206

7 **Statistical Theory of Heredity** **209**
A Theory Based on Pangenesis 210

"Typical Laws of Heredity" (1877) 211
An Experiment with Sweet Peas 212
Solution of the Problem 215
Johannsen's Experiments with Beans 218
The Inheritance of Human Height 224
The Advantages of Height 225
The Regression of Offspring on Mid-Parent 229
Kinship 231
Fraternal Regression 233
Variability in Fraternities and Co-Fraternities 235

8 The Law of Ancestral Heredity 238
Galton's Formulation of the Ancestral Law 239
Galton's Derivation of the Law in 1885 241
Derivation of the Law in 1897 244
Galton's Law As It Should Have Been 247
Karl Pearson's Interpretation of the Ancestral Law 250
The Ancestral Law and Mendelism 257
Weldon and Mendelism 259
Pearson and Mendelism 261
Yule's Reconciliation of the Law with Mendelism 266
Appendix: The Regression on Mid-Ancestral Values 272

9 Discontinuity in Evolution 275
Galton's Theory of Discontinuous Evolution 276
Stability of Type 277
Perpetual Regression 281
Selection Experiments 284
The Fallacy of Perpetual Regression 285
"Discontinuity in Evolution" (1894) 288
Speciation and Saltation 292
De Vries and *The Mutation Theory* 294
Punctuated Equilibria 297

10 Biometry **299**
The Demonstration of Natural Selection 300
The Career of W. F. R. Weldon 301
The Common Shrimp 302
The Shore Crab 303
Stabilizing Selection in Snails 308
Bumpus's Sparrows 309
Multivariate Selection 312
Quantitative Genetics 315
The Multiple Factor Hypothesis 316
The Hardy-Weinberg Law 318
Mendelian Theory of Quantitative Genetics 321
The Response to Selection 324
Coda 327
Appendix: Multivariate Selection Theory 329
Selection Differentials and Selection Gradients 329
The Response to Selection 331

References 333
Index 351

Acknowledgments

I am grateful to University College London Library for access to the Galton and Pearson papers, and for permission to use extracts from them; to the Biometrika Trust and to Oxford University Press for permission to quote extracts and to reproduce three figures from *The Life, Letters and Labours of Francis Galton* by Karl Pearson; to the librarians of the Radcliffe Science Library and the Bodleian Library at Oxford University for their help in locating references; to Professor Anthony Edwards of Gonville and Caius College, Cambridge and to an anonymous referee for valuable comments on the manuscript; to David Burbridge for helpful correspondence about Galton's theory of heredity; to Gavan Tredoux for use of his excellent website www.galton.org; to my copy editor Alice Calaprice of Princeton, New Jersey; to Trevor Lipscombe, Julie McCarthy, and Martha Sewall at the Johns Hopkins University Press, for their help in improving the manuscript and in preparing it for publication; and to Mark Boardman for making the scanned images.

Chronology

1822	Francis Galton born, 16 February
1828–38	At school
1838–40	Medical education in Birmingham and London
1840	Journey to Eastern Europe
1840–44	Trinity College, Cambridge, reading mathematics
1844	His father died, and he abandoned medicine
1845–46	Journey to Near East
1846–49	Hunting and shooting in England and Scotland
1850–52	Journey to South West Africa
1853	Married Louisa Butler
	Tropical South Africa
1855	*The Art of Travel*
1857	Bought 42 Rutland Gate, London
1864	Burton-Speke dispute at British Association, Bath
1869	*Hereditary Genius*
1872	Stanley's talk at British Association, Brighton
1874	*English Men of Science: Their Nature and Nurture*
1883	*Inquiries into Human Faculty and Its Development*
1889	*Natural Inheritance*
1892	*Finger Prints*
1897	Louisa Galton died
1908	*Memories of My Life*
1909	*Essays in Eugenics*
1911	Francis Galton died, 17 January

Honors and awards

1853	Gold Medal, Royal Geographical Society
1855	Member of Athenaeum Club by special election
1860	Fellow, Royal Society
1886	Royal Medal, Royal Society
1894	Hon. D.C.L., Oxford University
1895	Hon. D.Sc., Cambridge University
1901	Huxley Medal, Anthropological Institute
1902	Hon. Fellow, Trinity College, Cambridge
	Darwin Medal, Royal Society
1908	Darwin-Wallace Medal, Linnean Society
1909	Knighted
1910	Copley Medal, Royal Society

Introduction

Francis Galton made important contributions in many areas of science in the nineteenth century. He explored South West Africa, discovered and named the anticyclone, and wrote a book on fingerprints, in addition to his work on anthropometry, psychology, and photography. But most important were his pioneering studies of heredity, in the course of which he invented the statistical tools of regression and correlation, earning him the title of the father of biometry, the application of statistical methods to evolutionary biology.

Galton's work on heredity was closely linked to the evolutionary theory of his half-cousin Charles Darwin. Darwin proposed that species evolve through descent with modification, and that modification is predominantly due to the operation of natural selection on small inherited variations. The first part of Darwin's thesis was quickly accepted by contemporary scientists, but his theory of the mechanism of evolution by natural selection was not, because he did not have a convincing theory of heredity to explain how selection acted and because he had no direct evidence of natural selection. Thus the explanation of variability and heredity and their relationship to natural selection have been key questions in the history of Darwinism from the publication of *The Origin of Species* in 1859 until today. The work of Galton and his followers played a central role in the early history of Darwinism, which has been increasingly recognized by historians of science (Provine 1971, Olby 1985, Gayon 1998). Their work in the application of statistical methods to the biological sciences was equally important (Porter 1986, Stigler 1986 and 1999). This book is an attempt to clarify and explain the development of Galton's ideas on heredity and biometry, and to place them in their scientific context.

Ernst Mayr divides scientists into romantics and classics: "The romantic is bubbling over with ideas that have to be dealt with quickly to make room for the next one. Some of these ideas are superbly innovative; others are invalid if not silly.... The classic, by contrast, concentrates on the perfection of something that already exists. He tends to work over a subject exhaustively. He also tends to defend the status quo" (1982, 831). Galton was an archetypal romantic, an innovator with the gift of seeing problems in statistical terms, but lacking the mathematical ability and the inclination to push his ideas to their logical conclusion. Relying on intuition rather than mathematical expertise, he made mistakes and was sometimes confused or self-contradictory. He was also rather careless arithmetically; obvious numerical errors have been corrected in what follows without comment.

Thus, Galton was a pioneer who left it to others to perfect his ideas. It has been a challenge to present a clear account of these sometimes unclear ideas without distorting them. Apart from the first, biographical chapter, this is an internalist account of the development of scientific ideas, based on the premise that scientific knowledge is not just a subjective social construct (Ruse 1999), though scientists are fallible men and women who are constrained by their peculiar abilities, personalities, and social environment. Social constructionist historians, coming from a background in the social sciences, tend to stress the social context of science history and have little sympathy with "the old internalists who wrenched science from its social context and wrote ghostly histories of disembodied ideas" (Desmond 1989, 21). As a retired scientist, with a lifelong interest in population genetics, evolutionary biology, and biometry, I am more interested in the development of scientific ideas than in their social context, though I hope that I do not present these ideas as entirely disembodied nor wrenched completely out of context. I also hope that this book complements the excellent recent biography by Gillham (2001).

The first chapter briefly describes Galton's life and his position as a member of the Victorian intellectual aristocracy, an aristocracy of talent rather than of birth (Annan 1955). His first work on heredity, which was inspired by reading *The Origin of Species*, tried to demonstrate the inheritance of human mental abilities and to distinguish between the effects of nature and nurture (chapter 2). This work showed a hereditarian bias, but it laid the foundations of the study of human inheritance by statistical methods. He was also inspired by *The Origin of Species* to extend the idea of artificial selection of plants and animals to man, a subject that he later called eugenics (chapter 3).

Galton's thoughts about the mechanism of heredity were stimulated by the publication of Darwin's theory of pangenesis in 1868. He tried to verify the latter by transfusing blood between different varieties of rabbits, but the failure of these experiments led him to reject that part of it which required, as he thought, the transportation of hereditary particles in the blood and to develop his own theory, based largely on the other parts of Darwin's theory (chapter 4). Galton's theory of the "stirp" partially foreshadowed August Weismann's theory of the continuity of the germ-plasm. Though Galton was not a naturalist, he was also stimulated to apply Darwinian ideas of natural selection to several problems of evolutionary importance (chapter 5).

Perhaps dissatisfied with his physiological theory of the stirp, Galton then tried to develop a purely statistical theory of heredity based on the properties of the normal distribution (chapters 6 and 7). This work led to his invention of the techniques of regression and correlation and culminated in the law of ancestral inheritance (chapter 8). It was of dual significance in the history of Darwinism. The phenomenon of regression confirmed Galton in the view that selection on small variations was ineffective, so that evolution must proceed discontinuously in jerks (chapter 9). On the other hand, the early biometricians were inspired to develop his statistical methods and to use them in the empirical study of natural selection. After the rediscovery of Mendelism in 1900, a bitter dispute broke out between the Mendelians, who believed in discontinuous evolution through the occurrence of sports, and the biometricians, who stressed the Darwinian concept of continuous evolution through selection on small inherited variations and who distrusted Mendelism because of its apparent conflict with the law of ancestral inheritance; both sides appealed to Galton's work for support. After the resolution of this double conflict, biometrical methods led to the development of quantitative genetics based on Mendelian principles, which underpins our understanding of evolution today (chapter 10).

1

A Victorian Life

> There was no-one who has possessed in purer essence than he [Galton] the spirit of universal scientific curiosity.... It was not the business of his particular kind of brain to push anything far. His original genius was superior to his intellect, but his intellect was always just sufficient to keep him just on the right side of eccentricity.
>
> J. M. Keynes, 1937 (in M. Keynes 1993)

The life of Francis Galton (1822–1911) spanned the reign of Queen Victoria (1837–1901), and he was in many ways a typical Victorian. This chapter provides a broad picture of his life and work to place his studies of heredity and biometry in context. The main sources of information about his life are his autobiography *Memories of My Life* (1908), the *Life, Letters and Labours* by Karl Pearson (1914, 1924, and 1930a,b), the biographies by Forrest (1974) and Gillham (2001), and the Galton and Pearson papers at University College London. For the Galton family, see Pearson (1914, chap. 2) and Smith (1967).

Family Background and Education

Francis Galton's paternal grandfather, Samuel Galton, entered his father's gunmaking business in Birmingham, becoming a successful man of business and a contractor for the supply of muskets to the army during the Napoleonic wars. The Galton family had been Quakers for many generations; Samuel was formally disowned by them in 1795 "for fabricating and selling instruments of war" (Pearson 1914, 45) but he continued to attend Quaker worship. He retired from the gunsmith business in

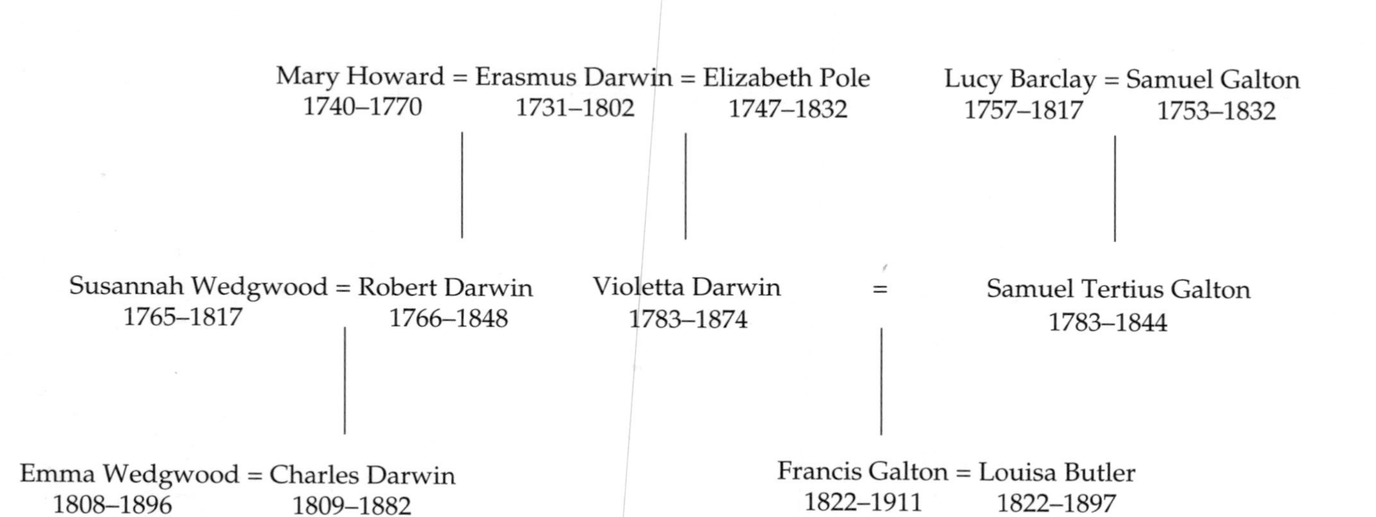

Fig. 1.1. Genealogy of Francis Galton

1804 and founded the Galton bank (he had married into the Barclay banking family) in collaboration with his son, Samuel Tertius Galton. The latter married Violetta Darwin, the daughter of Erasmus Darwin by his second wife, of whom her son Francis wrote: "I have heard ... many charming stories of her as a young bride. She, as I understand, had nothing of the Quaker temperament, but was a joyous and unconventional girl. In her later life she formed the centre of our family during thirty years of widowhood, after my father's comparatively early death at the age of sixty" (1908, 10). She persuaded her husband to join the established church, the Church of England, and her son Francis was baptized into this church. He lost his faith in the 1860s, but he kept much of the Quaker temperament. As he wrote to Karl Pearson, also of Quaker stock, during his last foreign holiday in the Pyrenees in 1906: "The Basque orderliness, thorough but quiet ways, and their substantial clean-looking houses, tug at every Quaker fibre in my heart" (Pearson 1930a, 279). In another letter to Pearson, he wrote, probably recognizing the same quality in himself: "They—the Quakers—were grandly and (simply) stubborn" (Pearson 1914, 29).

His maternal grandfather, Erasmus Darwin, was a highly regarded physician, to the Galton family among others, with a wide range of interests that made him a leading figure of the Enlightenment (King-Hele 1998). He devised a steering mechanism for carriages that was adopted for most modern automobiles until the 1940s; he published a classic paper defining the adiabatic expansion of gases and explaining the formation of clouds; and he advocated the idea of biological evolution, that all living things are descended from a single microscopic ancestor, in *Zoonomia* and in a long poem, "The Temple of Nature." Opinion turned against his radical ideas in the anti-Jacobin reaction to the French Revolution; his evolutionary ideas were attacked as materialistic and subversive of established religion, and his long poem, "The Loves of the Plants," was brilliantly satirized by the Tory minister George Canning in "The Loves of the Triangles." He had five children by his first wife, including Robert, the father of Charles Darwin; two daughters by his mistress; and seven children by his second wife, including Violetta.

Thus, Francis Galton came from a middle-class and comparatively wealthy background. There was an interest in science on both sides of his family. Samuel Galton was a keen amateur scientist who made an early contribution to the theory of color vision, and the scientific achievements of Erasmus Darwin are described above. Both men were Fellows of the Royal Society and members of the Lunar Society of

Birmingham, a scientific club whose members met at each others' houses on the day and night of the full moon, when it was easier to see to drive home (Schofield 1963, Uglow 2002). Other members were the manufacturers Matthew Boulton and James Watt, the potter Josiah Wedgwood, the chemist Joseph Priestley, and William Withering, physician, botanist, and discoverer of digitalis.

Of his father, Galton wrote that he also had scientific interests but that "he became a careful man of business, on whose shoulders the work of the Bank chiefly rested in troublous times. Its duties had cramped much of the joy and aspirations of his early youth and manhood, and narrowed the opportunity he always eagerly desired, of abundant leisure for systematic study. As one result of this drawback to his own development, he was earnestly desirous of giving me every opportunity of being educated that seemed feasible and right" (1908, 8). His son took this lesson to heart and eagerly seized the opportunity of abundant leisure for systematic study that his financial independence gave him.

Francis was the youngest of the seven surviving children (four girls followed by three boys) of Samuel Tertius and Violetta Galton. He was born in 1822 at the family home, The Larches, a large Georgian house in three acres of garden in Sparkbrook, which was then practically in the country but would later be engulfed by the encroaching suburbs of Birmingham. (It had been built by Withering on the site of Priestley's house, which had been burned down in the Birmingham riots of 1791.) He was a precocious child; a visitor to the Larches at Christmas 1828 wrote: "The youngest child Francis is a prodigy. He is 7 next February, and reads and enjoys *Marmion*, *The Lady of the Lake*, Cowper's, Pope's and Shakespeare's works for pleasure, and by reading a page twice over, repeats it by heart. He writes a beautiful hand, is in long division, has been twice through the Latin grammar, all taught by Adèle" (Pearson 1914, 66).

He was educated at home until he was five by his invalid sister, Adèle. He then went to a number of schools, including one in Boulogne, where he was sent to learn French but which he hated because of the harsh discipline, and King Edward's School in Birmingham, which he disliked because of the narrow classical curriculum: "The character of the education was altogether uncongenial to my temperament. I learnt nothing, and chafed at my limitations. I had craved for what was denied, namely, an abundance of good English reading, well-taught mathematics, and solid science. Grammar and the dry rudiments of Latin and

Greek were abhorrent to me, for there seemed so little sense in them" (1908, 20).

His mother was keen for her cleverest son to enter the medical profession, following the examples of her father, Dr. Erasmus Darwin, and her half-brother, Dr. Robert Darwin. Accordingly, he was removed from King Edward's School in 1838 at the age of 16, by no means against his will, and sent for one year to Birmingham General Hospital and for a second year to King's College Medical School in London. As an indoor pupil at the Birmingham General Hospital he was given practical experience and considerable responsibility, helping in the dispensary, accompanying the surgeons on their rounds, attending in the accident room, helping to dress wounds, and being present at operations and postmortem examinations. The medical conditions were primitive in those days before chloroform or asepsis, and even "the stethoscope was generally considered to be new-fangled; the older and naturally somewhat deaf practitioners pooh-poohed and never used it" (1908, 29). But he soon got used to the cries of patients under operation without anesthetic and the sight of suppurating wounds, and showed pride in developing the skills of neat bandaging, setting broken limbs, and reducing dislocations.

The following year he went to King's College, London, to obtain more theoretical instruction in chemistry, anatomy, physiology, and similar subjects. He did well academically, coming near the top of his class, and also enjoyed the social life of London. In particular, he visited his half-cousin Charles Darwin, who was living in Gower Street with his new bride Emma (née Wedgwood).

Before Francis left Birmingham, the idea had been discussed that he might interrupt his medical education to read mathematics at Cambridge. This scheme was encouraged by his London tutor and by Charles Darwin, and his father arranged for him to enter Trinity College, Cambridge, in October 1840, after a short tour of eastern Europe, which is described below. His academic career at Cambridge was undistinguished. He obtained a second class in the "Little Go" examination in 1842, and in his third year he had a breakdown and settled for taking a pass degree rather than reading for honors in his final year of 1843. His later work showed flair for innovation and for geometrical arguments but weakness in the analytic side of mathematics; after the Little Go examination, Galton told his father that his tutor had complimented him on his performance in mechanics, but he added, "I wish though that I were a better analyst" (Pearson 1914, 166). His decision to take a pass degree left him more time in his final year to spend on literary and social pursuits; during his time at Cambridge, he helped to found the Historical Society, a debating society set up in opposition to the Cambridge Union, which had become rather rowdy, and an English Epigram Society. He took advantage of the social opportunities that Cambridge offered, and he made many friends there who were useful to him in later life.

In 1844 he continued his medical studies in Cambridge and London, but in October his father died, leaving him financially independent. He immediately abandoned the idea of practicing medicine, which he had only pursued out of deference to his father's wishes, and in 1845 he set out for the Near East to satisfy his desire to travel and to sow his wild oats.

Travels

Galton was a keen traveler and he made three notable journeys as a young man. The first was a journey to eastern Europe in 1840 before going up to Cambridge, the second was an extended tour of Egypt and Syria in 1845–46 after the death of his father, and the third was an ambitious exploration of South West Africa in 1850–52. After his marriage in 1853, his travels were largely confined to vacation tours in the fashionable tourist resorts of Europe.

Eastern Europe, 1840

Galton wrote: "In the spring of 1840 a passion for travel seized me as if I had been a migratory bird. While attending the lectures at King's College I could see the sails of the lighters moving in sunshine on the Thames, and it required all my efforts to disregard the associations of travel which they aroused" (1908, 48).

He first planned a tour of Scandinavia, but abandoned it in favor of eastern Europe. A fellow student had arranged to spend the summer in Giessen, Germany, studying under the famous chemist Liebig, and asked Francis to accompany him. Francis's father agreed and sent him a liberal letter of credit. When he arrived in Giessen he soon found that his knowledge of chemistry was inadequate to profit from Liebig's teaching, and he determined instead to travel to Constantinople; reading Byron's poetry had given him a longing to see the East. He traveled by boat down the Danube past Vienna, Budapest, and Belgrade to a town from where he traveled overland to the Black Sea, and then by steamer through the Bosphorus to Constantinople, where he stayed for a week. He wrote whimsically to his father: "Egad the Bosphorus beats any thing in the way of a view I have ever set my peepers upon. The kiosks are so opera-scene-like, so white and so much trellis work about them, the mountains are so grand and the Bosphorus so broad and blue, that (I am stuck fast in the mud about how to finish the sentence being afraid of verging on the romantic)" (Pearson 1914, 137). He continued: "I saw the women's slave-market today—if I had had 50 pounds at my disposal I could have invested in an excessively beautiful one, a Georgian. Some of the slaves had their nails dyed in henna. Most of the black ones were fettered, but they seemed very happy dancing and singing and looking on complacently whilst a couple of Turks were wrangling about their prices."

From Constantinople he took a steamer through the Mediterranean to Venice, stopping off at several places and spending two extended periods in quarantine against bubonic plague, and then went by stagecoach via Milan and Geneva to Boulogne. He carried back in a bottle two specimens of the permanently gilled cave-dwelling amphibian *Proteus anguinis* which he had caught near Trieste, nursing them under his coat while crossing the Alps to prevent them from freezing, and gave them to King's College; one soon died, but the other was yearly lectured on until a cat ended its life. These were the first live specimens brought to England, though the bottled specimen in the Hunterian Museum had aroused much interest (Desmond 1989, 248).

The Near East, 1845–46

The death of his father in 1844 freed Galton from family responsibilities and from the need to earn a living, and enabled him to pursue his love of travel. He tried unsuccessfully to persuade a friend to join him on a shooting trip up the Nile, and so left London by himself in October 1845, bound for Egypt. On the steamer from Malta to Alexandria he met two Cambridge friends—Montague Boulton, a grandson of Matthew Boulton, and Hedwith Barclay, a distant relative—who were traveling with two servants (a butler and a cook). They agreed to join up, Galton engaging a dragoman (interpreter) called Ali when they reached Cairo. There they hired a lateen-sailed Nile sailing boat with its crew, plus an Arab boy Bob as coffee-bearer and general help. They sailed up the Nile to

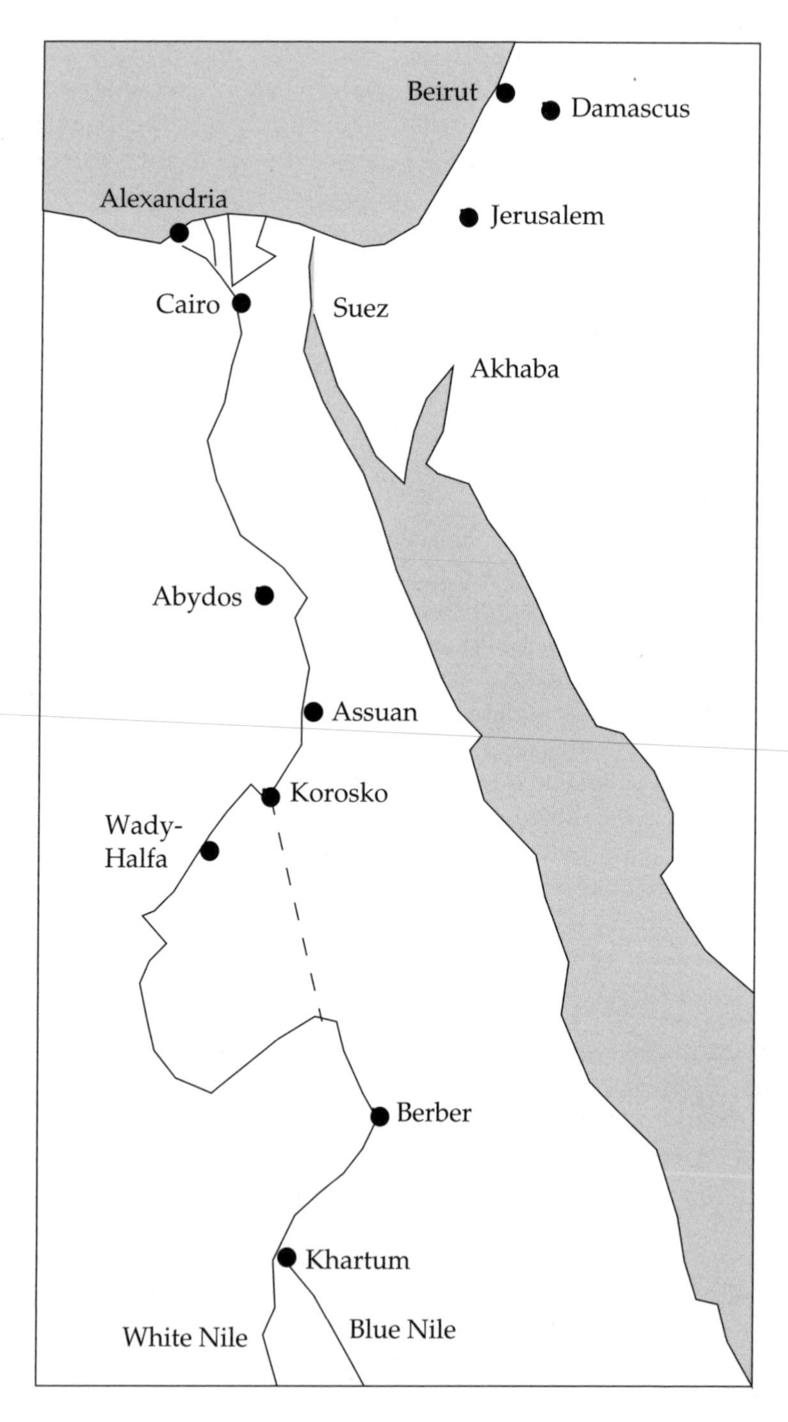

Fig. 1.2. Galton's travels in the Near East, 1845–46 (modified from Galton 1908)

Korosko beyond the first cataract, where they met an Egyptian who changed their plans. Galton tells the story against himself:

> A little above the first cataract, when near Korosko, the stream being swift, we went as usual by virtue of Barclay's firman to impress men to tug our boat, but found they had all been already impressed by the owner of a small and dirty looking Egyptian boat, who they told us was a Bey. We went to him and spoke impudently, like arrogant Britishers and discussed loudly in English together whether we should not pitch him into the river. He shortly astonished us by speaking perfect French and after a while [we] discovered he was a much more interesting person than we had dreamt of. He was Arnaud, a St. Simonian exile, in service of Mehemet Ali.... He then, after we had become friends, explained to us, that though he spoke English badly, he quite understood what we had said among ourselves when we first met him and made me feel very small indeed. However, we got on very well and made him talk of his travels and tell us of the country ahead, we had no map and knew nothing hardly. He said: "Why do you follow the English routine of just going to the 2nd cataract and just returning? Cross the desert and go to Khartoum." That sentence was a division of the ways in my subsequent life. We caught at the idea, he discussed it and said that the chief of the Korosko desert was then actually at the place with camels, that he knew him and would send for him to us that afternoon or evening, when we might finally settle matters. We asked Arnaud to dinner, received him in the grand style, Evard [the butler] doing his best, and gave our good friend and ourselves as much wine as was good for us. When in the midst of the carouse the door of the cabin opened, the cool air came in, and with the cool air, the dignified cold presence of the Sheikh, with the band of sand on his forehead, the mark of his having just prostrated himself in prayer. He did look disgusted, but we got over him and finally all was arranged. We were to start the very next afternoon. (Pearson 1914, 200)

It took their camel caravan eleven days to reach Berber. It was joined by many strangers, including a husband on foot with his wife and child mounted on a donkey, like pictures of the flight of the Holy Family into Egypt, and a mild-looking man who was on his way to Abyssinia to capture slaves in some fighting which was expected there. From Berber they took a boat to Khartoum, which was then just a group of huts with a large hall for the audiences of the Pasha. They met there a colorful Englishman, Mansfield Parkyns, who had left Trinity College, Cambridge, under a cloud and had since been traveling in Abyssinia. He acted as their guide to Khartoum, introducing them to some of the villainous slave dealers living there, and accompanied them on a boat trip up the

White Nile. Boulton and Parkyns shot a cow, mistaking it for a hippopotamus, and the party returned hastily to Khartoum to avoid any local unpleasantness. Leaving Mansfield Parkyns in Khartoum, they set off by boat and camel to find their own boat, which they had arranged to be sent to Wady Halfa, and returned down the Nile to Cairo, where they parted company; Galton sailed from Alexandria to Beirut with his dragoman, Ali.

It seems that Galton sowed his wild oats at this time. None of his letters from this period survive, and, a year after Galton's death, his great-niece and housekeeper Eva Biggs told his biographer Karl Pearson that his family "think it best not to take any notice of those blank years," adding, "I expect his mother tore up any letters of that date" (Kevles 1995, 11 and 304). One surviving piece of correspondence is a letter from Montague Boulton, who was still touring the Near East, to Galton: "What an unfortunate fellow you are, to get laid up in such a serious manner for, as you say, a few moments' enjoyment" (Forrest 1974, 33). Boulton continued that he was negotiating for the purchase of a pretty Abyssinian slave, and added: "The Han Houris are looking lovelier than ever, the divorced one has been critically examined and pronounced a virgin." Galton described Boulton as "an epicurean in disposition, that is to say a philosophical pleasure seeker and of sterling merit" (Pearson 1914, 204). He was killed shortly after at the siege of Multan in the Punjab.

Galton continued his tour of the Near East, visiting Damascus (where his dragoman died), the cedars of Lebanon, and Jerusalem, and leading an expedition down the valley of the Jordan by raft and on horseback. In Jerusalem he learned that the relatives of his late dragoman were pursuing him for money, and he received a letter from his sister Adèle asking for assistance in connection with the sudden death of her husband. He therefore returned to England, arriving in November 1846.

Galton spent the next few years in England and Scotland, shooting and fishing. He wrote: "I was also conscious that with all my varied experiences I was ignorant of the very ABC of the life of an English country gentleman, such as most of the friends of my family had been familiar with from childhood. I was totally unused to hunting, and I had no proper experience of shooting. This deficiency was remedied during the next three or four years" (1908, 110). He hunted in Warwickshire and joined a hunt club whose members were rather wild; several of them, including Jack Mytton (son of the more famous and even more extravagant father of the same name), ruined themselves by betting and gambling, though Galton said of himself that he was fortunate in being careless to the attraction of gambling. He was invited to shoot on a friend's grouse moor in Scotland and spent a summer in the Shetlands, seal shooting and bird nesting. He concluded his account of this period by stating that it was not really an idle life since he read a great deal, and digested what he read by much thinking about it. "It has always been my unwholesome way of work to brood much at irregular times" (1908, 119).

South West Africa, 1850–52

Galton now undertook a journey to southern Africa in search of adventure, big game, and discovery. At that time much of the interior of Africa had not been explored by Europeans; the Great Lakes had not been discovered and the sources of the White Nile and the Congo were unknown. The young medical missionary David Livingstone had recently crossed the Kalahari Desert to discover Lake Ngami and the well-watered country north of it, and Galton mentioned his plan to visit this region to his cousin Douglas Galton, who was a Fellow of the Royal Geographical Society and introduced him to its leading members. In consultation with them he drew up a plan, which they approved, to travel from Cape Town to Lake Ngami and to explore the surrounding area. Galton persuaded a Swedish amateur naturalist, Charles Andersson, to join him as second in command, and they sailed from Southampton on April 5 1850 on an old sailing vessel which could not beat to windward. The journey to the Cape took 86 days; Galton spent the time learning the Bechuana language and the practical skills he would need in establishing latitudes and longitudes and in surveying.

Galton was kindly received by the governor of the Cape, to whom he had an introduction from the colonial secretary, but learned that his original plan, of traveling north overland from Algoa Bay to Lake Ngami, was impracticable because he would be turned back by the Boers, who were asserting their independence. He decided instead to go by sea to Walfish Bay in South West Africa (now Namibia), and to explore the country inland from there. There were mission stations along the Swakop River east of Walfish Bay, which served him as starting points for a journey into the previously unexplored lands of the Damaras and the Ovampos. One or two highlights from his account of the journey in his autobiography (1908) and in *Tropical South Africa* (1853) illustrate his adventures.

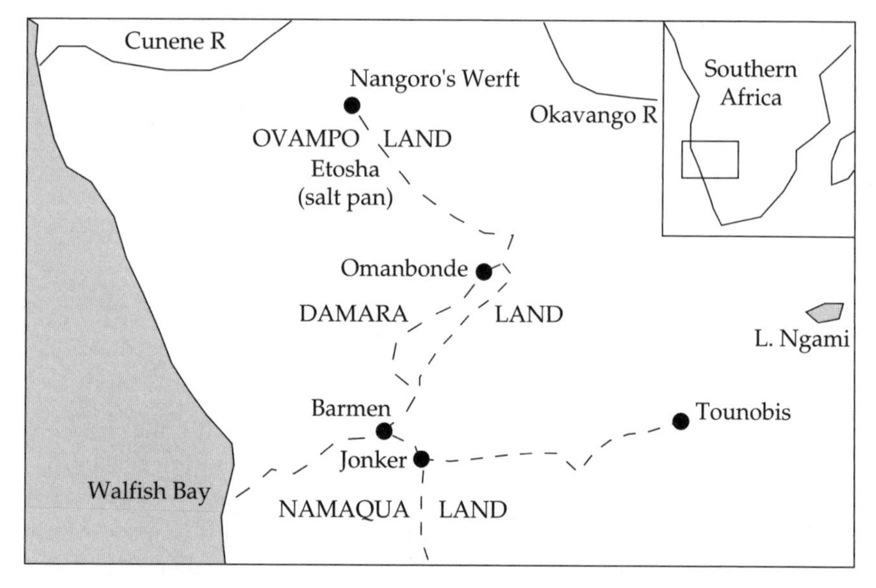

Fig. 1.3. Galton's travels in South West Africa, 1850–52 (modified from Galton 1853)

Galton arrived at Walfish Bay in August 1850 and spent the next six months in the Swakop River area, preparing his expedition to the unexplored regions to the north of it. They had several encounters with lions. On one occasion they had left the mules and horses to feed free at night by the river:

> The result was that a troop of lions dashed down upon them in the dark, killing one mule and one of my two horses.... I had no reserve food, so it was necessary to utilise the horse flesh, which I cut off and stored in an apparently safe hole in the side of a cliff. When I returned towards nightfall to remove it, one of my enemies had out-generalled me. He had clambered from behind and unseen to a ledge five or six yards above the hiding-place, and could be seen there by the party below, crouched like a cat above a mouse-hole. I got down safely, meat and all, and saw the head and the pricked ears of the brute as he kept his position. A shot struck the rock under his chin, and he decamped. (Galton 1908, 133)

While he was staying at the mission station of Barmen, he had occasion to use his sextant for an unusual purpose:

> The sub-interpreter was married to a charming person, not only a Hottentot in figure, but in that respect a Venus among Hottentots. I was perfectly

> aghast at her development, and made inquiries upon that delicate point as far as I dared among my missionary friends. The result is, that I believe Mrs. Petrus to be the lady who ranks second among all the Hottentots for the beautiful outline that her back affords, Jonker's wife ranking as the first; the latter, however, was slightly *passée*, while Mrs. Petrus was in full *embonpoint*. I profess to be a scientific man, and was exceedingly anxious to obtain accurate measurements of her shape; but there was a difficulty in doing this. I did not know a word of Hottentot, and could never therefore have explained to the lady what the object of my foot-rule could be; and I really dared not ask my worthy missionary host to interpret for me. I therefore felt in a dilemma as I gazed at her form.... The object of my admiration stood under a tree, and was turning herself about to all points of the compass, as ladies who wish to be admired usually do. Of a sudden my eye fell upon my sextant; the bright thought struck me, and I took a series of observations upon her figure in every direction, up and down, crossways, diagonally and so forth, and I registered them carefully upon an outline drawing for fear of any mistake; this being done, I boldly pulled out my measuring tape, and measured the distance from where I was to the place she stood, and having thus obtained both base and angles, I worked out the results by trigonometry and logarithms. (Galton 1853, 53–54)

One problem had to be resolved before Galton could travel north into the land of the Damaras. The local Hottentot tribes under the above-mentioned Jonker were in constant conflict with the Damaras and wanted to prevent white men from traveling into Damaraland. Galton tried to negotiate with Jonker with little success and finally decided to intimidate him. He rode to Jonker's village, dressed in hunting pink (red hunting-coat, jackboots, and hunting cap) and mounted on his favorite ox, Ceylon (the horses had all died of distemper), leaped a brook, trotted into Jonker's hut, and berated him. Jonker was suitably impressed, and he and the other chiefs later agreed to make peace with the Damaras so that it would be safe for white men to pass through their territory. They also agreed to abide by a simple code of law drawn up by Galton.

Galton now set out on his expedition through the unexplored region of Damaraland with a caravan consisting of two wagons, about ten Europeans and eighteen natives, ninety-four oxen, cows, and calves (for riding, for carrying packs, and for slaughter), and twenty-two sheep (for slaughter). (He called the animals destined for slaughter his "itinerant larder.") After a month they reached their first objective, Lake Omanbonde, but they found that it was completely dry that year due to exceptional lack of rain. Galton had a low opinion of the Damara people: "The

[Damara] vocabulary ... is not strong in the cardinal virtues; the language possessing no word at all for gratitude; but on looking hastily over my dictionary I find fifteen that express different forms of villainous deceit" (1853, 118); he was therefore keen to continue northwards into Ovampoland. One of the wagons now broke down, and he decided to leave both wagons with some of his party and to continue on ride-oxen with Andersson and three companions. They joined a caravan of Ovampo who had come south to Damaraland to trade beads, assegais, and other goods for cattle. After seeing the great salt pan of Etosha, they reached the boundary of Ovampoland: "Quite of a sudden the bushes ceased ... and the charming corn-country of the Ovampo lay yellow and broad as a sea before us. Fine dense timber-trees, and innumerable palms of all sizes were scattered over it.... The general appearance was that of most abundant fertility. It was a land of Goshen to us" (1853, 125). They were taken to meet the king, Nangoro, and invited to the ball he gave every night for the elite of Ovampoland.

Galton formed a high opinion of the Ovampo, who were an agricultural people more civilized than the nomadic Damaras, but it was not reciprocated. He and his party were treated with strict courtesy, but without friendliness, and Galton felt that there were ugly signs of an intention to allow their oxen to die of hunger, and then to make an easy

Fig. 1.4. Sketch by Galton of his ox Ceylon (from Pearson 1914)

end of them. He therefore abandoned the idea of traveling farther north to the Cunene River, which might have been used to open up Ovampoland to colonization, and returned to the mission station of Barmen the way he had come, rejoining his wagons and the rest of his party in Damaraland on the way. The lack of friendliness of the Ovampo may have been due to a natural suspicion of foreigners, compounded by several social gaffes on Galton's part: "I have reason to believe that I deeply wounded [Nangoro's] pride by the non-acceptance of his niece as, I presume, a temporary wife. I found her installed in my tent in negress finery, raddled with red ochre and butter, and as capable of leaving a mark on anything she touched as a well-inked printer's roller. I was dressed in my one well-preserved suit of white linen, so I had her ejected with scant ceremony" (1908, 143).

On arriving at Barmen in August 1851 he learned that a sailing vessel was expected to call at Walfish Bay in December on its way to England. He decided to take the ship, and in the meantime to make a journey eastward in the direction of Lake Ngami. He got as far as Tounobis, where he was told that there was a wagon route to Lake Ngami in the wet season, but that it was impassable in the prevailing very dry conditions. They stayed at Tounobis a week, shooting the abundant game:

> It is one of the most strangely exciting positions that a sportsman can find himself in, to lie behind one of these screens or holes by the side of a path leading to a watering-place so thronged with game as Tounobis. Herds of gnus glide along the neighbouring paths in almost endless files.... Now and then a slight pattering over the stones makes you start ... and a troop of zebras pass frolicking by. All at once you observe twenty or thirty yards off two huge ears pricked up high above the brushwood; another few seconds, and a sharp solid horn indicates the cautious and noiseless approach of the great rhinoceros. Then the rifle or gun is poked slowly over the wall.... The beast moves nearer and nearer.... Nearer, nearer still.... Bang! and the bullet lies well home under his shoulder....
>
> I like rhinoceros flesh more than that of any other wild animal. A young calf, rolled up in a piece of raw hide, and baked in the earth is excellent. I hardly know which part of the little animal is the best, the skin or the flesh. (Galton 1853, 168–169)

He clearly relished living rough. He recorded: "We rested a day, to have a really good breakfast and dinner. I have read in some old-fashioned books of fiction, entitled 'Natural History,' that an ostrich egg would feed six men; but I know that Stewartson, Andersson, and myself finished one very easily for breakfast, before beginning upon the giraffe"

(1853, 35). He was not in fact very interested in shooting, writing: "For my own taste, I should like to spend nights perched up in some tree with a powerful night glass watching these night frolics and attacks. I really do not much care about shooting the animals, though it makes a consummation to the night work, as the death of the fox does to a fox hunt, but it is the least pleasurable part of the whole" (1853, 171). But the idea of conservation was unknown to him, though he knew that enthusiasm for shooting had extirpated game in parts of Africa; he wrote when revising his essay on the domestication of animals for *Inquiries into Human Faculty*: "As civilisation extends they [animals that cannot be domesticated] are doomed to be gradually destroyed off the face of the earth as useless consumers of cultivated produce" (1883, 194).

The party then returned to Barmen. Andersson remained behind to investigate the natural history of the countries they had visited, and to explore Lake Ngami; he subsequently discovered the Okavango River. Galton went back to Walfish Bay, and took the ship home to England, where he arrived in April 1852 exactly two years after he had left. He had spent his time on the outward journey learning the Bechuana language and the use of the sextant. He never talked Bechuana because of his change of plans, but he put his knowledge of the sextant and of surveying techniques to good use. He triangulated the country through which he traveled, and made careful observations of latitude and longitude, the latter by the demanding method of lunar distances. He sent an account of this work from Africa to the Royal Geographical Society, which was published in their journal in 1852. The Society awarded him one of their two annual gold medals in 1854, "for having at his own expense and in furtherance of the expressed desire of the Society, fitted out an expedition to explore the centre of South Africa, and for having so successfully conducted it through the countries of the Namaquas, the Damaras, and the Ovampo (a journey of about 1700 miles), as to enable this Society to publish a valuable memoir and map in the last volume of the Journal, relating to a country hitherto unknown; the astronomical observations determining the latitude and longitude of places having been most accurately made by himself" (Galton 1908, 150).

On his return, Galton wrote a highly readable account of his journey, *Tropical South Africa*, which prompted a congratulatory letter from Charles Darwin: "I last night finished your volume with such lively interest, that I cannot resist the temptation of expressing my admiration at your expedition, and at the capital account you have published of it.... I should very much like to hear what your future plans are, and where you intend to settle.... I live at a village called Down near Farnborough in Kent, and employ myself in Zoology; but the objects of my study are very small fry, and to a man accustomed to rhinoceroses and lions, would appear infinitely insignificant" (Pearson 1914, 240–241). (Darwin was just finishing his eight years' study of barnacles.)

Galton then began to work on *The Art of Travel; Or, Shifts and Contrivances Available in Wild Countries*, in which he collected together a vast amount of practical information for intending travelers. It proved a very useful and popular book, running to eight editions. Some of the information seems rather quaint today: to prevent an ass from braying, lash a heavy stone to its tail; when an ass wants to bray he elevates his tail, and, if his tail be weighted down, he has not the heart to bray. Most of it is eminently practical: to prepare tea for a very early breakfast, make it overnight and pour it away from the tea leaves, into another vessel; in the morning it only needs to be heated to 140°F rather than to 212°. Some of it is more scientific: he presents a mathematical analysis of the optimum load for a pack animal, and concludes that an animal gets through most work in the day if he carries four-ninths of the greatest load he could just stagger under; in which case he will be able to travel one-third of the distance he could walk if he carried no load at all.

As a postscript, we note that South West Africa came under German control in 1884 during the scramble for Africa (Pakenham 1991, chap. 33). German settlement was at first accepted by the Namaquas and Damaras, but they were so humiliated by the settlers, who called them "baboons" to their face, that in 1904 the Damaras (= Hereros) rose in revolt, killing about one hundred German men. Kaiser Wilhelm II sent a punitive force from Germany to crush the revolt by fair means or foul. On arriving, its commander issued a *Vernichtungsbefehl* (extermination order): "The Herero are no longer considered German subjects.... Within the German boundaries, every Herero, whether found armed or unarmed, with or without cattle, will be shot. Signed: the Great General of the Mighty Kaiser, von Trotha" (Pakenham 1991, 611). He then surrounded them, leaving one escape route into the desert, and closed the wells behind them when they had passed through it. The population of Hereros fell from 80,000 in 1904 to 15,000 in 1907, and that of the Namaquas from 20,000 to 10,000. Pakenham (1991) calls this disgraceful episode the Kaiser's First War. During his Second War (1914–18) South West Africa became part of South Africa and gained independence as Namibia in 1990. The Ovampo were so warlike that no German dare set foot in

Ovampoland during the period of German control; they are today the dominant tribe in Namibia.

Vacation Tours

Galton spent Christmas 1852 with his mother and sister in Dover, where he met Louisa Butler, who was staying with her family next door. There was a rapid courtship and they became engaged in April 1853, though the marriage had to be postponed until August because of the death of her father. They left immediately on a honeymoon tour to Switzerland and then to Florence and Rome, where they spent the winter. On returning to London in the spring of 1854, they took temporary accommoda-

Fig. 1.5. Francis and Louisa Galton at the time of their marriage (from Pearson 1914)

tion until Galton bought a house at 42 Rutland Gate, South Kensington, in 1857, where he lived almost up to his death in 1911.

Louisa came from an intellectually distinguished family. Her father was headmaster of Harrow School before becoming dean of Peterborough; one of her brothers succeeded him as headmaster of Harrow and later became master of Trinity College, Cambridge; two other brothers were headmasters of well-known schools, and her fourth brother was a barrister. She herself was an intelligent woman with an interest in the arts, but she did not have enough to occupy her mind. The couple had no children, and she was not interested in domestic affairs and had trouble keeping servants. Her husband was busy with committee meetings and with his research and probably had little time to spend with her. She therefore looked forward to the opportunity to travel abroad with him on vacation. He also liked to take long vacations both in England and abroad since he recognized that overwork led to mental fatigue or even breakdown, which was curable by rest.

Thus, Louisa and her husband welcomed the opportunity of foreign travel for different reasons, she because she had too little to do in London, he because he had too much. Louisa kept an annual record of her life between 1830 and 1897 from which, together with his writings, a picture of their vacations abroad can be pieced together. The places visited form a roll-call of contemporary European resorts: Zermatt, Chamonix, St. Moritz, Grindelwald, Innsbruck, the Auvergne, the Pyrenees, Heidelberg, Berchtesgaden, the Bavarian and Italian lakes, the Black Forest, Cannes, Mentone, Sorrento, Rome, Florence, Venice, Vienna, and in later years, spas such as Vichy, Contréxeville, Royat, and Wildbad.

The Black Forest was a favorite with Louisa, who wrote in her annual record for 1879: "We went to the Black Forest where I had great enjoyment of the lovely scenery and left a bit of my heart behind." Francis was attracted by the mountainous scenery of the Alps and the Pyrenees, and for a few years took up the popular pastime of climbing. In 1860 he traveled with a party of astronomers to observe a total eclipse of the sun in Spain on July 18, and went on to meet Louisa for a tour of the Pyrenees, where "that remarkable madness of mountain climbing, to which every healthy man is liable at some period of his life … began to attack me with extreme severity" (1861, 236). He made several expeditions with one of the pioneers of Pyrenean exploration, Charles Packe, and advocated the sleeping bag, which he had observed in use by the officers who

Fig. 1.6. Daguerrotype of Galton in holiday garb at Vichy, August 1878 (from Pearson 1914)

patrolled the Franco-Spanish border; he was later toasted as the greatest "bagman" in Europe at an Alpine Club dinner (1908, 190).

Louisa died at Royat in the Auvergne in 1897, at the age of 75; Karl Pearson wrote of "his wonderful patience with an invalid wife, and after her death his splendid loyalty to her memory" (1930b, 441). Galton continued to travel abroad for some years, usually with his great-niece Eva Biggs, who became his companion. They spent the winter of 1899–1900 in Egpt, cruising down the Nile as far as Assuan; the highlight of the trip was a week spent with Professor Flinders Petrie at Abydos, where he and his team were excavating. Their last journey abroad was to the

Basque region of the Pyrenees in the winter of 1905–06. His physical health gradually deteriorated from then on, though he remained mentally alert, and he died in 1911.

Scientific Career

In 1853, the Royal Geographical Society (RGS) awarded Galton a gold medal for his geographical work in Southern Africa and next year elected him to its Council, on which he served almost continuously for forty years. This recognition gave him an established position in the scientific world, which led to his election in 1855 as a member of the Athenaeum Club and in 1860 as a Fellow of the Royal Society, his proposer for the latter being Charles Darwin. The Athenaeum was the leading London club in the arts and sciences, to which he had first been nominated for membership in 1840. He made much use of its facilities, and he recalled many years later that he had "a habit of spending an hour or two of the afternoon, during many years, in the then smoking room of the Athenaeum Club" (Pearson 1930b, 626). He frequently met Herbert Spencer there, as well as Richard Burton "generally to be seen with a folio volume before him" (Galton 1908, 171).

Galton was rather unsettled for a few years after his return to England in 1852, thinking of undertaking a fresh bit of geographical exploration (see below), or even of establishing himself in one of the colonies, but he mistrusted his health and eventually decided that there was an abundance of useful work at home. By 1857, when he bought 42 Rutland Gate, close to the main scientific institutions in London, he had committed himself to developing a scientific career in England. He played an active part in the scientific life of London for the rest of his life. Charles Darwin's son George, who knew Galton well, wrote in his obituary: "After their marriage Mr. and Mrs. Galton settled in London ... and went much into Society, especially in scientific and literary circles. His powers as a conversationalist and ready humour, seconded by Mrs. Galton's sympathetic nature, rendered them charming hosts and they were universally popular" (G. H. Darwin 1911, xii).

I now describe two aspects of his work in more detail. I first describe his work for the Royal Geographical Society and his involvement with British exploration in Central Africa. I then describe his wide-ranging research career by discussing the papers he read at meetings of the British Association.

The Royal Geographical Society

The London Geographical Society, later the Royal Geographical Society, was founded in 1830 to further scientific exploration overseas. After a period of mismanagement, it was revived in the 1850s under the leadership of Sir Roderick Murchison to become the leading institution underpinning scientific exploration. However, there was always a tension between its scientific aims and the need to attract the public support to finance those aims. Though all were Fellows, Lyell, Darwin, Hooker, Huxley, and Wallace shunned the Society's meetings because they believed Murchison's promotional efforts were devaluing science (Stafford 1989).

Galton was presented with the Society's Gold Medal in 1853, and was elected to its Council in 1854. He became enthusiastically involved in its activities, serving on several subcommittees, and was consequently invited to become one of its two honorary secretaries in 1857. The following pen-portrait of Galton at this time was made by his colleague, Clements Markham, in his history of the Society:

> He had intelligent rather eager blue eyes and heavy brows, a long straight mouth, bald head. He was very clever and perfectly straight in all his dealings with a strong sense of duty. Without an atom of vanity he held to his own opinions and aims tenaciously. His mind was mathematical and statistical with little or no imagination. He was essentially a doctrinaire not endowed with much sympathy. He was not adapted to lead or influence men. He could make no allowance for the failings of others and had no tact. (Forrest 1974, 69)

Galton's lack of tact led to a bitter quarrel with the executive secretary, Norton Shaw, whose efforts had been largely responsible for the flourishing state of the Society and with whom he had to work. Galton was eventually forced to resign his post as honorary secretary in 1863, though he kept his seat on Council. Markham's reference to Galton as a "doctrinaire" refers to his membership of the progressive wing of the Society which wanted to further geography as a science by encouraging its study in schools and universities, and by funding geographical research and scientific lectures; these aims were resisted by Markham as unpractical. The progressive "doctrinaires" won control of the Society for a short time in the 1880s and 1890s, but they split the Society from top to bottom by proposing the election of Lady Fellows. A general meeting of Fellows in 1893 defeated the proposal, a referendum of all Fellows resident in the British Isles brought a decision to admit women, but this decision was overturned at a second general meeting. The progressives, including Galton, resigned from the Society in 1894; Galton's resignation possibly reflected solidarity with his fellow-progressives rather than sympathy with women's right to Fellowship (Forrest 1974, 224–227).

Exploration in Central Africa

Galton's position on the expeditions committee during an exciting period of British exploration in Central Africa involved him in some well-known expeditions that are now briefly described. I have relied on these secondary sources: Moorehead 1960, Forrest 1974, Lovell 1998.

Information had reached London in the 1850s, Galton explained in his autobiography, of a snow-topped mountain called Kilimanjaro near the equator and of a vast Central African lake. He had been urged to investigate the neighborhood of Kilimanjaro himself by Sir Roderick Murchison, president of the RGS, but declined on grounds of health. But an expedition, for which Galton drafted the instructions, was set on foot in 1856, led by Richard Burton with Speke as second-in-command. They traveled inland along a slaving route from Zanzibar and became the first Europeans to see Lake Tanganyika in February 1858, exploring it briefly before returning part of the way to rest and recover their health. Speke now traveled north to investigate claims of an even larger lake, while Burton remained behind to continue his convalescence. Thus in August 1858, Speke became the first white man to see Lake Victoria, which he immediately guessed, without any evidence, to be the source of the White Nile. He hurried back to tell Burton of his discovery of the source of the Nile, but the latter was unconvinced: "The fortunate discoverer's conviction was strong. His reasons were weak" (Moorehead 1960, 38). Through lack of time to explore Lake Victoria more fully, the two men then returned to England. Speke arrived first, since Burton remained for a few weeks in Aden to convalesce, and immediately reported his discovery to the RGS, despite having promised Burton not to do so until the latter got back. They never spoke to each other again.

In discussing the Burton-Speke expedition, Galton remarked that the two men were incompatible in temperament: "Burton was a man of eccentric genius and tastes, orientalised in character and thoroughly Bohemian. . . . Speke, on the other hand, was a thorough Briton, conventional, solid and resolute. Two such characters were naturally unsympathetic" (1908, 199). He summarized Burton's character perceptively: "Burton had many great and endearing qualities, with others of

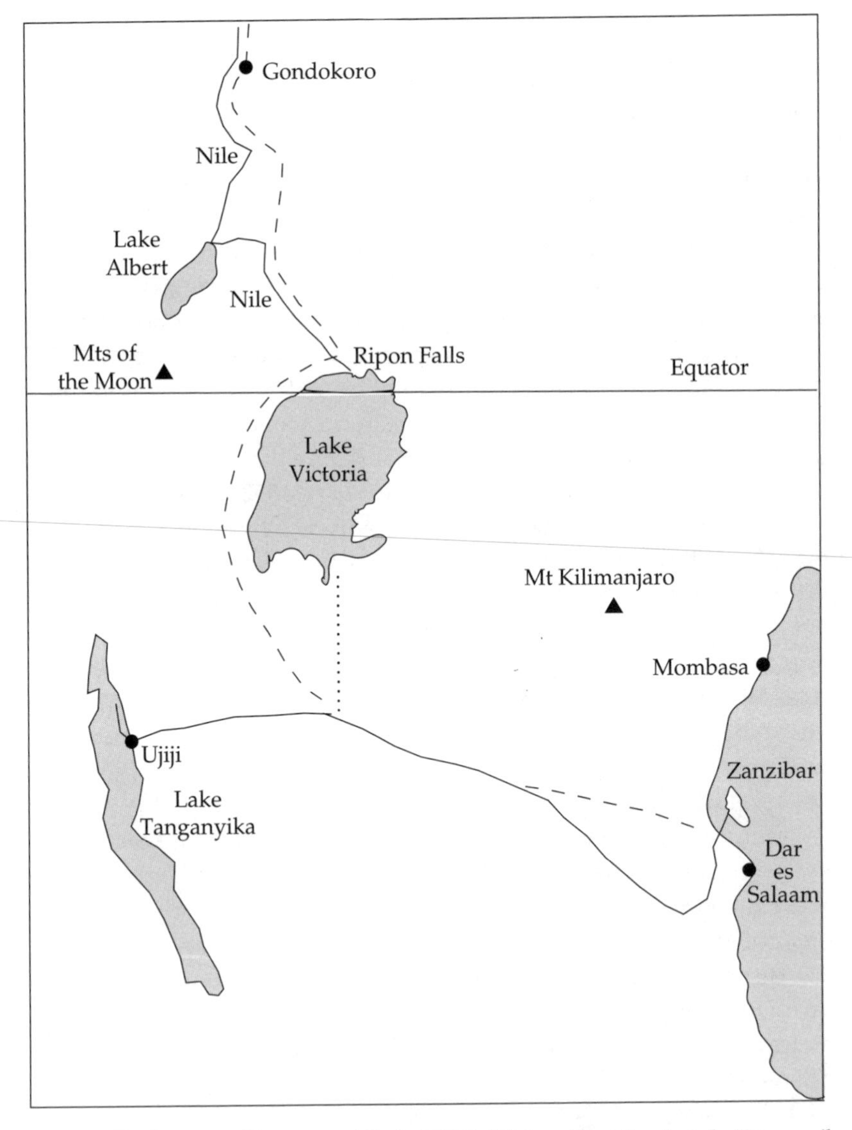

Fig. 1.7. The journey of Burton and Speke (1856–59) from Zanzibar to Lake Tanganyika and back (solid line); Speke's trip to Lake Victoria (1858) (dotted line); and the journey of Speke and Grant (1860–63) from Zanzibar to Gondokoro (dashed line).

which perhaps the most curious was his pleasure in dressing himself, so to speak, in wolf's clothing, in order to give an idea that he was worse than he really was" (202).

Speke's discovery of Lake Victoria was greeted with enthusiasm, and it was arranged for him to return there with Captain Grant. Galton describes their expedition (1860–63):

> They were to take up the quest at the point on [Lake] Victoria where Speke had reached it, and to travel onwards. This was done, and I may say that the attachment of Grant to Speke was most remarkable for its loyalty and intensity. They were fine manly fellows, and I can see them in my mind's eye, as they came to take a final leave, when I knocked two nails into the side of a cupboard as they stood side by side with their backs to it, to mark their respective heights and as a memento of them when away. As is well known, they followed the Nile, not however without a break, from the Lake into Egypt. This break, and the hypothetical placement of the 'Mountains of the Moon,' whose position Speke saw reason to modify in a second map, gave an opening to criticism of which bitter use was made. (Galton 1908, 200–201)

After discovering the source of the Nile at the Ripon Falls, Speke and Grant were short of supplies. They proceeded on foot to Gondokoro rather than following the course of the river, and from there down the Nile to Khartoum and Cairo. On their way Speke cabled home: "Inform Sir Roderick Murchison that all is well, that we are in latitude 14° 30' upon the Nile, and that the Nile is settled" (Moorehead 1960, 60). Speke was feted on his return to England in 1863, but it soon became clear, due to a number of geographical errors and assumptions he had made, that the Nile was not settled. A group of geographers, led by James MacQueen and including Galton, questioned Speke's claim, and the latter wrote to a friend in February 1864: "I have answered Galton in full and shown him he is wrong in *all* his conclusions. Moreover I have asked him to reconsider his resolution of not attending the meeting, as I wish to have all the pros and cons fairly argued out ... and I think no man more capable of finding cons" (Lovell 1998, 437).

On his return to England from West Africa in 1864, Burton joined the attack on Speke, arguing, among other things, that he had no evidence that the river flowing out of the Ripon Falls was the Nile, since he had not traced its course. (In fact, Baker would soon show that this river flowed from the Ripon Falls into Lake Albert, which Speke had not seen, and thence into the Nile. This left open whether Lake Albert was fed by another river farther south.) A confrontation between the two men was arranged at a meeting of the British Association at Bath in 1864, at which Livingstone would also be present. Burton was to read a paper severely criticizing Speke's work, and Speke, who was staying in the neighborhood with a shooting party, had been invited to take part in the discussion. Galton was present at the committee meeting just before Burton's paper when a message was received to say that Speke had accidentally shot himself dead, by drawing his gun after him while getting over a hedge. It has been suggested that Speke committed suicide, but this seems unlikely.

Galton's view of the roles of explorers and of London-based geographers like himself is shown in a letter he wrote to Grant just after Speke's death: "I should earnestly recommend your not burning your fingers with meteorological theorising. Poor Speke's notions on these things were so crude and ignorant that his frequent allusions to them did great harm to his reputation. What he could have done and what you can do, is to state accurately what you saw—leaving the stay-at-home men of science to collate the data" (Lovell 1998, 457).

The RGS now commissioned Livingstone to settle the question of the source of the Nile. He set out in 1865 with instructions from Galton and from the cartographer John Arrowsmith to make careful observations of latitude and longitude, which he found difficult to comply with because of the dense cloud conditions. "Put Arrowsmith & Galton into a hogshead," he complained to Murchison in 1867, "and ask them to take bearings out of the bunghole. I came for discovery and not for survey, and if I don't give a clear account of the countries traversed, I shall return the money" (Jeal 1973, 315). Livingstone was unsuccessful in solving the Nile problem because he concentrated his attention on the area around Lake Tanganyika, which he believed to be its source.

In the meantime, the British public had become concerned about the absence of news from Livingstone, and the RGS formed a Search and Relief Committee with Galton as secretary in 1871. A relief expedition was mounted, but before it reached East Africa a rival expedition led by H. M. Stanley, a journalist for the *New York Herald*, had found Livingstone at Ujiji on Lake Tanganyika. Overcome by the occasion, Stanley could only think to ask, "Dr. Livingstone, I presume?" The RGS expedition broke up in some acrimony, not having been instructed what to do if Stanley got there first. Livingstone welcomed Stanley but declined to return to England with him. The two men explored Lake Tanganyika for a few months before Stanley returned on his own; Livingstone would die there in 1873.

When he reached England in 1872, Stanley was invited to talk at the Geographical Section of the British Association at Brighton, of which Galton happened to be president that year. Stanley was unpopular with the geographers, for reasons articulated by Galton: "Mr. Stanley had other interests than geography. He was essentially a journalist aiming at producing sensational articles, and it was feared from the newspaper letters he had already written that he might utilise the opportunity in ways inappropriate to the British Association. However, the meeting went off without more misadventure than a single interference on my part, but under some tension" (1908, 207).

The full story, as recounted by Forrest (1974, 116–118), is amusing (see also Gillham 2001). In introducing the speaker to an audience of 3000 people, including the exiled Emperor Napoleon III and Empress Eugénie, Galton rather gratuitously hoped that Stanley would be able to clear up certain mysteries about himself and his nationality. (He claimed to be an American born in New York, but was in fact the illegitimate son of a Welsh farmer.) Stanley gave a popular talk about his adventures and the greatness of Livingstone, aimed at the majority of the audience rather than the scientific geographers in it, and concluded by saying that his origins were no concern of Galton's. Galton lost his temper and burst out that the talk had been all "sensational geography." What they needed were facts, and they should examine Stanley's stories to discover those facts. In that spirit, he would ask the first question: "Was the water of Lake Tanganyika sweet or brackish?" Stanley thought that he was being made a fool of, and replied that the water was delicious if not the best in the world, being particularly good for making tea. This rejoinder was greeted with applause, but Stanley never forgave Galton. He said later in a speech to the Savage Club: "It was at the British Association where Mr Francis Galton, F.R.S., F.R.G.S. and God knows how many letters to his name, said 'We don't want sensational speeches.' That does stick in my throat. (Laughter) ... He wanted facts. I gave him facts. (Cheers) They required no gilding."

The British Association

The British Association for the Advancement of Science was founded in 1831 to promote science throughout the country by holding annual meetings in major cities outside London. It was an influential organization, attracting the important scientists of the day, and was often known as the "Parliament of Science." Galton was an enthusiastic supporter of the

Association and usually arranged or interrupted his holiday to attend its meeting in late August or early September. (This irked Louisa, who wrote in her record for 1882: "It was such a boon not to be kept by a Brit. Asstn. meeting this summer.") He was on its Council almost from his return from South Africa, and served as its general secretary from 1863 to 1867, and as president of the Geographical Section in 1862 and 1872, and of the Anthropological Section in 1877 and 1885. He summarized its work thus:

> The papers read there and discussions upon them are not the most important part of its work.... Perhaps the most useful function of the British Association lies in causing persons who are occupied in different branches of science, and who rarely meet elsewhere, to be jostled together and to become well acquainted.... The plan of one meeting is as like that of another as two Roman camps, ... so after the experience of a single year a member finds himself at home on every future occasion. But the sustained racket of it is great, and I found it too long continued for my own nerves. (Galton 1908, 214–215)

Galton's contributions to the British Association, published in their annual reports, are listed in table 1.1. They give a good idea of the range and development of his scientific interests. Most of these contributions are very short, being little more than abstracts, apart from the three presidential addresses and the reports of the two committees of which he was chairman. Some of them are now briefly discussed.

Inventions

Five of the papers illustrate Galton's lifelong fascination with devising mechanical and optical instruments. His first contribution to the British Association, in 1858, was a description and demonstration of an ingenious hand heliostat for transmitting signals, which was much more compact than existing instruments; it was later manufactured commercially under the name of "Galton's Sun-Signal," and was used in nautical surveys to enable shore parties to make their exact whereabouts visible to those on the ship. At the same meeting, the General Committee of the British Association passed a resolution, at Galton's suggestion, recommending that Kew Observatory be asked to extend its work in the improvement and standardization of scientific instruments to cover instruments used by geographers. As a result, Galton was invited to become a member of the Kew Management Committee, and was active in its work until 1901, busying himself particularly with the testing of

Table 1.1. Galton's Contributions to the British Association

Year	Title of Article
1858	A hand heliostat, for the purpose of flashing sun signals, from on board ship or on land, in sunny climates
1862	On the "Boussole Burnier," a new French pocket instrument for measuring vertical and horizontal angles
1862	European weather-charts for December 1861
1864	First steps towards the domestication of animals
1865	On spectacles for divers, and on the vision of amphibious animals
1866	On an error in the usual method of obtaining meteorological statistics
1866	On the conversion of wind-charts into passage-charts
1870	Barometric predictions of weather
1872	Presidential address to the section of geography
1877	Presidential address to the section of anthropology
1880	On determining the heights and distances of clouds by their reflexions in a low pool of water, and in a mercurial horizon
1880	On a pocket registrator for anthropological purposes
1880	Mental images [Unpublished popular lecture]
1881	Isochronic postal charts
1881	On the application of composite portraiture to anthropological purposes
1883	Final report of the anthropometric committee [Chairman]
1885	Presidential address to the section of anthropology
1889	On the advisability of assigning marks for bodily efficiency in the examination of candidates for the public services
1889	On the principles and methods of assigning marks for bodily efficiency
1889	Feasible experiments on the possibility of transmitting acquired habits by means of inheritance
1889	An instrument for measuring reaction time
1893	Recent introduction into the Indian army of the method of finger prints for the identification of recruits
1898	Photographic records of pedigree stock
1899	Report of the committee on pedigree stock records [Chairman]
1899	The median estimate
1899	Finger prints of young children

sextants, thermometers, and watches (Pearson 1924, 49–50; Galton 1908, chap. 16).

At the 1865 meeting, he described a provisional attempt to construct spectacles to enable divers to see clearly under water. The convexity of the eyeball in contact with water creates a concave water lens which blurs vision, and he tried to compensate for this by spectacles with an equal and opposite convex lens. These spectacles were successful at a particular fixed distance, but left the eye with little power of accommodating to other distances. He recalled in his autobiography that he had had spectacles made for him with which he could read the print of a newspaper perfectly under water, when it was held at the exact distance of clear vision, but that the range of clear vision was small. He continued: "I amused myself very frequently with this new hobby, and being most interested in the act of reading, constantly forgot that I was nearly suffocating myself, and was recalled to the fact not by any gasping desire for breath, but purely by a sense of illness, that alarmed me" (Galton 1908, 186).

In 1880, he exhibited a "pocket registrator" for recording the number of individuals of different kinds in a crowd without attracting attention, and he also drew attention to the ease with which registers may be kept by pricking holes in paper in different compartments with a fine needle. He used the latter technique to construct a "Beauty-Map" of the British Isles, classifying the girls he passed in the street or elsewhere as attractive, indifferent, or repellent. He found London to rank highest for beauty, and Aberdeen lowest. He also made a practice of counting the number of fidgets at meetings of the Royal Geographical Society as an index of the boredom of the audience (Galton 1885a).

Meteorology

Another group of five papers reflects his early interest in meteorology, which arose from his geographical interests. This work was important in the development of his understanding of statistical concepts that would later be applied to heredity.

In 1862, he presented a series of weather charts of observations made at eighty stations in Europe on the morning, afternoon, and evening of each day in December 1861. The observations, which he had compiled from rather patchy responses to a circular sent to meteorologists throughout Europe, consisted of cloud cover, wind force and direction, temperature of wet and dry thermometers, and barometric pressure;

they were represented on the charts partly in symbols and partly in figures.

Subsequent analysis of these charts led him to discover and name the anticyclone (Galton 1863; 1908, 229–231). It was known that winds rotated in an anticlockwise direction (in the northern hemisphere) round an area of low pressure due to the rotation of the earth, forming what was called a cyclone. By analogy, Galton supposed that winds should rotate in a clockwise direction (in the northern hemisphere) round an area of high pressure, forming what he called an anticyclone. By good fortune, the weather over Europe changed from cyclonic to anticyclonic during December 1861, enabling him to make this discovery.

In the course of this work he developed a facility for interpreting maps and charts, in particular by drawing isobars, isotherms, and so on through points of equal barometric pressure, temperature, and other physical quantities. For example, in 1881 he read a paper on isochronic postal charts, in which he displayed maps of the world on which all places within ten days' journey of London were colored green, those between ten and twenty orange, between twenty and thirty red, between thirty and forty blue, and those beyond forty days brown. He was thus well prepared to discover the properties of the bivariate normal distribution by drawing ellipses through points of equal frequency of occurrence on a diagram of the joint frequency distribution of the heights of parents and offspring. He described this discovery in his presidential address to the Section of Anthropology in 1885 (see below).

Another meteorological contribution shows Galton's intuitive understanding of statistics. In 1866, he pointed out a statistical error in obtaining meteorological data at sea. These data were obtained by sailing ships, which spent less time in an area in which they found favorable winds than where they were becalmed or had unfavorable winds. Hence there was a bias toward reports of calms or head winds, for which allowance should be made.

In 1870, he read a paper in which he discussed the problem of predicting wind velocity from measurements on barometric pressure, temperature, and humidity from an formula that would now be called a linear multiple regression equation. Though his method of estimating the formula is very crude, it shows that he was accustomed through his meteorological work to think in statistical terms that he would later apply to problems in heredity.

Heredity and Evolution

Galton was deeply impressed by reading Darwin's *The Origin of Species* in 1859, which helped turn his interests toward heredity and evolution. His first paper in this area was a discussion at the British Association in 1864 of the domestication of animals, which is described in chapter 5. The brevity of papers read to the British Association made it an unsuitable place for Galton to present his more substantial contributions to heredity, but he devoted his presidential address to the Section of Anthropology in 1885 to his recent work on "Regression towards mediocrity in hereditary stature," which is considered in chapters 6 and 7. There was a discussion on the inheritance of acquired characters at the 1889 meeting, where Galton proposed some possible experiments to settle the question, but they were never implemented. He was from the first skeptical about the inheritance of acquired characters.

Psychology

Galton was a pioneer in analyzing mental processes by introspection. He had observed that some people had much stronger visual memory than others, and he devised a questionnaire to investigate the matter further. He discussed the results in a popular lecture at the Swansea meeting of the British Association in 1880. The questionnaire asked the respondent to think of the breakfast table as he or she had sat down to it that morning, and to consider the picture that rose to the mind's eye: Was the image dim or clear? Were the colors of the china, the toast, mustard, meat, parsley distinct and natural? and so on. The replies showed a large range in the vividness of mental imagery. In many people the faculty was highly developed: "I can see my breakfast-table or any equally familiar thing with my mind's eye, quite as well in all particulars as I can do if the reality is before me." Many others, in particular the great majority of scientists, were almost completely defective in the faculty, and often doubted its existence: "These questions presuppose assent to some sort of a proposition regarding the 'mind's eye,' and the 'images' which it sees.... This points to some initial fallacy.... It is only by a figure of speech that I can describe my recollection of a scene as a 'mental image' which I can 'see' with my 'mind's eye'." Galton remarked that these people had no more idea of the nature of mental imagery than a color-blind man, who has not discerned his defect, has of the nature of color.

He explained the absence of the faculty in most scientists by saying that the perception of sharp mental pictures is antagonistic to highly generalized and abstract thought, especially when carried on by verbal argument. As an experiment, he asked his acquaintances what idea the word "boat" called up, when he said to them, "I want to tell you about a boat," with this result: "One person, a young lady, said that she immediately saw the image of a rather large boat pushing off from the shore, and that it was full of ladies and gentlemen, the ladies being dressed in white and blue.... Another person, who was accustomed to philosophise, said that the word 'boat' had aroused no definite image, because he had held his mind in suspense. He had exerted himself not to lapse into any one of the special ideas that he felt the word boat was ready to call up, such as a skiff, wherry, barge, launch, punt, or dingy." This was precisely the sort of result that he had anticipated.

In his questionnaire, Galton encouraged his correspondents to describe other types of visual imagery, and so discovered the existence of number forms. Persons with strong visual imagination tend to see numbers; if the idea of *six* occurs to them, the figure 6 rises before their mental eye. In some people who visualize numbers, a particular number always appears in the same characteristic position in the visual field, so that the numbers 1, 2, 3, and so on form a pattern or number form. For example, the barrister George Bidder, son of a well-known rapid calculator, saw numbers in a form which began with the face of a clock, numbered I to XII, and then tailed off, much like the tail of a kite, into an undulating curve, having 20, 30, 40, and so on, at each bend. The physicist Professor Schuster saw numbers in the form of a horseshoe, with the open end toward him, with 0 at the bottom right, 50 at the top, and 100 at the bottom left.

Galton found that about 1 in 30 men and 1 in 15 women saw number forms. He concluded his talk to the British Association by saying, "Now, will every person in this large meeting who is conscious of seeing a Number-Form, hold up his hand?" There was a dead silence, since those who should have responded were too shy to do so, and not a hand was raised. He therefore told them a humorous story to overcome their shyness, and called on them again to put up their hands, at the same time naming one person on the platform (probably Professor Schuster) whom he knew to perceive number forms. The appeal succeeded, and a multitude of scattered hands went up all over the hall.

This account is based on Galton (1880; 1883, 57–105; 1908, 270–272) and Pearson (1924, chap. 11). See also Burbridge (1994).

Photography

In his presidential address to the section of anthropology in 1877, Galton discussed, among other topics, an investigation in progress about a possible correlation between physical and mental characteristics. Through the assistance of the Surveyor-General of Prisons, he had a large number of photographs of criminals classified into three groups: (1) murder, manslaughter, and burglary; (2) felony and forgery; (3) sexual crimes. By visual inspection he thought that the photographs could be sorted into certain natural classes, and that the three groups of criminals contributed in very different proportions to the different physiognomic classes. In other words, there seemed to be distinct physical types with differing propensities for different types of crime. To test this idea objectively, he developed a method for obtaining a composite portrait from photographs of several persons, by copying them on the same photographic plate, giving each of them a fraction of the full exposure. The results revealed little difference between the composite portraits of the three groups of criminals (Galton 1878).

Galton used composite portraiture in many other contexts. In 1881, he exhibited at the British Association a composite picture of eight skulls of male Andaman Islanders. In the same paper he referred to numerous composites illustrating the physiognomy of disease made in collaboration with Dr. Mahomed, a physician at Guy's Hospital. They photographed 442 patients with tuberculosis, with 200 patients suffering from other diseases as controls. Their conclusions were negative: "The results lend no countenance to the belief that any special type of face predominates among phthisical patients, nor to the generally entertained opinion that the narrow ovoid or 'tubercular' face is more common in phthisis than *among other diseases*. Whether it is more common than among the rest of the *healthy* population we cannot at present say" (Forrest 1974, 141).

Much later, in 1898, he presented composite portraits of race horses and proposed that the technique could be combined with his law of ancestral heredity (chapter 8). To predict the appearance of the offspring of a particular mating, he suggested that a composite should be made from photographs of the two parents and the four grandparents, with the parents receiving four times as many units of exposure as the grandparents, to reflect their relative weight in the ancestral law. He presented composites based on this principle, but the idea was not used by breeders, who were more interested in the speed than the appearance of the offspring.

Fingerprints

The use of fingerprints for the identification of criminals had been advocated in *Nature* in 1880 by both Henry Faulds and Sir William Herschel, but was put on a scientific basis only in Galton's book on *Finger Prints*, published in 1892 (Hauser 1993, Beavan 2002). Two papers to the British Association concerned fingerprints. The first, in 1893, reported that fingerprinting had recently been introduced into the Indian army to identify recruits. The second, more interesting paper, in 1899, concerned their persistence over time and the first age at which they can be identified. Galton stated that he had been approached by the police authorities in an unnamed country who had received information that a baby who was heir to a great title and estate might be kidnapped for the sake of extorting ransom. He was asked whether prints of the fingers of a baby would serve forever afterwards to identify him, and to prove that he was not a changeling. He had shown from a series of longitudinal studies in his book that fingerprints persisted unchanged from 30 months to 80 years old, but he had no information on younger children. An American lady volunteered to collect prints on her daughter from birth, and he had sets of prints on this girl at the following ages: fourteen days, one month, six months, seventeen months, two and a half years, and four and a half years. The last two sets were identical, but the first four sets were more difficult to deal with; this was partly because of the difficulty of taking the prints (it was only possible to make a mere dab of the finger rather than a rolling impression, even when the baby did not make a blur by closing her fists), and partly because the pattern was less distinct. However, after careful examination he had little doubt that the prints of all ten fingers of a baby would suffice for identification after examination by an expert. The correspondence between Galton and the American lady about the fingerprinting of her baby daughter survives (Pearson 1930b, 496–499), but there is no mention of its use in identifying a kidnapped baby; the latter may have been embroidery designed to add interest to his talk.

Characterization

Annan (1955) has drawn attention to an aristocracy of intellect that began to form at the beginning of the nineteenth century. A particular type of middle class family (usually prosperous, and often from an evangelical, Quaker, or Unitarian background) then started to intermarry and produced children who became scholars and teachers. When these sons came to marry, it was natural for them to choose a wife from a similar background, so that the same names recurred as professors and headmasters. Members of these intellectual families became the leaders of the new intelligentsia and of the professional civil service, criticizing the assumptions of the ruling class above them and forming the opinions of the upper middle class to which they belonged. They valued intellectual freedom, campaigned for entry to the professions to be open to talent, irrespective of religious belief or class, and represented many of the progressive forces of Victorian society.

Francis Galton was a typical member of this intellectual elite. Both his Galton and his Darwin ancestry provided predisposing factors for him to join it, and his cousin Charles Darwin a positive exemplar, though it should be noted that both his brothers became country gentlemen. He deliberately married into another intellectual family, the Butlers. He wrote of

> the far greater importance of being married into a family that is good in character, in health, and in ability, than into one that is either very wealthy or very noble, but lacks these primary qualifications.... I protest against the opinions of those sentimental people who think that marriage concerns only the two principals; it has in reality the wider effect of an alliance between each of them and a new family. (Galton 1908, 158)

It is sad that he and Louisa had no children to continue the line.

Galton had intellectual ambitions from his youth; when he was four he announced that he was saving his pennies to buy honors at university, and he arranged to be nominated for the Athenaeum when he was eighteen (Pearson 1914, 114 and 145). He settled down to pursue the career of an independent scientist in London in his early thirties when he had satisfied his urge to travel and when his financial position and his reputation as a geographer allowed him to do so. From that time, he found his identity as a member of the scientific elite, reading and socializing in the Athenaeum, allying himself with the progressive wing in the Royal Geographical Society, and assiduously attending the annual meetings of the British Association, the "Parliament of Science," which was much more important in the nineteenth century than it is now. His identification with this progressive scientific elite contrasts with his social and political conservatism.

Galton shared most of the opinions and characteristics of the intellectual aristocracy described by Annan. "Their Glorious Revolution was achieved in 1870–1 when entry to public service by privilege, purchase of army commissions and the religious tests were finally abolished. Then it was ordained that men of good intellect should prosper through open competitive examination" (Annan 1955, 247). Galton emphasized competitive examinations in his eugenic proposals, and at the British Association discussed the assignment of marks for bodily efficiency in choosing candidates for the public services (see table 1.1). "Most of them . . . had become town-dwellers. But they had not lost touch with Nature whom they sought mountaineering in the Alps or on forty-mile tramps" (249). Galton's enthusiasm for climbing in the Alps and the Pyrenees has already been described. "Their experience of the visual arts was meagre" (251). Galton displayed some talent for sketching in his youth (see fig. 1.4), but in his old age he relieved the boredom of sitting for his portrait by counting the number of brush-strokes made by the painter. Discussing the results in a letter to *Nature* in 1905, he wondered whether painters had mastered the art of getting the maximum result from their labor, but he admitted making this remark as a confessed Philistine. "Beautiful objects and elegant rooms were not to them necessities: their comfortable ugly houses, in Kensington, Bayswater and North Oxford, rambling, untidy, full of glory-holes and massive furnishings and staffed by two or three despairing servants, were dedicated to utility, not beauty" (251). When he took 42 Rutland Gate, South Kensington, partly furnished, Galton carefully designed additional furniture, but his interest reflected his practicality rather than his aestheticism. For example, the lavatory was situated on the half-landing, and he had one of the wooden door panels replaced by frosted glass and fixed a vertical rod to the bolt. This rod could be seen from the bottom of the stairs when the lavatory was occupied and the bolt across, thus obviating a futile climb and the disturbance of the occupant (Forrest 1974).

Fancher (2001) suggests that Galton's abandonment of his orthodox religious faith in the 1860s was due as much to his interaction with scientists such as T. H. Huxley, Herbert Spencer, G. H. Lewes, John Tyndall, and John Lubbock as to reading *The Origin of Species*. These men, with whom Galton interacted closely in the 1860s (for example, in founding the weekly journal *The Reader*, the precursor of *Nature*), were leading figures in promoting the doctrine of scientific naturalism, that supernatural modes of explanation must be rejected in favor of explanations based on empirical science. This led them to reject orthodox Christianity, though they regarded themselves as profoundly religious in a broad sense. Galton's rejection of Christianity dates to the time of his association with these scientific naturalists. In 1860, the third edition of *The Art of Travel* contained a passage speculating about how Adam, Eve, and Cain might have learned to make fire, showing his acceptance of the literal interpretation of the Bible at that time, but this passage was deleted from the fourth edition in 1867. *Hereditary Genius*, published in 1869, attacked orthodox Christianity but attributed religious significance to the Darwinian worldview in rather extravagant language: "We may look upon each individual as something not wholly detached from its parent source,—as a wave that has been lifted and shaped by normal conditions in an unknown, illimitable ocean. There is decidedly a solidarity as well as a separateness in all human, and probably in all lives whatsoever; and this consideration goes far, as I think, to establish an opinion that the constitution of the living Universe is a pure theism, and that its form of activity is what may be described as cooperative" (Galton 1869, 376). From these sentiments he developed his eugenic program as a kind of secular religion (chapter 3).

In assessing Galton's character, it may be useful to see him from the viewpoint of his opponents. We have already quoted the opinion of Clements Markham. Another opponent was Richard Burton, whom Galton characterized with obvious disapproval as "a man of eccentric genius and tastes, orientalised in character and thoroughly Bohemian" (1908, 199). Burton wrote politely of Galton in his published works, but his real opinion is revealed in a private letter written in 1872: "The RGS has as usual put its foot into the wrong hole but what can you expect of a body which owns as one of its heads Mr Galton? The creature is Grundy, knows Grundy and owes all his strength to Grundy. He hates with a harsh and frustrated (almost Xstian) hatred all who take the position that ought to have been, but has not been, taken by himself—Galton. For years he inflicted his corvine voice upon every meeting simply for the same petty vanity" (Lovell 1998, 828). Grundy is the personification of conventional propriety. Burton's assessment contains some truth, if one ignores his characteristic exaggeration and his contempt for those who disagreed with him; he was defending his belief that the Nile rose from Lake Tanganyika.

Galton was socially and politically conservative, with little sympathy for liberal causes. He wrote of the Chartist disturbance in 1839 during his year at the Birmingham General Hospital: "Some riots connected with the 'Charter' occurred at this time, and many people were hurt. It was curious to observe the apparent cleanness of the cuts that were made through the scalp by the blow of a policeman's round truncheon" (1908, 30–31). This comment reveals a detached intellectual interest, a tendency to see people as cases to be studied or examples to be counted rather than as fellow humans to be sympathized with, which remained part of his character (Fancher 1985, MacKenzie 1981). In similar vein, he showed no emotional reaction when he visited the women's slave market in Constantinople in 1840, in contrast to Charles Darwin's strong antipathy to slavery. Fancher (1998) suggests that Galton's emotional detachment was a defense mechanism adopted, at least in part, to cope with his experience of suffering at the Birmingham General Hospital, but it can just as plausibly be attributed to nature rather than to nurture.

His political views were formed in his youth. While on vacation from Cambridge in 1841 he wrote to his father: "I had a very entertaining fellow-traveller; he had a hooked nose, gold spectacles, was a member of the Reform Club, and a *ne plus ultra* radical.... We had a red hot argument on politics, which I firmly believe neither of us knew anything about, but he would talk about them, and as I must answer yes or no, even Bessy will excuse my not assenting to a radical's views" (Pearson 1914, 155). In later life he tended, more often than not, to vote Conservative (Cowan 1977, quoting his nephew Frank Butler). Toward the end of his life he joined the Committee of the Anti-Suffrage League, which distressed the women working in the Eugenics Laboratory which he was supporting financially (Pearson 1930a, 359).

Galton's conservative views can also be inferred from a comment in a letter to George Darwin in 1886: "Josephine Butler joined her husband at [Positano] last night.—Well, well! one can't talk to her about her favourite topics, holding as I do most diametrically opposed views in nearly every particular of faith, morals, and justifiable courses of action; but for all that she is, or was, very charming and keenly alive and sympathetic" (Pearson 1930b, 475). Josephine Butler was the daughter of John Grey, a strong advocate of social reform who played a significant role in the campaigns for the 1832 Reform Act, the repeal of the Corn Laws, and the abolition of slavery. She shared her father's views and, after her marriage to Louisa's brother George Butler, she campaigned for the provision of higher education for women and for women's suffrage. She was best known for her successful campaign for the repeal of the Contagious Diseases Acts, which attempted to reduce venereal disease in the armed forces by allowing the police in garrison towns and seaports to arrest women they believed to be prostitutes and to require them to have a medical examination. Despite his opposition to her views, Galton had a friendly social relationship with his sister-in-law (Jordan 2001).

Galton's conventionality and his dislike of emotionalism may have been reinforced by his shyness. Though outwardly reserved, he showed the real warmth of his personality among his circle of close friends and relatives. One of his great-nieces wrote: "I expect we all see our friends differently; if I were to write a memoir of Uncle Frank I should just say what a pet he was, and how good tempered and full of delightful naive sayings, and that everyone wanted to kiss him! I should not bother about his intellect, which did not come my way" (Pearson 1914, 26). George Darwin's daughter, Gwen Raverat, who was no respecter of persons, has this account in her charming book of childhood recollections, *Period Piece*:

> We must have seen a good many Great Men in our youth, but most of them seemed to me very uninteresting. There was Lord Kelvin; he looked very fine, but he seemed to be always absorbed in his own thoughts, and never opened his mouth, except once.... But Francis Galton was both pleasant and impressive, with his bushy, twitching eyebrows. We went to his house once to have our fingerprints taken for some experiment in the classification of fingerprints, on which he was working. He did not provide us with any means of washing off the printers' ink, and we had to go about all day in London with sticky black hands. (Raverat 1952, 273–274)

A lengthier pen-portrait is extracted from an appreciation of Galton by his niece Millicent Lethbridge (daughter of his sister Adèle):

> I have an amusing recollection of a little trip to Auvergne which he and I took together in the summer of 1904.... The heat was terrific, and I felt utterly exhausted, but seeing him perfectly brisk and full of energy in spite of his 82 years, dared not, for very shame, confess to my miserable condition. I recollect one terrible train-journey, when, smothered with dust and panting with heat, I had to bear his reproachful looks for drawing a curtain forward to ward off a little of the blazing sun in which he was revelling. He drew out a small thermometer which registered 94°, observing: "Yes, only 94°. Are you aware that when the temperature of the air exceeds that of blood-heat, it is apt to be trying?" I could quite believe it!—By and by he asked me whether it would not be pleasant to wash our face and hands? I certainly thought so, but did not see how it was to be done. Then, with

> perfect simplicity and sublime disregard of appearances and of the astounded looks of the other occupants of our compartment, a very much "got-up" Frenchman and two fashionably dressed Frenchwomen, he proceeded to twist his newspaper into the shape of a washhand-basin, produced an infinitesimally small bit of soap, and poured some water out of a medicine bottle, and we performed our ablutions—I fear I was too self-conscious to enjoy the proceeding, but it never seemed to occur to him that he was doing anything unusual!
>
> He had ordered rooms at Royat [where he had gone to visit his wife's grave], insisting that they should have a southern aspect. On arriving at the Hotel it was found that they looked due north. Then, for the first and only time since I had known him, he was guilty of a very forcible and by no means parliamentary expression. A minute or two later he turned round and saw me. He appeared exceedingly uncomfortable, and at last could stand it no longer: "Er—er—did you hear what—er—I said just now?" I could not resist the temptation of declaring myself extremely pained and shocked, but he was so genuinely distressed I had to hasten and assure him I was only talking nonsense. (Pearson 1930b, 447)

The year after this trip Galton sent Millicent Lethbridge a letter while on holiday for the last time in Biarritz, including a riddle that illustrates his typically Victorian sense of humor: "'Explain the relationship between (1) a gardener, (2) a billiard-player, (3) an actor, (4) a verger.' The gardener attends to his p's (peas), the billiard player to his q's (cues), the actor to his p's and q's, and the verger to his keys and pews" (Pearson 1930b, 555). He posed an even worse riddle in a letter to his father in 1844: "*Riddle.* If a man wants to obtain a vegetable time piece at what hour should he rise? *Answer.* He must get up at eight o'clock (must get a potato clock)" (Pearson 1914, 188).

2

Hereditary Ability

> The publication in 1859 of the *Origin of Species* by Charles Darwin made a marked epoch in my own mental development, as it did in that of human thought generally.... I was encouraged by the new views to pursue many enquiries which had long interested me, and which clustered round the central topics of Heredity and the possible improvement of the Human Race.
>
> Galton, *Memories of My Life*

The Origin of Species marked a turning point in Galton's intellectual life. After its publication in 1859, he wrote to Darwin: "Pray let me add a word of congratulation on the completion of your wonderful volume.... I have laid it down in the full enjoyment of a feeling that one rarely experiences after boyish days, of having been initiated into an entirely new province of knowledge which, nevertheless, connects itself with other things in a thousand ways" (Pearson 1924, plate 18). When Darwin congratulated Galton on the publication of *Hereditary Genius* in 1869, Galton replied: "It would be idle to speak of the delight your letter has given me.... I always think of you in the same way as converts from barbarism think of the teacher who first relieved them from the intolerable burden of superstition. I used to be wretched under the weight of the old-fashioned arguments from design, of which I felt, though I was unable to prove to myself, the worthlessness. Consequently the appearance of your *Origin of Species* formed a real crisis in my life; your book drove away the constraint of my old superstition as if it had been a nightmare and was the first to give me freedom of thought" (Pearson 1914, plate 2). The quotation at the beginning of this chapter shows Galton's mature view of its influence on him. He was probably also influenced by interacting with the scientific naturalists in the 1860s (Fancher 2001), as discussed in chapter 1.

In *The Origin of Species*, Darwin showed that species became adapted to their environment through natural selection acting on heritable variations. In chap. 1, on "Variation under Domestication," he drew a direct analogy with the effect of artificial selection by man in producing improved varieties of domestic species. He did not, at this time, extend his conclusions to man except for a brief passage near the end of the book: "In the distant future I see open fields for far more important researches. Psychology will be based on a new foundation, that of the necessary acquirement of each mental power and capacity by gradation. Light will be thrown on the origin of man and his history" (Darwin 1859, 488).

Galton had little interest in natural history or in the evolution of plants and animals, but he focused on the implications of Darwin's theory for man. It seemed likely that human mental abilities had evolved through natural selection, in which case they must show heritable variation and would be susceptible to artificial selection. This was the source of Galton's interest in "heredity and the possible improvement of the human race." He published his first paper on heredity in 1865, which he expanded in 1869 into his book on *Hereditary Genius*. (He stated in the second edition in 1892 that he regretted not having called it *Hereditary Ability* since the original title was apt to mislead [1892a, viii–ix].) He described in the preface how he had come to work on this topic:

> The idea of investigating the subject of hereditary genius occurred to me during the course of a purely ethnological inquiry, into the mental peculiarities of different races; when the fact, that characteristics cling to families, was so frequently forced on my notice as to induce me to pay especial attention to that branch of the subject. I began by thinking over the dispositions and achievements of my contemporaries at school, at college, and in after life, and was surprised to find how frequently ability seemed to go by descent. Then I made a cursory examination into the kindred of about four hundred illustrious men of all periods of history, and the results were such, in my own opinion, as completely to establish the theory that genius was hereditary, under limitations that required to be investigated. Thereupon I set to work to gather a large amount of carefully selected biographical data. (Galton 1869, v)

(The "four hundred illustrious men" are presumably the 605 notable persons discussed in "Hereditary talent and character.")

In these investigations he tried to demonstrate the inheritance of ability by showing that close relatives of distinguished men were themselves much more likely to be distinguished than average members of their class. The main weakness of this statistical approach is that the effects of inheritance and family background, or in Galton's words of nature and nurture, may be confounded. He tried to distinguish between these factors in several ways, culminating in his study of the life history changes of twins, from which he concluded that the effect of nurture was very weak compared with that of nature.

Though he was fully aware of the distinction between nature and nurture and tried to distinguish between them, Galton showed a hereditarian bias in interpreting his results and tended to exaggerate the importance of nature. Nevertheless, his work laid the foundations of the study of human inheritance by statistical methods. He justifiably claimed, in the introduction to *Hereditary Genius*, to be the first person to treat the subject in a statistical way.

"Hereditary Talent and Character" (1865)

> The power of man over animal life, in producing whatever varieties of form he pleases, is enormously great. It would seem as though the physical structure of future generations was almost as plastic as clay, under the control of the breeder's will. It is my desire to show, more pointedly than—so far as I am aware—has been attempted before, that mental qualities are equally under control. (Galton 1865a, 157)

Galton began his first paper on heredity with the above words that owe an obvious debt to the discussion of selective breeding in the first chapter of *The Origin of Species*. He continued that he could not show directly the success of breeding for mental qualities since no one had tried to select for general intelligence in animals, but that he could show that "talent and peculiarities of character are found in the children, when they have existed in either of the parents, to an extent beyond all question greater than in the children of ordinary persons" (1865a, 158).

To show this, he first examined a reference book which listed 605 notable persons with original minds who lived between 1453 and 1853, of whom 102, or 1 in 6, had at least one close relative in the list. For example, the list included John Adams, president of the United States, his son Samuel Adams, and his nephew John Quincy Adams, president, the family contributing 3 individuals to the total of 102 notable individuals with at least one notable close relative.

Galton recognized that in some professions, particularly statesmanship and generalship, the son of an eminent father will be in a more favorable social position for advancement than the son of an ordinary person, so that the figure of 1 in 6 may exaggerate the effect of heredity. To counter this objection, he argued that science and literature were more open fields in which social position offers no favor beyond the advantage of a good education. There were 330 literary or scientific people in the list, and he calculated that there must have been more than a million students educated in Europe in the previous four centuries, so that the chance of an educated person achieving literary or scientific distinction was less than 1 in 3000. Yet, in the sublist of 330 distinguished literary people, 51, or 1 in 6.5 had a distinguished literary relative.

Galton examined other biographical lists with similar results (table 2.1). He concluded that "intellectual capacity is so largely transmitted by descent that, out of every hundred sons of men distinguished in the open professions, no less than eight are found to have rivalled their fathers in eminence" (318). (To obtain the figure of 8 per cent, he has implicitly assumed that each distinguished man had, on average, one son. Sweeney

Table 2.1. Frequency with Which Distinguished Men Have a Distinguished Relative

Type of distinction	Distinguished near relative	Distinguished son	Distinguished brother
Men of original minds[a]	17%	6%	2%
Living notabilities[b]	29%	7%	7%
Painters of all dates[c]	17%	5%	4%
Musicians[d]	10%	6%	3%
Lord chancellors[e]	33%	16%	4%
Senior classics of Cambridge[f]	29%	Too recent	10%
Average	16%	8%	5%

Source: After Galton 1865a

[a] Sir Thomas Phillips, *The Million of Facts;* 605 cases between 1453 and 1853.
[b] Walford, *Men of the Time,* letter A; 85 cases.
[c] Bryan, *Dictionary of Painters,* letter A; 391 cases.
[d] Fétis, *Biographie Universelle des Musiciens,* letter A; 515 cases.
[e] Lord Campbell, *Lives of the Chancellors;* 54 cases.
[f] 41 cases.

[2001] has accused Galton of gross carelessness and/or dishonesty in his use of *Men of the Time*, but his comments depend on a misidentification of the edition used by Galton [David Burbridge, personal communication].) He gave three reasons that the son of a very distinguished father is likely to be less distinguished than him. First, great success implies the simultaneous inheritance of many qualities (good health, industry, ambition) in addition to intellectual capacity. Second, the proportion arrived at ignores one-half of the hereditary influences on the child since it takes no account of the mother. Third, even though both parents were very distinguished, their sons would tend to revert to more remote ancestors, who would on average be less distinguished. He concluded that, in view of these impediments, "eight per cent is as large a proportion as could have been expected on the most stringent hypothesis of hereditary transmission" (319). The second and third reasons introduce two principles, biparental inheritance and reversion, which run through his work on heredity; they are discussed in chapter 4.

Hereditary Genius (1869)

Galton consolidated and expanded these results in *Hereditary Genius* for many groups of professional people. To justify his methodology, he began by comparing the classification of men according to their reputation and according to their natural gifts. He was about to consider groups of men with an eminent reputation in their profession, and to ask how many eminent relatives they had. He wanted to interpret the results as evidence of the inheritance of ability, which would only be justified if reputation were a good surrogate measure of ability. He therefore asked: "Is reputation a fair test of natural ability? It is the only one I can employ—am I justified in using it? How much of a man's success is due to his opportunities, how much to his natural power of intellect?" (1869, 37). He explained that by reputation he meant the reputation of a leader of opinion, of an originator, disregarding high social or official position; and that by natural ability he meant "those qualities of intellect and disposition, which urge and qualify a man to perform acts that lead to reputation."

Under this definition, Galton believed that "if the 'eminent' men of any period had been changelings when babies, a very fair proportion of those who survived and retained their health up to fifty years of age, would, notwithstanding their altered circumstances, have equally risen to eminence" (38). He applied this fostering test of the heredity of eminence by considering the adopted sons of Popes, who were in fact their nephews or more remote relatives. He argued that, if social help were really of the highest importance, the nephews of the Popes would attain eminence as frequently as the sons of other equally eminent men; otherwise they would not. The evidence was that "individuals whose advancement has been due to nepotism are curiously undistinguished. The very common combination of an able son and an eminent parent, is not matched, in the case of high Romish ecclesiastics, by an eminent nephew and an eminent uncle. The social helps are the same, but hereditary gifts are wanting in the latter case" (42). But he did not provide statistical evidence to back this claim.

He added three arguments in respect to literary and artistic eminence. First, many men of high ability but humble rank rise to eminence; the mathematician D'Alembert is quoted as an example. ("He was a foundling (afterwards shown to be well bred as respects ability), and put out to nurse as a pauper baby, to the wife of a poor glazier. The child's indomitable tendency to the higher studies, could not be repressed by his foster-mother's ridicule and dissuasion, nor by the taunts of his schoolfellows, nor by the discouragements of his schoolmaster, who was incapable of appreciating him" [43–44].) Second, in countries such as America where there were fewer social hindrances than in England to a poor man rising in life, there was a much larger proportion of persons of culture, but not of eminent men. Third, socially advantaged men are unable to achieve eminence unless they are endowed with high natural gifts; men belonging to great county families might become influential members of Parliament, but when they died, "there is no Westminster Abbey and no public mourning for them—perhaps barely a biographical notice in the columns of the daily papers" (41). However, he accepted that large qualifications were required in applying these arguments to statesmen and commanders, and that only the most illustrious men in these groups should be considered to have high natural ability. With this proviso he concluded that "no man can achieve a very high reputation without being gifted with very high abilities; and I trust I have shown reason to believe, that few who possess these very high abilities can fail in achieving eminence" (49).

Having satisfied himself that reputation was a satisfactory measure of ability, he examined in great detail the family histories of English judges, followed in less detail by those of other professional people, to determine whether ability was hereditary.

English Judges

Instead of taking each judge in turn and counting his eminent relatives, as he had done in "Hereditary talent and character," Galton chose to do his calculations through families. There were 286 judges, who fell into 262 families. He first determined the most eminent member of each of these families, and counted the numbers of fathers, brothers, and so on, of each of them who were eminent. The results are shown in column A of table 2.2.

He also estimated (see the second column, headed C) the average total number of relatives of different types, so that he could express the number of distinguished relatives of each type as a percentage of the total number of relatives of that type. He found, from very crude data, that each judge had on average one adult son and one daughter, and that adult families consisted on average of 2.5 sons and 2.5 daughters. Consequently, Galton argued, each judge has one father, and on average one

Table 2.2. Eminent Relatives of Judges

Relationship	A	C	E
Father	22	1	8
Brother	30	1.5[a]	8[a]
Son	31	1	12
Grandfather	13	2	2
Uncle	15	4	1
Nephew	16	4	2
Grandson	16	2	3
Great-grandfather	2	4	0.2
Great-uncle	3	8	0.1
First-cousin	9	8	0.4
Great-nephew	15	8	0.7
Great-grandson	5	4	0.5

Source: After Galton 1869

Note: A = number of eminent relatives of this type of the most eminent member of each of the families; C = estimated number of relatives of this type per person; E = percentage chance that this type of relative of the most eminent member of a family with at least one judge will himself be eminent.

[a] Galton later estimated that C = 2 for brothers with a consequent reduction of E from 8 to 6 (see appendix).

son, 1.5 brothers and 2.5 sisters; the nephews consist of the brothers' sons and the sisters' sons, comprising 1.5 + 2.5 = 4 nephews, and so on. These estimated figures are clearly crude and unreliable, but they are accepted for the want of better. Galton later attempted to improve these estimated numbers of relatives of different types, and he concluded that an adult male has on average 2 adult brothers and 2 sisters, rather than 1.5 and 2.5. The other numbers in column C are approximately correct, with the proviso that, for example, the 4 nephews are composed of 2 brothers' sons and 2 sisters' sons. (See the appendix to this chapter for further discussion.)

He then calculated three numbers from these data:

$$\begin{aligned} B &= 100A/F^* \\ D &= 100A/(C \times F^*) \\ E &= 100A/(C \times F), \end{aligned} \qquad [1]$$

where F is the total number of families and F* is the number of families with two or more eminent members.

B is the number of eminent fathers, brothers, sons of the most eminent member of 100 families containing at least one judge and another eminent member. It has little intrinsic significance, but it is useful for comparing results for different professions.

D is the percentage chance that a father, brother, son of the most eminent member of a family containing at least one judge and another eminent member is himself eminent.

E has been standardized to the total number of families, including those with no eminent members other than the judge through whom it entered the sample. It is the percentage chance that a father, brother, son of the most eminent member of a family containing at least one judge, is himself eminent.

Galton laid considerable emphasis on B and D, but E has a more meaningful interpretation. It is shown in the third column in table 2.2. (It seems more natural today to use the direct method of taking each of the judges in turn and counting their eminent fathers, brothers, sons, and so on, say A′, and then calculating E′ = 100 A′/(C x *n*), where *n* (= 286) is the total number of judges. E′ is the percentage chance that a father, brother, son, etc., of a judge is eminent. I have calculated E′ from the original data for judges in *Hereditary Genius*; the results are similar to E. Galton does not explain why he chose to do his calculations through families;

perhaps he thought that the direct method would give rise to double counting.)

Galton reasoned that the results in table 2.2 provided strong evidence of the inheritance of ability. He argued that a judgeship is a guarantee of its possessor being exceptionally gifted, and that evidence of inheritance is provided both by the high frequency of eminent first-degree relatives in column E, and by the sharp fall in this frequency in second- and third-degree relatives; if ability were randomly distributed all relatives would have the same chance of eminence, so that these frequencies would be constant.

He also noted that among the families with two or more eminent members, 39 had two, 32 had three or four, and 15 had five or more such members. He argued that the large numbers of families with several eminent members was clear evidence that ability is not distributed at haphazard, but that it clings to certain families.

Finally, he noted that the highest, and presumably most able, legal officers, the Lord Chancellors, had a higher percentage of eminent relatives than the rest of the judges, and he rejected the view that this was due to their having more opportunities of thrusting their relatives into eminence, by jobbery, than other judges. For all these reasons, he concluded that the data provided strong evidence of the inheritance of ability. The relatives of judges were often themselves judges, showing inheritance of the peculiar type of ability required in a judge, but there were also bishops and archbishops, poets, novelists, physicians, admirals, and generals.

At the end of this section, he gave data on the birth order of 72 judges which had been recorded in his notes. The judge was an only son in 11 percent of the cases, the eldest in 17, second in 38, third in 22, and fourth or later in 12 percent of the cases. Comparing the figure of 17 percent for eldest sons with 38 percent for second sons, he concluded that the eldest sons did not succeed as judges half as well as the cadets, presumably because social influences were against their entering, or against their succeeding at the law. Thus he accepted the effect of some social factors on success at the law, but the main thrust of his argument was that such success was predominantly due to inherited ability.

Comparison of Results for All Professions

Galton obtained similar data from different biographical sources for statesmen, commanders, literary men, scientists, poets, musicians, painters, and divines. He treated them in less detail than the judges, though he gave brief details of all the men in the samples and their eminent relatives. He took care for the statesmen and commanders to select only men of great eminence who had obtained their positions through ability rather than influence.

For the 196 evangelical divines extracted from Middleton's *Biographica Evangelica*, he first examined whether statistical evidence supported the idea that fertility, material well-being, health, or longevity were dependent on godliness, as suggested by religious writers and in the Psalms. He concluded that they were not. He conducted this inquiry because "if an exceptional providence protects the families of godly men, it is a fact that we must take into account. Natural gifts would then have to be conceived as due, in a high and probably measurable degree, to ancestral piety, and in a much lower degree than I might otherwise have been inclined to suppose, to ancestral natural peculiarities" (158). This analysis was the forerunner of his "Statistical inquiries into the efficacy of prayer" (1872a), in which he showed that royal persons, who are frequently prayed for, and clergymen, who belong to a prayerful class, have the same longevity as lawyers and medical men, who are much less prayerful. Galton's lack of sympathy for religion ensured his book a hostile reception in the religious press.

After presenting the data, he compared all the results for the statistic B, defined in eq.[1], which shows the numbers of eminent men related in different ways to the most eminent man of each family, expressed as a percentage of the number of families with at least two eminent members. These numbers, shown in table 2.3, have little meaning in themselves, but they are comparable with one another.

Galton drew attention to the general uniformity of the data among the kinsmen in the different groups, which he thought demonstrated the existence of a law of distribution of ability in families. This uniformity makes it meaningful to calculate the average value of B, and then to divide this average value by C, the estimated number of kinsfolk, to estimate D in the last column of table 2.3. The main feature is a sudden dropping off of the numbers in going from first-degree to second-degree relatives, which is conspicuous in column D, and again in going from second- to third-degree relatives, which he took as evidence of inheritance. He attributed the excess of eminent sons over eminent fathers to assortative mating: "Able men take pleasure in the society of intelligent women, and, if they can find such as would in other respects be suitable, they will marry them in preference to mediocrities" (326).

Table 2.3. Eminent Relatives of Different Groups (B)

Relationship	Jud	St	Com	Lit	Sc	Po	Art	Div	Average B	Average D
Father	26	33	47	48	26	20	32	28	31	31
Brother	35	39	50	42	47	40	50	36	41	27
Son	36	49	31	51	60	45	89	40	48	48
Grandfather	15	28	16	24	14	5	7	20	17	8
Uncle	18	18	8	24	16	5	14	40	18	5
Nephew	19	18	35	24	23	50	18	4	22	5
Grandson	19	10	12	9	14	5	18	16	14	7
Great-grandfather	2	8	8	3	0	0	0	4	3	1
Great-uncle	4	5	8	6	5	5	7	4	5	1
First-cousin	11	21	20	18	16	0	1	8	13	2
Great-nephew	17	5	8	6	16	10	0	0	10	1
Great-grandson	6	0	0	3	7	0	0	0	3	1

Source: After Galton 1869

Note: Jud = judges; St = statesmen; Com = commanders; Lit = literary men; Sc = scientists; Po = poets; Art = artists (painters and musicians); Div = divines. B and D are defined in eq.[1] and explained in the text.

He also tried to explain a few exceptions to these rules. He explained the small number of eminent sons of commanders by suggesting that commanders had small families because "they usually begin their active careers in youth, and therefore, if married at all, they are mostly away from their wives on military service" (319). He attributed the very large number of eminent sons of scientists to a favorable family environment: "It is, I believe, owing to the favourable conditions of their early training, that an unusually large proportion of the sons of the most gifted men of science become distinguished in the same career. They have been nurtured in an atmosphere of free enquiry, and observing as they grow older that myriads of problems lie on every side of them, simply waiting for some moderately capable person to take the trouble of engaging in their solution, they throw themselves with ardour into a field of labour so singularly tempting" (197). He accounted for the large number of eminent sons of artists in the same way: "The remarks I made about the descendant of a great scientific man prospering in science, more than his ancestor, are eminently true as regards Artists, for the fairly-gifted son of a great painter or musician is far more likely to become a professional celebrity, than another who has equal natural ability, but is not especially educated for professional life" (320–321). However, he did not use this opportunity to discuss the role of family environment, but went on: "The large number of artists' sons who have become eminent, testifies to the strongly hereditary character of their peculiar ability" (321).

Galton now proceeded to work out what he called the final and most important result: if nothing else is known about a person than that he is a father, brother, son, or other relation of an illustrious man, what is the chance that he is or will be eminent? He pointed out that the answer to this question was given for judges by column E in table 2.2, and that it remained to discover what it was for illustrious men generally. From eq.[1], E can be calculated as D x (F*/F). Galton found that the average value of F*/F for all the groups in table 2.3 was 0.5, so that E could be calculated by halving the values of D in the last column. However, this calculation conceals the very wide variability in the value of F*/F between different groups. It is 0.32 for judges, 0.70 for scientists, and 0.13 for divines; in other words, most of the scientists and few of the divines had an eminent relative. A more complete picture is revealed by calculating E separately for each group, showing a high chance of eminence

Table 2.4. E = (B/C) x (F/F) for Different Groups*

Relationship	Judges	Scientists	Divines
Father	8	18	4
Brother[a]	8	22	3
Son	12	43	5
Grandfather	2	5	1
Uncle	1	3	1
Nephew	2	4	0
Grandson	3	5	1
Great-grandfather	0.2	0.0	0.1
Great-uncle	0.1	0.4	0.1
First cousin	0.4	1.4	0.1
Great-nephew	0.7	1.4	0.0
Great-grandson	0.5	1.2	0.0

Note: Calculated from Galton's data

[a] If C = 2, E is reduced by 25 percent.

among relatives of scientists, compared with judges, and a low chance among those of divines (table 2.4). The high chance of eminence among relatives of scientists may reflect the rather low standard of eminence he adopted in determining them.

Transmission through Male and Female Lines

Finally, Galton contrasted the power of the male and female lines of kinship in transmitting ability. Table 2.5 shows the numbers of eminent second-degree relatives in the male and female lines for the different groups. The lines are defined as:

Male line = G + U + N + P, where G = father's father, U = brother's father, N = brother's son, and P = son's son;
Female line = g + u + n + p, where g = mother's father, u = sister's father, n = sister's son, and p = daughter's son.

It will be seen that the ratio of transmission through male and female lines is nearly constant at about 2:1 in the first five groups (Judges, Statesmen, Commanders, Literary Men, and Scientists). The figures for the different types of relative for these five groups combined were:

21 G	23 U	40 N	26 P	=	110 in all
21 g	16 u	10 n	6 p	=	53 in all.

Galton's first idea was that this reflected an ascertainment bias: "The relative smallness of the numbers in the lower line appears only in those kinships which are most difficult to trace through female descent, and the apparent inferiority is in exact proportion to that difficulty. Thus the parentage of a man's mother is invariably stated in his biography; consequently, an eminent g is no less likely to be overlooked than a G; but a u

Table 2.5. Number of Eminent Second-Degree Relatives of Different Groups through the Male and Female Lines

	Jud	St	Com	Lit	Sc	Po	Art	Div	Total
Male line	41	19	12	18	20	12	13	4	139
Female line	19	10	6	9	9	1	3	16	73

Source: After Galton 1869

Note: See text for definition of male and female lines; see table 2.3 for the column headings.

is more likely to be overlooked than a U, and an n and a p much more likely than an N and P" (327–328). However, he rejected this explanation "because the differences appear to be as great in the well-known families of the Statesmen and Commanders, as in the obscure ones of the Literary and Scientific men" (328). His later explanation was that the aunts, sisters, and daughters of eminent men do not marry, on the average, so frequently as other women due to what may be called the bluestocking effect; they might be more selective in their choice of husband because they were accustomed to a higher form of culture in their family circle, or they might be less attractive to men because of belonging to a dogmatic and self-assertive type or because of having shy, odd manners. He pointed out that this model also accounts for g being as large as G, because every man has one maternal and one paternal grandfather, but he admitted that he must leave the question unresolved in the absence of hard information about the numbers of kinsfolk. His first explanation of an ascertainment bias seems to me more plausible.

The apparent weakness of transmission through the female line is even more marked in the case of poets and artists, and he thought the idea of a bluestocking effect would be even more appropriate in these groups. Among the divines, however, the situation is reversed, for which he suggests two reasons. First, the bluestocking effect does not operate since their female relatives "consider intellectual ability and a cultured mind of small importance compared with pious professions, and religious society is particularly large; . . . therefore the necessity of choosing a pious husband is no material hindrance to the marriage of a near female relation of an eminent divine" (329). But this does not explain the reversal of the roles of the male and female lines. To account for this, Galton postulated a real biological cause:

> The female line has an unusually large effect in qualifying a man to become eminent in the religious world.... It requires unusual qualifications, and some of them of a feminine cast, to become a leading theologian. A man must not only have appropriate abilities, and zeal, and power of work, but the postulates of the creed that he professes must be so firmly ingrained into his mind, as to be the equivalent of axioms. The diversities of creeds held by earnest, good, and conscientious men, show to a candid looker-on, that there can be no certainty as to any point on which many of such men think differently. But a divine must not accept this view; he must be convinced of the absolute security of the groundwork of his peculiar faith,—a blind conviction which can best be obtained through maternal teachings in the years of childhood. (Galton 1869, 276)

This would today be called a heritable maternal effect, a characteristic in the mother which has a direct effect on her sons, and which is inherited by her daughters, thus causing the predominance of the female line in leading to prominence in the religious world.

Galton also thought that there was a strong maternal influence on scientists, due to the antithesis of the maternal characteristic shaping divines. He noted that many scientists had distinguished mothers, aunts or grandmothers, from which he concluded:

> It therefore appears to be very important to success in science, that a man should have an able mother. I believe the reason to be, that a child so circumstanced has the good fortune to be delivered from the ordinary narrowing, partisan influences of home education. Our race is essentially slavish; it is the nature of all of us to believe blindly in what we love.... We are inclined to look upon an honest, unshrinking pursuit of truth as something irreverent.... Women are far more strongly influenced by these feelings than men: they are blinder partisans and more servile followers of custom. Happy are they whose mothers did not intensify their naturally slavish dispositions in childhood, by the frequent use of phrases such as, "Do not ask questions about this or that, for it is wrong to doubt"; but who showed them, by practice and teaching, that inquiry may be absolutely free without being irreverent. (Galton 1869, 196)

Since it does not increase the influence of the female line, this is presumably a non-heritable maternal factor (another aspect of family environment), but Galton did not articulate this distinction. However, he abandoned this idea later in *English Men of Science*, after analyzing the replies from scientists to a question about the origin of their taste for science:

> Attention should be given to the relatively small encouragement received from the mother.... In many respects the character of scientific men is strongly anti-feminine; their mind is directed to facts and abstract theories, and not to persons or human interests. The man of science is deficient in the purely emotional element.... In many respects they have little sympathy with female ways of thought. It is a curious proof of this, that in the very numerous answers which have reference to parental influence, that of the father is quoted three times as often as that of the mother. (Galton 1874a, 206–208)

The characterization of the man of science might almost be a self-portrait!

The Reception of *Hereditary Genius*

Louisa Galton's record for 1869 contains the entry: "Frank's book 'Hereditary Genius' published in November, but not well received, but liked by Darwin and men of note" (Pearson 1924, 88). A recent analysis of its reception in the Victorian periodical press concludes that it was favorably received by scientists, in particular by Darwin and Wallace, that (not surprisingly) it had a hostile reception in the religious press, and that it had a mixed reception from "neutral" reviewers in political and literary journals, who praised the sincerity, ingenuity, and intelligence of the author, but were skeptical of his exclusively hereditarian interpretation of the interesting information he had collected (Gökyigit 1994).

Galton observed in his autobiography that he was particularly pleased to receive a letter from Darwin, which began: "I have only read about 50 pages of your book (to Judges), but I must exhale myself, else something will go wrong in my inside. I do not think I ever in all my life read anything more interesting and original—and how well and clearly you put every point!" (Galton 1908, 290). Darwin expressed his mature view in his own autobiography: "I am inclined to agree with Francis Galton in believing that education and environment produce only a small effect on the mind of any one, and that most of our qualities are innate" (Barlow 1958, 43).

Of greater interest to us are some of the more critical "neutral" reviews, which articulate our reaction to reading *Hereditary Genius* today. In a long, fair-minded review, Herman Merivale (1870) began by saying that the doctrine of the influence of heredity on genius (understood in a loose sense as meaning ability) was generally accepted; it was as acceptable a mode of expression to say that someone belonged to a clever family as to say that he belonged to a tall family or a fair family. Galton's idea that he had to overcome a popular prejudice against the inheritance of ability was a misconception that had led him to overstate his case by ignoring the effects of family environment and family influence. He considered the case of two children of equal abilities, born from an inferior and a superior couple in point of intellect. "In such a case we may be quite sure that the latter—the child of clever parents—has a much better chance of being well instructed, and through such instruction of becoming 'eminent,' and filling a place in statistical lists after Mr. Galton's fashion, than the child of the other pair" (Merivale 1870, 109). In addition, Merivale argued that "in a great number of cases a father who has made his way in the world has advantages for bringing forward his sons and other relatives in the career of life beyond what are possessed by others who have not thriven in the same way" (110).

Merivale concluded this general argument:

> The truth is that the success in life which leads to distinction is due to two causes, the one consisting in natural aptitude or ability, the other in surrounding circumstances. Even if it be possible to refer the former condition to the laws of descent, who shall attempt to calculate the variations of the latter? Who shall say how often talents of a high order are repressed by penury, by the want of education, by the drudgery of life? We cannot agree with Mr. Galton that men endowed with a certain amount of genius always force their way to the front ranks of society. For one who succeeds, a hundred, perhaps not inferior in natural gifts, fail and perish by the way.... It is not less certain that many of those whose names are rescued from oblivion owe their celebrity to favourable opportunity, to patronage or family influence, or to what is termed good fortune, quite as much as to their natural gifts. (Merivale 1870, 110–111)

To illustrate his argument, Merivale pointed out that, of the 250 clever relations of judges in Galton's list, more than 100 had been lawyers themselves, which seemed strange "unless we are to assume, not only that talent is hereditary, but that the special talent of the lawyer is hereditary also" (111). He drew attention to the family of Atkyns in Galton's list of judges: "There have been four judges of the name and (let us note in passing) nobody, except a law student or a painstaking county antiquary, ever heard more than the name of any of them. These Atkynses are credited with seven or eight remarkable relatives, but of these there is only one who was not a lawyer, and he was reader of Lincoln's Inn. The whole list has the unmistakeable character of a snug little family party of jobbers, rather than that of a galaxy of genius" (112).

Similar points were made by W. F. Farrar, then a master at Harrow School, and later headmaster of Marlborough and dean of Canterbury. (He knew Galton quite well, probably through Galton's brother-in-law Montagu Butler, who was headmaster of Harrow. He cooperated with Galton in trying to introduce geographical education in schools, and in collecting anthropometric data at Marlborough.) In his review in 1870, he accepted the role of heredity in determining ability, but rejected Galton's claim that it was exclusively responsible. In particular, he found that Galton applied the title of "eminent" to many men of average ability, helped forward by incidental advantages, and that he was reluctant to allow for the effects of "family tradition, and surrounding circumstances, and early education." He agreed that there was a tendency of sons to follow in their father's footsteps, which he illustrated by the facts that the sons of Aeschylus and Sophocles were very moderate tragedians and that the son of Linnaeus was a tolerable naturalist; but he pointed out that this could be explained as easily by family influence as by heredity. He reinforced this argument by citing two examples from Roman history of adopted children (Marius the Younger and P. Scipio Africanus Minor) who as adults reflected the characteristics of the family into which they had been introduced.

Galton paid most attention to a book published by the Swiss botanist Alphonse de Candolle in 1873. He was the son of the eminent botanist Augustin de Candolle, and the two men appeared in the list of eminent related scientists in *Hereditary Genius*. While Galton was collecting information for *Hereditary Genius*, de Candolle had been writing his book on the effect of social factors on scientific productivity. He examined the foreign membership of the scientific academies of London, Berlin, and Paris in the four years 1750, 1789, 1829, and 1869, to discover which countries were most scientifically productive. He identified eighteen factors correlated with the flourishing of science, such as a significant number of persons of independent means, well-organized facilities for scientific education and work, public opinion favorable to science, the absence of clerical opposition to scientific pursuits, a small independent country or a union of such countries, and a northern or temperate climate. In particular, he attributed the thriving state of science in his native Geneva and in the neighboring canton of Basle to these social factors.

De Candolle had been working on this thesis for forty years, and his book was nearly complete when *Hereditary Genius* appeared. In considering this work, he maintained that Galton had exaggerated the effect of heredity at the expense of social factors. He considered that general intellectual ability was heritable, and that an individual with high inherited ability could go far in any profession requiring such ability when helped by favorable family and social circumstances; he did not think that there were specific inherited abilities, for example for a particular science, or for the law, except for mathematics.

His viewpoint is expressed well in a letter to Galton written in 1873:

> I still believe that there is, not a conflict but a substantial *difference* in our assessment of the causal factors which have determined the facts. You as a rule rely on heredity as the main factor. When you speak of other factors they are examined cursorily and without trying to distinguish how much effect they have either individually or together.... As for me, I have had

> the advantage of coming after you. It was not difficult for me to confirm by new facts the influence of heredity, but I never lost other factors from sight, and the result of my researches has convinced me that they are in general more important than heredity, at least among men of the same race.... The effect of tradition, example and advice within the family has seemed to me more influential than heredity strictly speaking. In addition there is education outside the family, public opinion, institutions etc. I have tried to distinguish the relative influence of all these factors, varying between countries and periods, and favouring or impeding the effects of heredity. (Pearson 1924, 281–282; my translation)

This is from a letter forming part of a lengthy and civilized correspondence between Galton and de Candolle. But Galton did not accept that de Candolle's work undermined the conclusions of *Hereditary Genius*, and in a short paper published in 1873 he referred to

> a volume written by M. de Candolle ... in which my name is frequently referred to and used as a foil to set off his own conclusions. The author maintains that minute intellectual peculiarities do not go by descent, and that I have overstated the influence of heredity, since social causes, which he analyses in a most instructive manner, are much more important.... The author, however, continually trespasses on hereditary questions, without, as it appears to me, any adequate basis of fact, since he has collected next to nothing about the relatives of the people upon whom all his statistics are founded. (Galton 1873, 346)

Nature and Nurture

Nevertheless, de Candolle's work stimulated Galton to try to assess the relative importance of heredity and environment, or, as he called them, nature and nurture, resulting in his book on *English Men of Science* and in his pioneering studies of twins.

English Men of Science: Their Nature and Nurture (1874)

Galton wrote in the preface that, when he read de Candolle's book, he was engaged on a leisurely investigation which would supplement his work in *Hereditary Genius*. The object of that book had been to assert the claims of heredity, the importance of which had previously been overlooked, but he recognized that this was only one of what he called the "preefficients," or causes that had gone to the making of eminent men, and he was trying to work out the relative efficacy of environmental factors such as education, tradition, fortune, and opportunity, compared with heredity, in determining ability. Stimulated by de Candolle's work and by his criticism of the importance attributed to heredity in *Hereditary Genius*, Galton had undertaken his own investigation of the roles of nature and nurture in the history of men of science.

Galton first defined the distinction between nature and nurture:

> The phrase "nature and nurture" is a convenient jingle of words, for it separates under two distinct heads the innumerable elements of which personality is composed. Nature is all that a man brings with himself into the world; nurture is every influence from without that affects him after his birth.... Neither of the terms implies any theory; natural gifts may or may not be hereditary; nurture does not especially consist of food, clothing, education or tradition, but it includes all these and similar influences whether known or unknown. (Galton 1874a, 12)

In other words, nature includes any prenatal influence, whether hereditary or not; natural in this sense is synonymous with congenital. He probably adopted this as a pragmatic definition because he saw no way of distinguishing hereditary from environmental factors before birth. He may have borrowed the phrase "nature and nurture" from the passage in Shakespeare's play *The Tempest* in which Prospero complains about his adopted son Caliban: "A devil, a born devil, on whose nature / Nurture can never stick."

To evaluate the roles of nature and nurture among the "preefficients" of scientists, Galton sent a detailed questionnaire to about 190 selected Fellows of the Royal Society, and received just over 100 replies, a remarkably good response. His book is a detailed analysis of the replies to the questionnaire. The four chapters consider the antecedents of the scientists, their own personal qualities, the origin of their taste for science, and their education. All four chapters contain information of great interest to historians of nineteenth-century science, but only the first and third need be discussed here.

In chap. 1, on the antecedents of the scientists, Galton first discussed their race and birthplace, and the occupation and physical peculiarities of their parents. He then considered their birth order, and found that the scientists in his sample were twice as likely to be the eldest as the youngest son, in families with more than one son. He concluded that elder sons had decided advantages of nurture over younger sons for three reasons: They are more likely to become possessed of independent means, and therefore able to follow pursuits to their taste; they are treated more as companions by their parents, and have earlier responsibility, developing independence of character; and probably, in less well-to-do families, the first-born child would have more attention and better nourishment in his infancy, than his younger brothers and sisters. (Sulloway [1996] notes that the tendency for firstborns to be overrepresented among eminent scientists has been confirmed in several recent studies. Galton had previously found the opposite in judges.)

He then considered the role of heredity in the same way as in *Hereditary Genius*, but the results add little to those obtained there. His main contribution to the nature–nurture issue is contained in chap. 3 on "Origin of Taste for Science," in which he analyzed the replies to the questions: "Can you trace the origin of your interest in science in general, and in your particular branch of it? How far do your scientific tastes appear to have been innate?" Ninety-one replies were received, which were classified by Galton as shown in table 2.6. This classification is not exclusive, many individuals being classed under several heads; for example, Charles Darwin is classed under *a* since he had replied "certainly innate," and also under *h* because of the voyage in the *Beagle*.

From the 91 replies to these questions, Galton found that there were 56 cases in which the taste for science was decidedly innate, 11 in which it was decidedly not innate (including 4 of the 7 medical men in the sample), with 24 doubtful cases. He concluded that a strong and innate taste for science is a prevailing characteristic among scientific men: "As a rough numerical estimate, it seems that 6 out of every 10 men of science were gifted by nature with a strong taste for it; certainly not 1 person in 10, taken at haphazard, possesses such an instinct; therefore I contend that its presence adds five-fold at least, to the chance of scientific success" (Galton 1874a, 195). However, this instinctive taste is not necessarily hereditary, and he concluded that "instinctive tastes for science are, generally speaking, not so strongly hereditary as the more elementary qualities of the body and mind. I have tabulated the replies, and find the proportion to be 1 case of inheritance to 4 that are not inherited from either parent. There is no case in which the correspondent speaks of having inherited a love of science from his mother, though, of course, she may, and probably has, often transmitted it from a grandparent" (196–197).

Table 2.6. Classification of Origin of Taste for Science

Symbol	Cases	Classification
a	59	Innate tastes (not necessarily hereditary)
b	11	Fortunate accidents (which generally testify to an innate taste)
c	19	Indirect opportunities and indirect motives
d	24	Professional influences to exertion
e	34	Encouragement at home of scientific inclinations
f	20	Influence and encouragement of private friends and acquaintances
g	13	Influence and encouragement of teachers
h	8	Travel in distant regions
z	3	Residual influences, unclassed

It has been argued, in a very thorough appraisal of *English Men of Science* by Hilts (1975), that the two questions Galton asked about the origin of the taste for science were not phrased to elicit whether it was innate, and if so hereditary. If a scientist could not trace the origin of his interest in science, it was only natural for him to assume that it was innate, whereas it might well have been due to an influence at an early age which he could not remember. Furthermore, if one of the parents had scientific tastes, Galton took this as evidence of heredity, but he had no way of distinguishing heredity from encouragement at home, whose importance was acknowledged by one-third of the respondents (table 2.6). Indeed, it is difficult to see what questions could have been asked to throw further light on this problem. Thus, *English Men of Science*, though it provides some fascinating information as the first sociological questionnaire concerned with the scientific community, does little to resolve the question of the relative importance of nature and nurture which it was designed to address. But in the introduction to the book, Galton briefly mentioned identical twins as evidence of the predominance of nature over nurture and related some anecdotes to show that twins who were identical at birth retained their near identity in adult life despite differences of nurture. He subsequently elaborated this anecdotal approach with a more systematic enquiry on "The history of twins, as a criterion of the relative powers of nature and nurture," which was his major contribution to this subject.

"The History of Twins" (1875)

The opening paragraph of the paper recognized the inadequacy of statistical evidence for proving the inheritance of mental ability and suggested that twins might be used to determine the relative importance of nature and nurture:

> The exceedingly close resemblance attributed to twins has been the subject of many novels and plays, and most persons have felt a desire to know upon what basis of truth those works of fiction may rest. But twins have many other claims to attention, one of which will be discussed in the present memoir. It is, that their history affords means of distinguishing between the effects of tendencies received at birth, and of those that were imposed by the circumstances of their after lives; in other words, between the effects of nature and of nurture. This is a subject of especial importance in its bearings on investigations into mental heredity, and I, for my part, have keenly felt the difficulty of drawing the necessary distinction whenever I tried to estimate the degree in which mental ability was, on the average, inherited. The objection to statistical evidence in proof of its inheritance has always been: "The persons whom you compare may have lived under similar conditions and have had similar advantages of education, but such prominent conditions are only a small part of those that determine the future of each man's life. It is to trifling accidental circumstances that the bent of his disposition and his success are mainly due, and these you leave wholly out of account—in fact, they do not admit of being tabulated, and therefore your statistics, however plausible at first sight, are really of very little use." (Galton 1875a, 391)

His method of using twins to estimate the effect of nurture was to track their life history changes, to see whether twins who were similar at birth diverged in dissimilar environments or whether twins who were dissimilar at birth converged in similar environments. He sent a questionnaire about their life history to a number of twins or their close relatives, expanding the sample by asking for the addresses of other twins known to them. He only considered those who gave detailed replies, and he concentrated his attention on two extreme groups. He first identified 35 pairs of like-sexed twins who had been very similar as children, most of them so similar that they were difficult to tell apart. They were reared alike until adulthood, but since then they had led separate lives. What effect had the differences in their environment as adults had upon them? The records showed that "in some cases the resemblance of body and mind had continued unaltered up to old age, notwithstanding very different conditions of life; and they showed in the other cases that the parents ascribed such dissimilarity as there was wholly, or almost wholly to some form of illness" (401). He concluded that nature was far stronger than nurture within the limited range of environmental differences encountered, but that illness, usually infectious, in one twin could have a large effect.

Galton then identified a group of 20 pairs of like-sexed twins who were very dissimilar as young children, and examined how far an identity of nurture in childhood and youth tended to assimilate them. The answer was clearcut: not at all. A typical answer from a parent was: "The twins have been perfectly dissimilar in character, habits, and likeness from the moment of their birth to the present time, though they were nursed by the same woman, went to school together, and were never separated till the age of fifteen" (404). Galton drew this conclusion:

> The impression that all this evidence leaves on the mind is one of some wonder whether nurture can do anything at all, beyond giving instruction and professional training. It emphatically corroborates and goes far beyond the conclusions to which we had already been driven by the cases of similarity. In these, the causes of divergence began to act about the period of adult life, when the characters had become somewhat fixed; but here the causes conducive to assimilation began to act from the earliest moment of the existence of the twins, when the disposition was most pliant, and they were continuous until the period of adult life. There is no escape from the conclusion that nature prevails over nurture when the differences of nurture do not exceed what is commonly to be found among persons of the same rank of society and in the same country. (Galton 1875a, 404)

Galton is often credited with inventing the twin method for distinguishing the roles of heredity and environment, but this is misleading (Rende, Plomin, and Vandenberg 1990). He did invent a method of using twins to investigate the effect of the environment, but it was not the classical twin method that compares the resemblance of monozygotic and dizygotic twins, on the assumption that the former are genetically identical while the latter are not. In fact, he could not have used this method since he believed that all twins were genetically identical, as were siblings. To explain this I anticipate his discussion of the reasons for the resemblances between siblings and between twins, which is fully described in chapter 4.

Galton thought that the zygote contains a large number of hereditary elements, which collectively form the "stirp." Some of these elements are patent, developing into the cells of the embryo and hence determining the appearance of the adult, while others remain latent but can be transmitted to future generations. He thought that parents transmitted the same elements to all their offspring, so that brothers and sisters had identical stirps, and he attributed phenotypic differences between siblings to variability in the choice of patent elements from the stirp; in other words, to developmental variability.

He also knew from contemporary work that there are two types of twins, monozygotic twins derived from the division of a single fertilized ovum and dizygotic twins derived from the independent fertilization of two ova, that monozygotic twins are always of the same sex, and that many of them are very similar in appearance (Spaeth 1860 and 1862, Kleinwächter 1871). However, he believed that all twins, dizygotic as well as monozygotic, had identical stirps, just as siblings did, and he attributed the fact that many monozygotic twins are almost identical to the similarity of their developmental environment. Furthermore, he thought that the group of twenty twins who were markedly dissimilar were also monozygotic, and he supposed that their dissimilarity was due to late division of the ovum. His reason for thinking these twins to be monozygotic was that they were all like-sexed, but he acknowledged that he had no direct evidence that these twins were monozygotic, and it seems likely that they were not. In Galton's view, the dizygotic twins were all contained in the group of twins of intermediate similarity, including all the unlike-sexed twins, which he did not use in his analysis.

Thus Galton did not use the classical twin method of comparing the similarities of monozygotic and dizygotic twins because he did not know that dizygotic twins only share half their genes while monozygotic twins are genetically identical. According to Rende, Plomin, and Vandenberg (1990) the classical twin method was developed simultaneously in America and Germany in the 1920s. It came into some disrepute because of the studies on twins in Auschwitz by Josef Mengele, who did cruel experiments and murdered some of his subjects to obtain postmortem samples; but this is an indictment not of the twin method when properly conducted but of unethical human experimentation. The classical twin method, of comparing the similarities between monozygotic and dizygotic twins, is one of the most powerful tools of human genetics.

Galton's twin method was to track the life history changes of twins to see whether twins who were similar at birth diverged in dissimilar environments or whether twins who were dissimilar at birth converged in similar environments. He recognized that the first comparison was rather weak because the similar twins did not experience different environments until they became adult; the modern extension of this method is the study of twins separated at birth and reared apart in foster homes (see below). His study can also be faulted for relying on anecdotal replies to a questionnaire, but it was the first study of its kind. As usual, Galton was the pioneer who left his methods and his results to be perfected by others (Burbridge 2001).

Galton's Hereditarianism

In the preface to *Hereditary Genius,* Galton explained that the idea of investigating this subject occurred to him during the course of a purely ethnological inquiry into the mental peculiarities of different races. He briefly described the results of this inquiry in "Hereditary talent and character," in which he contrasted the American Indian, who is "naturally cold, melancholic, patient, and taciturn," with the West African Negro, who "has strong impulsive passions, and neither patience, reticence, nor dignity. He is warm-hearted, loving towards his master's children, and idolised by the children in return. He is eminently gregarious, for he is always jabbering, quarrelling, tom-tom-ing, or dancing" (1865a, 321). He warned that care must be taken before inferring that these characters are innate because of the exceeding docility of man: "His mental habits in mature life are the creatures of social discipline, as well as of inborn aptitudes, and it is impossible to ascertain what is due to the latter alone, except by observing several individuals of the same race, reared under different influences, and noting the peculiarities of character that invariably assert themselves" (320). But he concluded that the character of the American Indian, at least, was innate because it was the same over an enormous area, through every climate from the frozen North, through the equator, down to the inclement regions of South America, and under a great variety of political systems.

Having convinced himself that inter-racial differences in mental characters were innate, it was reasonable to ask if the same were true for intra-racial differences. He began by thinking over the dispositions and achievements of his contemporaries and was surprised to find how frequently ability seemed to go by descent; he was thus led into his full-scale investigation of the relatives of distinguished men reported in "Hereditary talent and character" and in *Hereditary Genius*. He showed without any doubt that the tendency of ability to run in families was not just a matter of chance, but he was less successful in demonstrating his conviction that this tendency was largely due to heredity rather than to family background and influence, to nature rather than to nurture. Galton accepted the role of environmental factors, such as birth order and parental example and encouragement, in determining career success, but in the absence of other evidence he preferred the hereditarian over the environmental explanation of the tendency of "eminence" to run in families. It clearly seemed self-evident to him that descent usually demonstrated heredity. In his autobiography, for example, he wrote that the Myttons were an unquestionable instance of a very peculiar hereditary temperament. He had known Jack Mytton junior in the late 1840s (see chapter 1), and one might wonder today whether he had inherited his wild character from his father, or whether he was wild because he copied his father. Galton was in no doubt.

Herman Merivale, in his review of *Hereditary Genius*, suggested that Galton had been led to overstate his case by his misconception that there was a popular prejudice against the inheritance of ability. Galton was probably familiar with the passage in the well-known *History of Civilisation in England* (1857) in which Henry Buckle expressed his skepticism about the inheritance of human mental and moral characters:

> We often hear of hereditary talents, hereditary vices, and hereditary virtues; but whoever will critically examine the evidence will find that we have no proof of their existence. The way in which they are commonly proved is in the highest degree illogical; the usual course being for writers to collect instances of some mental peculiarity found in a parent and in his child, and then to infer that the peculiarity was bequeathed. By this mode of reasoning we might demonstrate any proposition; since in all large fields of inquiry there are a sufficient number of empirical coincidences to make a plausible case in favour of whatever view a man chooses to advocate. But this is not the way in which truth is discovered; and we ought to inquire not only how many instances there are of hereditary talents, &c. but how many instances there are of such qualities not being hereditary. Until something of this sort is attempted, we can know nothing about the matter inductively; while, until physiology and chemistry are much more advanced, we can know nothing about it deductively. (Buckle 1857, quoted in Olby 1985, 169–170)

In this passage, it is implied that the empirical case for heredity would be proved by showing that family patterns were not just due to coincidence. If Galton set out to counter Buckle's argument, he might have been initially led to think that disproving chance would prove heredity; it was certainly a necessary, though not a sufficient, condition for heredity.

Galton's hereditarianism may also have been a reaction against the environmentalism of Victorian social reformers. In a well-known passage in *Hereditary Genius* he wrote:

> I have no patience with the hypothesis occasionally expressed, and often implied, especially in tales written to teach children to be good, that babies are born pretty much alike, and that the sole agencies in creating differences between boy and boy, and man and man, are steady application and moral effort. It is in the most unqualified manner that I object to pretensions of natural equality. The experiences of the nursery, the school, the University, and of professional careers, are a chain of proofs to the contrary. (Galton 1869, 14)

This passage is a clear rejection of the extreme environmentalist view, originally advocated by John Locke, that the mind is at birth a tabula rasa, a clean slate, whose development is determined by postnatal environmental influences.

Fancher (1985) has contrasted two factors that combined to make John Stuart Mill an environmentalist and Francis Galton a hereditarian. The first was their educational experience. Mill was successfully educated at home by his distinguished father, who encouraged him to believe that any intellectual superiority he might have was due not to natural ability but to his unusual educational advantage; he was thus predisposed to attribute success to environmental rather than innate factors. Galton was intellectually precocious, but his experience at Cambridge taught him that he was not as mathematically gifted as many of his contemporaries, despite his ambition to succeed; he attributed his lack of success to lack of innate mathematical ability and concluded that there were large inherited differences in natural ability. The passage quoted above continued:

> I acknowledge freely the great power of education and social influences in developing the active powers of the mind, just as I acknowledge the effect of use in developing the muscles of a blacksmith's arm, and no further. Let the blacksmith labour as he will, he will find there are certain feats beyond his power that are well within the strength of a man of herculean make, even although the latter may have led a sedentary life.... There is a definite limit to the muscular powers of every man, which he cannot by any education or exertion overpass.
>
> This is precisely analogous to the experience that every student has had of the working of his mental powers. (Galton 1869, 14–15)

The second factor that distinguished Mill from Galton was their social attitude. Mill, like his father, was a liberal social reformer who believed that environmental explanations ought to take precedence over hereditarian ones on moral grounds; if people in power believe that the poor have an innate and natural inferiority, they will have little reason to improve their environment, so that politicians have a moral obligation to prefer the environmentalist explanation. He expressed this view in his *Autobiography*: "I have long felt that the prevailing tendency to regard all the marked distinctions of human nature as innate, and in the main indelible, and to ignore the irresistible proofs that by far the greater part of these differences, whether between individuals, races, or sexes, are such as not only might, but naturally would be produced by differences in circumstances, is one of the chief hindrances to the rational treatment of great social questions, and one of the greatest stumbling blocks to human improvement" (Mill 1873, 203). Galton had little sympathy with radical social reform and was perhaps predisposed toward a hereditarian position in reaction against the environmentalism of the social reformers.

Thus several factors may have predisposed Galton to take a hereditarian rather than an environmentalist position. His experience at Cambridge convinced him of the importance of innate differences in ability. He had little sympathy with radical social reform or with the environmentalism that it presupposes. He tended to see the world in black and white and was therefore predisposed to take a definite stance one way or the other rather than to sit on the fence. Having started on his statistical investigations of how ability ran in families, he was committed to interpret them as due primarily to heredity, since otherwise he could draw few conclusions from them. Lastly, he was deeply committed to Darwin's theory of evolution through natural selection, which depends on the existence of adequate heritable variation. Darwinism does not of course exclude the existence of environmental variability, but its supporters are more interested in heritable variability, and it is understandable that an early supporter like Galton should have been biased toward a hereditarian position. Nevertheless, he did make an honest attempt to discriminate between the effects of nature and nurture.

Another explanation of Galton's hereditarian bias has been put forward by social constructionist historians of science, who consider that Galton's enthusiasm for eugenics had priority over his scientific interests. In their view, his belief in the heredity of mental characters was determined by the fact that this was a necessary condition for eugenics to work. For example, Ruth Cowan considers that "Hereditary talent and character" was not a scientific treatise but "an exercise in political propaganda," and she continues: "Galton's attachment to the idea of mental heredity went far beyond what was warranted by the scientific evidence he adduced. He had convinced himself of the validity of mental heredity, not because he thought it was a solution to a great scientific problem, but because he was fascinated by the social programs that could be built around it" (1977, 140). Donald MacKenzie, discussing Galton's contributions to statistical theory in a similar vein, suggests that "it is reasonable to see Galton's eugenics not merely as providing the motive for his statistical work, but also as conditioning the content of it" (1981, 68).

Galton was the founder of eugenics, which provided one of the main motives for his hereditary and statistical work. But it does not follow that eugenics conditioned the content of his work. This assessment ignores the origin of his interest in both heredity and eugenics, Darwin's *The Origin of Species*, and it is at variance with his lifelong commitment to the advancement of science. It is more logical, and more plausible, to think that his enthusiasm for eugenics was determined by his belief in mental heredity rather than the other way around. This view is supported by his impeccably scientific treatment of Darwin's theory of pangenesis, discussed in chapter 4, and by his adherence to the plausible but mistaken idea of perpetual regression, discussed in chapter 9. If he had only wanted to construct a theory favorable to eugenics, why should he have saddled himself with perpetual regression, which was an obstacle to eugenic progress?

Epilogue

The relative importance of heredity and environment, of nature and nurture, has been the subject of research and controversy since Galton's time. The most important developments have been the concentration on the inheritance of measurable, quantitative characters, and in particular of the intelligence quotient (IQ) as a measure of intellectual ability; the use of the correlation coefficient, invented by Galton in 1888, to measure resemblance between relatives; understanding of the predicted correlations between relatives based on Mendelian genetics (Fisher 1918), and of the genetic relatedness of monozygotic and dizygotic twins; and the accumulation of data on cross-fostered children and on twins raised apart which allows the effect of heredity to be distinguished from that of shared family environment.

Table 2.7 shows data for a number of characters on monozygotic (genetically identical) twins reared together and apart. The first column shows the correlation between twins reared together, r_{MZT}, which reflects the effects of heredity and of their shared environment. The second column shows the correlation between twins reared apart, r_{MZA}, which reflects only the effect of heredity and thus measures the proportion of the total variance that is of genetic origin (the broad heritability). The difference between these two correlations shown in the third column therefore measures the proportion of the total variance due to the shared environment between twins reared in the same family. The quantity $1 - r_{MZT}$ in the fourth column measures the proportion of the total variance due to nonshared environmental factors. The contribution of shared environment (column 3) is modest for all characters except IQ. The broad heritability (column 2) shows a substantial genetic contribution to all the characters listed, but there is also an appreciable environmental contribution to all the characters except fingerprint ridge count. This brief summary represents just the tip of a rapidly growing iceberg. For further information, see Bouchard and Propping 1993, Brody 1992, Falconer and Mackay 1996, Lynch and Walsh 1998, Mascie-Taylor 1993, Plomin et al. 1997.

Thus Galton has been largely justified in his conclusion that there is a strong hereditary component in determining human mental ability,

Table 2.7. Correlation Coefficients for Monozygotic Twins Raised Together (r_{MZT}) and Raised Apart (r_{MZA})

Character	r_{MZT}	r_{MZA}	$r_{MZT} - r_{MZA}$	$1 - r_{MZT}$
Fingerprint ridge count	0.96	0.97	–0.01	0.04
Height	0.93	0.86	0.07	0.07
Blood pressure	0.70	0.64	0.06	0.30
Heart rate	0.54	0.49	0.05	0.46
IQ	0.88	0.69	0.19	0.12
Extraversion	0.51	0.38	0.13	0.49
Neuroticism	0.46	0.38	0.08	0.54

Sources: Lynch and Walsh 1998, Plomin et al. 1997

Note: r_{MZA} measures the proportion of the total variance that is of genetic origin (the broad heritability); $r_{MZT} - r_{MZA}$ measures the proportion due to the shared environment between twins reared in the same family; $1 - r_{MZT}$ measures the proportion due to nonshared environmental factors.

though he may have exaggerated its importance. He also believed, based on much cruder evidence, that there were substantial hereditary differences in ability between races, and in particular that the Negro race both in America and Africa was substantially inferior in inherited ability to the Anglo-Saxon race. Modern research has confirmed that the black population of the United States scores, on average, one standard deviation (15 IQ points) lower than the white population on various tests of intelligence; there has been heated controversy whether this difference reflects genetic inferiority or whether it is due to environmental or cultural poverty, or to cultural bias in the tests (Herrnstein and Murray 1994, Gould 1996, Rushton 1997, Harris 1998). The question can be answered only from cross-fostering data.

One study was of a sample of children whose biological fathers were black soldiers serving in the Army of Occupation in Germany at the end of World War II (Brody 1992). The children were reared by their biological mothers, who were white German women. The children were compared to a sample of white children whose fathers had been white soldiers in the same army in Germany. The mean IQ of the children of white fathers was 97.2, that of the children of black fathers was 96.5; the difference between the two groups is negligible, suggesting that there was no genetic difference in IQ between the two groups of fathers. However, there was no information about the actual IQ of the fathers, which may have happened to be similar in the two groups, perhaps through selection for army service.

Another study compared natural and adopted children raised by white parents (Weinberg, Scarr, and Waldman 1992). The average IQ was 109 for their natural children, 106 for adopted children with two white parents, 99 for adopted children with one black and one white parent, and 89 for adopted children with two black parents. This seems to provide strong evidence that the difference in IQ between blacks and whites has a substantial genetic component, but doubt has been cast on this conclusion because the data were not corrected for preplacement history (Cross 1996); many of the children with two black parents were placed for adoption at a later age than the other foster children, which has a deleterious effect on their development. It does not seem possible to draw a firm conclusion about the magnitude of inter-racial differences in IQ from either of these studies.

Appendix: Number of Kinsfolk

Galton had some difficulty in estimating the total number of kinsfolk of different degrees (see column C in table 2.2) in order to find the relative frequency of distinguished relatives of a distinguised person. In comparing the results for different professions in *Hereditary Genius*, he wrote:

> Little dependence can be placed on the entries in C.... I did indeed try to obtain real and not estimated data for C, by inquiring into the total numbers of kinsmen in each degree, of every illustrious man, as well as of those who achieved eminence. I wearied myself for a long time with searching biographies, but finding the results very disproportionate to the labour, and continually open to doubt after they had been obtained, I gave up the task, and resigned myself to the rough but ready method of estimated averages. (Galton 1869, 318–319)

He explained how he obtained these estimated averages for 286 judges, which he used for all other professions:

> I find that 23 of the Judges are reported to have had "large families," say consisting of four adult sons in each; 11 are simply described as having "issue," say at the rate of 1.5 sons each; and that the number of the sons of others are specified as amounting between them to 186; forming thus far a total of 294. In addition to these, there are 9 reported marriages of judges in which no allusion is made to children, and there are 31 judges in respect to whom nothing is said about marriage at all. I think we are fairly justified, from these data, in concluding that each judge is father, on average, to not less than one son who lives to an age at which he might have distinguished himself, if he had the ability to do so. I also find the (adult) families to consist on an average of not less than 2.5 sons and 2.5 daughters each, consequently each judge has an average of 1.5 brothers and 2.5 sisters.
>
> From these data it is perfectly easy to reckon the number of kinsmen in each order. Thus the nephews consist of the brothers' sons and the sisters' sons: now 100 judges are supposed to have 150 brothers and 250 sisters, and each brother and sister to have, on the average, only one son; consequently the 100 judges will have (150 + 250, or) 400 nephews. (Galton 1869, 82)

This passage reveals the crude nature of the data on which column C in table 2.2 was based. It is certainly true that every individual has one father and one mother. It is also plausible to assume that the population size is approximately stationary, which implies that every adult individual has, on average, one adult son and one adult daughter, in accordance with Galton's calculation. But this needs to be reconciled with the conclusion that an average completed family contains five adult children; and, as he later showed, this does not imply that a man has on average 1.5 brothers and 2.5 sisters.

When Galton designed the questionnaire for Fellows of the Royal Society which underpinned *English Men of Science* in 1874, he asked them to state the total number of relatives of different kinds who had attained 30 years of age as well as naming those who had achieved distinction. He therefore had direct information about the numbers of relatives of the scientists:

> It appears from my returns, which are rather troublesome to deal with, owing to incompleteness of information, that 120 scientific men have certainly not more than 250 brothers, 460 uncles, and 1200 male cousins who reach adult life. They have somewhat *less* than 120 fathers and 240 grandfathers, because the list contains brothers and cousins. (Galton 1874a, 64)

He went on:

> My data afford an approximate estimate of the ratio, according to which effective ability (hereditary gifts *plus* education *plus* opportunity) is distributed throughout the different degrees of kinship.... It is therefore only requisite ... to add the returns together, and to compare the number of distinguished kinsmen in the various degrees with the total number of kinsmen in those degrees, to obtain results whose *ratio to one another* is the one we are in search of. (Galton 1874a, 70–71)

It is strange to find that he does not give these results, but instead presents some completely garbled figures at the end of chap. 1.

Galton sent a similar questionnaire to 464 Fellows of the Royal Society in 1904, receiving 207 useful replies. He published a preliminary report in the same year (1904a) and a full report in 1906 in *Noteworthy Families* by Galton and Schuster. (Galton contributed a full analysis of the data as a preface to this volume; the Galton Research Fellow, Edgar Schuster, compiled biographical details of the most noteworthy families.) Just over a hundred of the replies contained complete returns of the total number of adult kinsfolk, and Galton selected the 100 most reliable of these for the analysis shown in table 2.8. (I do not discuss the data on the numbers of noteworthy kinsfolk.) It will be seen that each scientist had on average the same number of just over 2 adult brothers and 2 sisters. In the previous generation their parents each had on average about 2.25 adult brothers and 2.25 sisters. The latter brothers and sisters had on average just over one adult son and one daughter (2.30/2.24 = 1.03 sons and 2.56/2.23 = 1.15 daughters).

After presenting these results, Galton continued:

> It may seem at first surprising that a brother and a sister should each have the same average number of brothers. It puzzled me until I had thought the matter out, and when the results were published in "Nature," it also seems to have puzzled an able mathematician [G. H. Bryan 1904], and gave rise to some newspaper controversy, which need not be recapitulated. The essence of the problem is that the sex of one child is supposed to give no clue of any practical importance to that of any other child in the same family. Therefore, if one child be selected out of a family of brothers and sisters, the proportion of males to females in those that remain will be, *on the average*, identical with that of males to females in the population at large. It makes no difference whether the selected child be a boy or a girl. Of course, if the conditions were "given a family of three boys and three girls," each boy would have only two brothers and three sisters, and each girl would have three brothers and two sisters, but that is not the problem. (Galton and Schuster 1906, xxxi)

I do not consider Galton's detailed calculations in *Nature* (1904b),

Table 2.8. Average Number of Relatives Who Survived Childhood in 100 Families

Relationship	Average number	Relationship	Average number
Bro	2.06	Si	2.07
Fa Bro	2.28	Fa Si	2.07
Me Bro	2.19	Me Si	2.38
Uncles[a]	2.24	Aunts[a]	2.23
Fa Bro Son	2.65	Fa Bro Da	3.02
Fa Si Son	1.84	Fa Si Da	2.08
Me Bro Son	2.36	Me Bro Da	2.66
Me Si Son	2.37	Me Si Da	2.46
Nephews[b]	2.30	Nieces[b]	2.56

Source: After Galton and Schuster 1906

Note: Bro = brother; Si = sister; Fa = father; Me = mother; Da = daughter; e.g. Me Bro means Mother's Brothers.

[a] Mean of previous two values.

[b] Mean of previous four values.

which are not illuminating, but instead I reconsider the problem from first principles. Consider a completed family with a fixed number of n adult children, of whom x are male and $n - x$ female. If equal numbers of boys and girls are born and if they have the same childhood mortality, x follows a binomial distribution with probability 1/2:

$$P(x) = \frac{n!}{x!(n-x)!}\left(\tfrac{1}{2}\right)^n, \qquad x = 0, 1, \cdots, n. \qquad [2]$$

If we select males at random from families of size n, this amounts to choosing families with probability proportional to x, so that the distribution of x in the selected families after standardization is

$$P^*(x) = \frac{xP(x)}{\mathrm{E}(x)}. \qquad [3]$$

($\mathrm{E}(x)$ denotes the Expected value of x.) The average number of brothers of these males is

$$\sum (x-1)P^*(x) = \frac{\sum (x-1)xP(x)}{\mathrm{E}(x)} = \tfrac{1}{2}(n-1) \qquad [4]$$

from the formulas for the first two moments of the binomial distribution. Since there are $n - 1$ brothers and sisters, the average number of sisters must also be $(n - 1)/2$.

This result was obtained by Galton (1904b) by direct enumeration for the binomial distribution with different values of n, but he did not adequately explain why the average number of brothers (or sisters) should be about 2. To do this, we must regard the family size n as a random variable with a probability distribution, which we denote $\Pi(n)$. If we select males from families of all sizes, this amounts to choosing family size with probability proportional to n, so that the distribution of n in the selected families after standardization is

$$\Pi^*(n) = \frac{n\Pi(n)}{\mathrm{E}(n)}. \qquad [5]$$

Hence the average number of brothers (or sisters) is

$$\sum \tfrac{1}{2}(n-1)\Pi^*(n) = \frac{\sum \tfrac{1}{2}(n-1)n\Pi(n)}{\mathrm{E}(n)} = \frac{1}{2}\left[\mathrm{E}(n) - 1 + \frac{\mathrm{Var}(n)}{\mathrm{E}(n)}\right], \qquad [6]$$

where Var(n) denotes the variance. If the population size is stationary so that E(n) = 2, the average number of brothers (or sisters) is

$$\frac{1}{2}\left[1+\frac{\mathrm{Var}(n)}{\mathrm{E}(n)}\right]. \qquad [7]$$

Under the simplest model, family size n follows a Poisson distribution with a mean of 2, which would imply that the variance to mean ratio, Var(n)/E(n), should be unity. Under this assumption it follows from eq.[7] that an individual should on average have one brother and one sister; but if the variance is greater than the mean, the average number of brothers and sisters of an individual selected at random is larger than this because selection of an individual provides information about the size of the family from which he or she comes. To obtain an average value of 2 brothers (or sisters) requires that the variance to mean ratio is about 3.

The variance to mean ratio of fertility is also important in population genetics in the theory of effective population size, and Crow and Morton (1955) have calculated it as 1.88, after adjustment to stationary population size, from the data of Pearson and Lee (1899) on British upper-class women. These data only included women whose marriage had lasted at least 15 years, or until the death of husband or wife. If the substantial number of women who did not marry, and so had no children, had also been included, it is likely that the variance would have been substantially increased. It is therefore plausible that the variance to mean ratio of fertility among all women who had reached adulthood and had completed their reproductive life (either through age or death) was of the order of 3 in the nineteenth century.

In conclusion, it is certain that everyone has one father and one mother; it is plausible in a stationary population that every adult person has on average one adult son and one daughter; and it can be estimated from empirical data (see table 2.8), and justified from theoretical considerations, that an adult person selected at random has, on average, about two adult brothers and two sisters.

3

Eugenics

> If a twentieth part of the cost and pains were spent in measures for the improvement of the human race that is spent on the improvement of the breed of horses and cattle, what a galaxy of genius might we not create!
>
> Galton, "Hereditary talent and character"

Francis Galton is best known to the general public as the founding father of eugenics, the science of the hereditary improvement of the human race by selective breeding. He was enthusiastic about the subject from 1865 to the end of his life, and he coined the word *eugenics* to describe it in *Inquiries into Human Faculty* in 1883. It was an obvious extension of the improvement of domestic animals by man which was discussed in the first chapter of *The Origin of Species*. It played a central role in motivating Galton's work, but it was argued in chapter 2 that it was secondary to his ideas on heredity and evolution, which are our main theme in this book; in other words, his eugenic ideas were derived from his ideas on heredity and evolution rather than the other way around. It is logically unnecessary to discuss eugenics in considering the latter, but some discussion of the subject is justified by the strong popular interest in it today.

Galtonian Eugenics

Most of Galton's eugenic ideas were formulated in 1865 in "Hereditary talent and character." In this paper, he first discussed the consequences of dysgenic practices: "Many forms of civilization have been peculiarly unfavourable to the hereditary transmission of rare talent. None of them were more prejudicial to it than that of the Middle Ages, where almost every youth of genius was attracted into the Church, and enrolled in the ranks of the celibate clergy" (1865a, 164). Another hindrance was a costly tone of society, which discouraged an ambitious and talented man from encumbering himself with domestic expenses until he could afford them: "Here also genius is celibate, at least during the best part of manhood" (164).

He then advocated the opposite idea, of improving the human race by encouraging early marriage between talented men and women. In a fanciful passage, he suggested that this objective could be brought about by a system of endowment based on examination results: "Let us, then, give reins to our fancy, and imagine a Utopia—or a Laputa, if you will—in which a system of competitive examination for girls, as well as for youths, had been so developed as to embrace every important quality of mind and body, and where a considerable sum was yearly allotted to the endowment of such marriages as promised to yield children who would grow into eminent servants of the State" (165). Ten young men and ten girls would be chosen each year, and the Sovereign herself would give away the brides in Westminster Abbey at any marriages between them that might be agreed; each of the couples would be given £5000 as a wedding present, and the state would defray the expenses of maintaining and educating their children.

He argued that hereditary improvement was necessary because civilization was advancing more rapidly than our ability to cope with it: "The natural qualifications of our race are no greater than they used to be in semi-barbarous times, though the conditions amid which we are born are vastly more complex than of old. The foremost minds of the present day seem to stagger and halt under an intellectual load too heavy for their powers" (166).

Finally, he met the objection that merely encouraging marriages between gifted individuals would not in itself lead to racial improvement: "If we divided the rising generation into two castes, A and B, of which A was selected for natural gifts, and B was the refuse, then, supposing marriage was confined within the pale of the caste in which each individual belonged, it might be objected that we should simply differentiate our race—that we should make a good and a bad caste, but we should not improve the race as a whole" (319). His reply was that it was also necessary to increase the fertility of the A's (positive eugenics) and to decrease that of the B's (negative eugenics): "Any agency, however indirect, that would somewhat hasten the marriages in caste A, and retard those in caste B, would result in a larger proportion of children being born to A than to B, and would end by wholly eliminating B, and replacing it by A" (319).

He concluded his treatment of this subject in "Hereditary talent and character":

> I hence conclude that the improvement of the breed of mankind is no insuperable difficulty. If everybody were to agree on the improvement of the race of man being a matter of the very utmost importance, and if the theory of the hereditary transmission of qualities in men were as thoroughly understood as it is in the case of our domestic animals, I see no absurdity in supposing that, in some way or other, the improvement would be carried into effect. (Galton 1865a, 319–320)

Galton elaborated his discussion of the effect of age at marriage in *Hereditary Genius* in a chapter on "Influences That Affect the Natural Ability of Nations." He attacked the Malthusian maxim of delaying marriage to avoid over-population on the grounds that it was dysgenic; it was put forward as a rule of conduct for the prudent part of mankind to follow, while the imprudent were left free to disregard it. He also renewed his attack on the Church: "The long period of the dark ages under which Europe has lain is due, I believe in a very considerable degree, to the celibacy enjoined by religious orders on their votaries. Whenever a man or woman was possessed of a gentle nature that fitted him or her to deeds of charity, to meditation, to literature, or to art, the social condition of the time was such that they had no refuge elsewhere than in the bosom of the Church [which] chose to preach and exact celibacy. The consequence was that ... the Church brutalized the breed of our forefathers.... No wonder that club-law prevailed for centuries over Europe" (1869, 357). He concluded the chapter with a passage that illustrates the meritocratic nature of his proposals:

> The best form of civilization in respect to the improvement of the race, would be one in which society was not costly; where incomes were chiefly derived from professional sources, and not much through inheritance; where every lad had a chance of showing his abilities, and, if highly gifted, was enabled to achieve a first-class education and entrance into professional life, by the liberal help of the exhibitions and scholarships which he had gained in his early youth; where marriage was held in as high honour as in ancient Jewish times; where the pride of race was encouraged (of course I do not refer to the nonsensical sentiment of the present day, that goes under that name); where the weak could find a welcome and a refuge

> in celibate monasteries or sisterhoods, and lastly, where the better sort of emigrants and refugees from other lands were invited and welcomed, and their descendants naturalized. (Galton 1869, 362)

In 1883 Galton published his book *Inquiries into Human Faculty and Its Development* in which he brought together many of his researches. The intention of the book was "to touch on various topics more or less connected with that of the cultivation of race, or, as we might call it, with 'eugenic' questions, and to present the results of several of my own separate investigations" (17). His ideas for implementing eugenics remained rather vague. He suggested that a scheme of marks for family merit should be devised, so that ancestral qualities as well as personal qualities could be taken into account. Individuals with a high composite score should then be encouraged to marry early by the provision of endowments as well as by a sense of pride: "The attitude of mind that I should expect to predominate among those who had undeniable claims to rank as members of an exceptionally gifted race, would be akin to that of the modern possessors of ancestral property or hereditary rank.... A man of good race would shrink from spoiling it by a lower marriage" (216). (Galton used the word "race" in a very loose way.) He pointed out that medieval endowments to Oxford and Cambridge had been dysgenic, since college statutes forbade Fellows of Colleges to marry, under the penalty of losing their Fellowships: "It is as though the winning horses at races were rendered ineligible to become sires, which I need hardly say is the exact reverse of the practice" (215). These ancient statutes had recently been repealed, and he had no doubt that the number of Englishmen naturally endowed with high scholastic faculties would in consequence be sensibly increased in future generations. Endowments directed to promoting early marriages in the classes to be favored would have a corresponding beneficial effect in improving the race: "The stream of charity is not unlimited, and it is requisite for the speedier evolution of a more perfect humanity that it should be so distributed as to favour the best-adapted races" (219).

With regard to the other side of the coin, he wrote: "I have not spoken of the repression of the rest, believing that it would ensue indirectly as a matter of course; but I may add that few would deserve better of their country than those who determine to live celibate lives, through a reasonable conviction that their issue would probably be less fitted than the generality to play their part as citizens" (219). He later hardened this rather naive attitude, writing in his autobiography: "I think that stern compulsion ought to be exerted to prevent the free propagation of the stock of those who are seriously afflicted by lunacy, feeble-mindedness, habitual criminality, and pauperism.... How to restrain ill-omened marriages is a question by itself, whether it should be effected by seclusion, or in other ways yet to be devised that are consistent with a humane and well-informed public opinion" (1908, 311). He wrote to Karl Pearson in 1907: "Except by sterilization I cannot yet see any way of checking the produce of the unfit who are allowed their liberty and are below the reach of moral control" (Kevles 1995, 94).

Galton stated that in 1865 he had overrated the speed with which a great racial improvement might be made because he had not then made out the law of regression (1908, 318). By this he meant that selection of small deviations from the racial type (individual differences) would not lead to permanent change since they would be subject to regression back to the central type; only selection of large deviations (sports) would be effective since they represent a new focus of regression. He expressed this idea in *Inquiries into Human Faculty*:

> So long as the race remains radically the same, the stringent selection of the best specimens to rear and breed from, can never lead to any permanent result. The attempt to raise the standard of such a race is like the labour of Sisyphus in rolling his stone uphill; let the effort be relaxed for a moment, and the stone will roll back. Whenever a new typical centre appears, it is as though there were a facet upon the lower surface of the stone, on which it is capable of resting without rolling back. It affords a temporary sticking-point in the forward progress of evolution....
>
> Whenever a low race is preserved under conditions of life that exact a high level of efficiency, it must be subjected to rigorous selection. The few best specimens of that race can alone be allowed to become parents, and not many of their descendants can be allowed to live. On the other hand, if a higher race be substituted for the low one, all this terrible misery disappears. The most merciful form of what I ventured to call "eugenics" would consist in watching for the indications of superior strains or races, and in so favouring them that their progeny shall outnumber and gradually replace that of the old one. (Galton 1883, 198–200)

In other words, eugenics should concentrate on selecting sports rather than small individual differences. His ideas on discontinuity in evolution are discussed in chapter 9.

Galton spent the last decade of his life expounding eugenics as a social, political and religious creed, writing: "It must be introduced into the national conscience, like a new religion. It has, indeed, strong claims to become an orthodox religious tenet of the future, for Eugenics cooperates with the workings of Nature by securing that humanity shall be represented by the fittest races. What Nature does blindly, slowly, and ruthlessly, man may do providently, quickly, and kindly.... The improvement of our stock seems to me one of the highest objects that we can reasonably attempt" (1904c, 42).

Later History of Eugenics

Eugenic movements based on Galtonian principles developed in more than thirty countries in the first few decades of the twentieth century. The most significant and best studied of these national movements were in Britain, America, and Germany, which are considered briefly here. I have used these secondary sources: Blacker (1952), Burleigh (2000), Carlson (2001), Kershaw (1998 and 2000), Kevles (1995), Kühl (1994), Lombardo (2002a,b,c), Mazumdar (1992), Peel (1998), Proctor (1988), Ridley (1999), Weindling (1989), and Weiss (1990).

Britain

In his paper on "Eugenics. Its definition, scope and aims," Galton concluded: "I see no impossibility in Eugenics becoming a religious dogma among mankind, but its details must first be worked out sedulously in the study. Over-zeal leading to hasty action would do harm ... and cause the science to be discredited" (1904c, 43). In accordance with his view that further research should precede action, he gave London University £500 a year in 1904 to support a Eugenics Research Fellow, and he left the university £45,000 in his will to endow a Chair at University College London. Karl Pearson was Galton Professor of Eugenics from 1911 to 1933; R. A. Fisher, statistician, geneticist, and keen eugenicist, from 1933 to 1944; and Lionel Penrose, an expert on mental deficiency and an opponent of eugenics, from 1944 to 1965. The department conducted biometric and genetic studies on human heredity, many of them published in its journal *Annals of Eugenics*, which was founded in 1925 and changed its title to *Annals of Human Genetics* in 1954.

The more activist Eugenics Education Society was founded in 1907 by a group of enthusiasts to educate the public about the importance of eugenics. Galton agreed, rather reluctantly, to become its honorary president in 1908, though he feared it might attract cranks; its president from 1911 to 1928 was Charles Darwin's son Leonard Darwin. The society, known as the Eugenics Society from 1926 and as the Galton Institute from 1989, was similar in its attitudes and membership to the meliorist societies that arose in Victorian and Edwardian England to improve the drains and to counter alcoholism and venereal disease, though it concentrated on improving the genetic quality of the nation rather than its sanitation. It published popular articles in the *Eugenics Review*, and it advocated practical measures for both positive and negative eugenics.

British eugenics was dominated by the problem of the differential fertility of the classes. Galton (1901) suggested that "civic worth" was approximately normally distributed and that it could be identified with the social categories used by Charles Booth in his social survey of London. At the bottom was the small number of criminals and loafers of little worth; in the middle was the great mass of the respectable working class of moderate worth; at the top were the independent professionals and large employers of high civic worth. Galton assumed that civic worth was heritable, from which he concluded that the human breed could be improved by encouraging the reproductive productivity of the upper at the expense of the lower classes.

It became clear at the beginning of the twentieth century that the upper social classes were in fact breeding less rapidly than the lower ones, presumably due to greater use of birth control methods. For example, a study at the Galton Laboratory compared the birthrates in the upper-class London suburb of Hampstead and the working-class suburb of Shoreditch. In the 1880s the rate was about 30 births per 1000 in both suburbs; it was about the same in Shoreditch in 1910, but had declined to about 17 births per thousand in Hampstead. An analysis of the census results of 1911 for England and Wales allowed a comparison to be made between the fertilities of the five social classes, with class 1 representing the upper and middle classes, class 3 skilled workers and class 5 unskilled workers. The average number of living children per family for the five classes, from class 1 to class 5, in 1911 were 1.68, 2.05, 2.32, 2.37, and 2.68, the unskilled workers having 1.6 times as many children as the upper and middle classes. If one equates civic worth with high social class and assumes it to be largely hereditary, the higher fertility of the lower classes implies rapid national deterioration.

To counter this problem, the Eugenics Education Society, led by its president Leonard Darwin and his close friend R. A. Fisher, proposed that the tax system should be modified to give larger family allowances to the better-off in order to encourage them to have more children. (See the last few chapters of Fisher 1930 and the letters from Fisher to Darwin in Bennett 1983). The society opposed flat-rate allowances as wholly dysgenic, since they were worth less to the better-off than to the worse-off. In Fisher's scheme the government would provide an allowance for each child proportional to the family's earned income; he defended this scheme by remarking that it replaced the principle of equal pay for equal work by that of an equal standard of living for equal work. Governments expressed sympathy with such arguments but never took them seriously; the idea of redistributing wealth from the poor to the rich was presumably politically unattractive. The Eugenics Society, under its next president, C. P. Blacker, tried to forge links with the birth control movement, presumably in the hope that this would lower the fertility of the lower classes, though Leonard Darwin and the old guard in the society always regarded birth control as dysgenic since it would be more effectively put into practice by the better educated.

The other plank of the British eugenics movement was to prevent feeble-minded people from reproducing, either by segregation or sterilization. A Royal Commission in 1908 concluded that feeble-mindedness was largely inherited and that such people needed to be segregated from society. After intensive lobbying, a bill was introduced into Parliament which would restrict procreation by feeble-minded people and would punish those who married mental defectives. It was an open secret that it could be amended to allow compulsory sterilization, and it had the enthusiastic support of Winston Churchill, who privately advocated the sterilization of the mentally unfit (Ridley 1999, 294). But it was fiercely opposed on libertarian grounds, and was only approved in a watered-down form in which the clauses regulating marriage and preventing procreation were dropped. The resulting Mental Deficiency Act of 1913 allowed the forcible commitment of mental and moral defectives in colonies where they were segregated from the public. This had the practical effect of limiting their chances of reproducing, and was claimed as a success by the eugenics movement.

But it did not go far enough. In 1929 the Wood Committee, on which the Eugenics Society was well represented, defined the lowest 10 percent on the social scale as the social problem group; it was associated not only with feeble-mindedness, but also with insanity, epilepsy, pauperism, crime, unemployability, and alcoholism. Only about 10 percent of this group could be certified under the 1913 act, but the committee thought the others probably carried hereditary defect and suggested that they should also be segregated to protect them from themselves and to protect society from their excessive fertility. Meanwhile, the Eugenics Society campaigned for the legalization of sterilization on eugenic grounds and persuaded the government to set up a committee under Sir Laurence Brock to examine the question. The Brock Committee, of which R. A. Fisher was an influential member, reported in 1934 in favor of allowing voluntary sterilization of the mentally unfit (though it glossed over the problem of obtaining informed consent from them). But no further legislation followed to implement the recommendations of the Wood and Brock committees. The climate of opinion had moved against eugenic ideas. There was growing recognition of the importance of environmental explanations of human behavior, of the complexity of genetic explanations, and of the slowness of eugenic measures in reducing the incidence of genetic disorders. There was hostility from libertarians, from the political left, which considered eugenics with its talk of "the residuum" and "the social problem group" as an attack on the working class, and from the Catholic Church, which condemned eugenics in general and sterilization in particular.

America

Charles B. Davenport (1866–1944) was trained in mathematics as well as biology and read Karl Pearson's papers on the mathematical theory of evolution in the 1890s. He was recruited by Pearson as the American editor of *Biometrika* when he founded the journal in 1901 (see chapter 10), and after visiting Galton, Weldon, and Pearson in England, he wrote to Galton in 1902 that he returned home with "renewed courage for the fight for the quantitative study of Evolution" (Kevles 1995, 45). In 1904 he obtained a handsome grant from the Carnegie Institution of Washington to establish a station for the experimental study of evolution at Cold Spring Harbor on Long Island, New York. He was an enthusiastic Mendelian, and did important early work on Mendelism in canaries and poultry, and on the inheritance of human eye color. (The latter work displeased Pearson, who wrote to Galton in 1909 that his assistant Heron "dealt with attacks on your eye-colour data from the side of Davenport and Hurst, who assert that two true blue eyed parents always have blue eyed children" [Pearson 1930a, 376]. Davenport and Hurst were of course correct in their assertion that blue eye color is a Mendelian recessive trait. The dispute between the biometricians and the Mendelians is discussed in chapter 8.)

Davenport's interest in eugenics was stimulated by his genetic work and by his interaction with Galton and Pearson. In 1910 he obtained a gift from the wealthy widow Mrs. E. H. Harriman to establish the Eugenics Record Office on land near the experimental station, which he described in a letter to Galton:

> As the enclosed printed matter will show in some detail, there has been started here a Record Office in *Eugenics*; so you see the seed sown by you is still sprouting in distant countries. And there is great interest in Eugenics in America, I can assure you.
>
> We have a plot of ground of 80 acres, near New York City, and a house with a fireproof addition for our records. We have a Superintendent, a stenographer and two helpers, besides six trained field-workers.... We ... have established very cordial relations with institutions for imbeciles, epileptics, insane and criminals. We are studying communities with high consanguinity also [the Amish in Pennsylvania]. (Pearson 1930b, 613)

The Eugenics Record Office became the most influential organization of the American eugenics movement through its collection and analysis of pedigree data on dysgenic traits and through its training of fieldworkers. Unfortunately, the superintendent whom Davenport appointed, Harry H. Laughlin (1880–1943), was rather overzealous in his enthusiasm for eugenics and rather careless in his use of data. This eventually led to disillusionment with the work of the Record Office and to its closure in 1939.

Eugenic legislation fell into three categories, compulsory sterilization of the unfit, the prevention of dysgenic marriages, and immigration control. The first two came under state control, the third under federal control. The following account of this legislation is based on Kevles (1995), Lombardo (2002a,b,c), and Carlson (2001).

The first law for compulsory eugenic sterilization was passed in Indiana in 1907, the preamble stating: "Whereas heredity plays a most important part in the transmission of crime, idiocy and imbecility" (Kevles 1995, 109). By 1917 sterilization laws were passed by 15 more states, most of them giving the power to sterilize habitual criminals as well as epileptics, the insane, and idiots in state institutions. However, these early laws were not widely enforced because of legal difficulties; by 1924 only about 3000 people had been involuntarily sterilized in America, most of them (2500) in California.

In 1924 Virginia decided to pass a law that would withstand legal challenge, based on a model eugenical sterilization law published by Harry Laughlin in 1914. Its preamble stated that "heredity plays an important part in the transmission of insanity, idiocy, imbecility, epilepsy and crime" (Lombardo 2002a), and it focused on defective persons whose reproduction represented a menace to society. An opportunity to test the law was presented when a seventeen-year-old girl named Carrie Buck was committed to the Virginia Colony for Epileptics and Feebleminded in Lynchburg. Carrie was found to have a mental age of nine years, and she had just had an illegitimate daughter, Vivian. Carrie's mother, Emma, was also certified to be feebleminded and had lived in the colony since 1920. The colony's board of directors ordered Carrie to be sterilized under the new law, and a court-appointed guardian appealed the order to establish its legality. Virginia officials consulted Harry Laughlin at the Eugenics Record Office, and he gave a deposition, without having examined her, that Carrie's feeblemindedness was primarily hereditary, and that she and her forebears "belong to the shiftless, ignorant, and worthless class of anti-social whites of the South" (Lombardo 2002a). Arthur Estabrook of the Eugenics Record Office gave Vivian a mental test for an infant, from which he concluded that she was below average intelligence for her age and he testified that the feeblemindedness in the Buck line conformed to the Mendelian laws of inheritance. The judge upheld the sterilization order, and his ruling was confirmed by the Virginia Supreme Court of Appeals. In 1927 the case of *Buck* v. *Bell* (John Bell was the colony's superintendent) was taken to the United States Supreme Court, which upheld the Virginia law by a vote of eight to one. The court's opinion was written by Justice Oliver Wendell Holmes:

> We have seen more than once that the public welfare may call upon the best citizens for their lives. It would be strange if it could not call upon those who already sap the strength of the state for these lesser sacrifices ... in order to prevent our being swamped with incompetents. It is better for the world, if instead of waiting to execute degenerate offspring for crime, or to let them starve for their imbecility, society can prevent those who are manifestly unfit from continuing their kind. The principle that sustains compulsory vaccination is broad enough to cover cutting the fallopian tubes. Three generations of imbeciles are enough. (Carlson 2001, 255)

This ruling provided legitimacy for sterilization throughout America. The average rate of about 400 sterilizations per year before 1930 increased to about 3000 sterilizations per year in the 1930s (Kühl 1994). It is now known that Carrie's illegitimate child was not the result of promiscuity, but that she had been raped by a relative of her foster parents.

Vivian died in 1932, but her school records show that she was an above average student.

Marriage laws were of two kinds, prohibiting the marriage of the mentally deficient and prohibiting the marriage of blacks and whites. Laws preventing the marriage of the mentally deficient were in force in about thirty states by 1914, and were justified both because they were incapable of making contracts and on eugenic grounds. For example, Indiana passed a law forbidding the marriage of the mentally deficient, persons having a transmissible disease, and habitual drunkards, and requiring a health certificate of those released from institutions.

Laws forbidding marriage between people of different races were common in America from the colonial period, and twenty-eight states made marriages between Negroes and white persons invalid. The arguments supporting such restrictions were reinforced by the eugenics movement, particularly by the extreme views expressed by Madison Grant in his influential book *The Passing of the Great Race* (1916). Grant maintained that the white race, and in particular its Nordic subrace, was genetically superior to other races. He also had an extreme view about the consequence of mixed marriages: "The cross between a white man and an Indian is an Indian; the cross between a white man and a negro is a negro.... When it becomes thoroughly understood that the children of mixed marriages between contrasted races belong to the lower type, the importance of transmitting in unimpaired purity the blood inheritance of ages will be appreciated at its full value" (Lombardo 2002b). Grant was a New York lawyer with little knowledge of biology but a strong interest in conservation, in particular the founding of the Bronx Zoo and the campaign to save the redwoods. His views on race were not considered controversial in the 1920s when many Americans feared the transition from an Anglo-Saxon to a multiethnic country (Carlson 2001, 263).

In Virginia, three local eugenicists consulted with Madison Grant and Harry Laughlin to frame Virginia's Racial Integrity Act of 1924. The three local men were respectively, the founder of the Anglo-Saxon Clubs of America (dedicated to maintaining Anglo-Saxon ideals and civilization in America), the author of *White America* (emphasizing white supremacy and the dangers of racial mixing), and the registrar of the Bureau of Vital Statistics (who included advice about racial interbreeding as the source of public health problems in pamphlets distributed to people planning to marry). The Racial Integrity Act provided that "it shall hereafter be unlawful for any white person in this state to marry any save a white person, or a person with no other admixture of blood than white and

American Indian.... The term 'white person' shall apply only to such person as has no trace whatever of any blood other than Caucasian; but persons who have one-sixteenth or less of the blood of the American Indian and have no other non-Caucasic blood shall be deemed to be white persons" (Lombardo 2002b). The clauses about American Indians were included to satisfy members of the Virginia General Assembly who were descendants of Pocahontas.

In 1958 a white man, Richard Loving, married a black woman in Washington, D.C., and the couple moved to Virginia. They were convicted under the 1924 act and were sentenced to one year in jail, suspended provided that they accepted banishment from the state. The judge declared: "Almighty God created the races white, black, yellow, malay and red, and He placed them on separate continents.... The fact that He separated the races shows that He did not intend for the races to mix" (Lombardo 2002b). The couple moved back to Washington, D.C., until 1967, when the United States Supreme Court, in *Loving* v. *Commonwealth of Virginia*, unanimously struck down the Racial Integrity Act and similar laws of fifteen other states as violating the Equal Protection Clause (the Fourteenth Amendment to the Constitution of 1868 which became the centerpiece for the civil rights movement after the Second World War).

The 1882 Act to Regulate Immigration prohibited entry to the country to "any person unable to take care of himself or herself without becoming a public charge" (Lombardo 2002c). By 1917 the definition had been expanded to include idiots, imbeciles, feebleminded persons, epileptics, insane persons, and mentally or physically defective persons. These provisions excluded immigrants who were likely to be costly to society as well as having an obvious eugenic justification. But by this time the immigration rate was increasing rapidly, and there was pressure to reduce it from several quarters, from nativists wanting to preserve American culture, from labor unions wanting to restrict the influx of cheap foreign labor, from employers fearful of the radical ideas of the new immigrants, and from eugenicists fearful that the new immigrants from southern and eastern Europe were below average intelligence and were racially inferior to the native population. Charles Davenport argued that selection should be on an individual basis, with no national or racial group being prioritized for acceptance or rejection, but by the 1920s this eugenic principle had been submerged under racial prejudice against southern and eastern Europeans, a prejudice shared by the superintendent of the Eugenics Record Office, Harry Laughlin, who was proud that his family could be traced back to the American Revolution. In 1920, Laughlin appeared before a House of Representatives Committee on Immigration and argued, from a survey of the number of foreign-born persons in jails, prisons, and reformatories, that the American gene pool was being polluted by the large number of intellectually and morally defective immigrants, primarily from southern and eastern Europe. The chairman, Albert Johnson, liked the message and appointed Laughlin as the committee's expert eugenics agent. In this capacity Laughlin did more research along the same lines, which he presented to Congress in support of a bill to restrict immigration, which became the Immigration Restriction Act of 1924. This law scaled the number of immigrants from each country in proportion to their percentage of the U.S. population in 1890, before large-scale immigration began. Under this law, the quota of southern and eastern Europeans was reduced from 45 percent to 15 percent.

Germany

The German eugenics movement was founded by Alfred Ploetz (1860–1940). As a young man he wanted to establish some kind of pan-Germanic utopian socialist commune and he spent six months in the cooperative known as Icarus in Iowa to study how such communes worked. He was appalled by the egotism and squabbling that he found there, and he concluded that "the plans we wished to execute would be destroyed as a result of the low quality of human beings.... For this reason I must direct my efforts not merely toward preserving the race but also toward improving it.... My views ... immediately led me to the field of medicine—which appeared to be relevant to the biological transformation of human beings." (Weiss 1990, 15). In 1895 he published a monograph on *Rassenhygiene*, or race hygiene. He used the word "race" in a loose sense for any interbreeding human population, including a small ethnic community, a nation, an anthropological race, or the entire human race, so that his term *Rassenhygiene* meant much the same as Galton's term *eugenics*. He argued that protection of the weak and underprivileged in a civilized society must lead to degeneration of the race unless steps were taken to counter it by selective breeding. He thought that Galton's solution of state controls on marriage was too elitist, and he advocated a system of "reproductive hygiene" which would be elaborated in a promised second volume. By this he meant the selection of reproductive cells, for which the technology would not be developed for a hundred years. The idea is similar to that of embryo selection, but he did not pursue it further.

In 1904 Ploetz founded the *Archiv für Rassen- und Gesellschafts-Biologie* (Archives for racial and societal biology), the first journal dedicated to eugenics in the world. It sought to attract articles bearing on the optimal preservation and development of the race, and included contributions from leading biologists such as Weismann, Plate, Correns, de Vries, and Johannsen. In the next year Ploetz was the major force underlying the foundation of the Gesellschaft für Rassenhygiene (Racial Hygiene Society), the world's first professional eugenics organization, of which Francis Galton became honorary vice-president in 1909. It offered membership to white people who were ethically, intellectually, and physically fit and who were likely to be economically prosperous.

Thus the society restricted its membership to a breeding elite of the eugenically fit, the criterion of fitness having a strong class bias toward the professional middle class. As another German eugenicist, Wilhelm Schallmayer (1857–1919), wrote in 1903: "In the meantime, it would not be incorrect to view highly socially productive individuals, especially the better educated, as being, on the average, more biologically valuable" (Weiss 1990, 21). Membership was restricted to whites, but this included any type of white person; almost all educated white people in Europe and America at that time accepted the racial and cultural superiority of Caucasians. This did not imply the idea of Aryan or Nordic superiority over other branches of the Caucasian race, though there was some ambivalence about this. Ploetz was himself mildly anti-Semitic and was sympathetic to pan-Germanism and to ideas of Nordic superiority. (He joined a secret Nordic cabal called The Bow, symbolizing Nordic vitality, within the Racial Hygiene Society.) But he was keen to maintain the scientific reputation of the archives and of the society, and he distanced himself from the wilder Nordic excesses of Gobineau and of Houston Stewart Chamberlain (see, for example, the extract from *Mein Kampf* below). Schallmayer, on the other hand, was a firm opponent of any form of Aryan racial theory, which he regarded as unscientific nonsense. Thus all members of the Racial Hygiene Society favored eugenic proposals that used a class-based definition of fitness, while some of them wished also to allow the possibility of a Nordic race hygiene. The hidden conflict between the racist and the nonracist members of the society became apparent in the Weimar years after the First World War as a conflict between the Berlin and Munich chapters of the society, with the more left-wing, nonracist members concentrated in Berlin and the more nationalist, racist members, headed by Fritz Lenz, in Munich. But the Munich chapter never tried to put any racist ideas into practice.

The German eugenics movement before the Nazis came to power was unsuccessful in implementing eugenic ideas. The Racial Hygiene Society was more a talking shop than a populist movement, though it gave birth under the Weimar Republic to a more populist daughter group, the Deutscher Bund. The society had monitored developments in America, particularly the sterilization laws, but did not press even for voluntary sterilization of the unfit under the empire because it was felt that it would not have had popular support. During the economic and psychological depression of the Weimar Republic, opinion moved in favor of sterilization, and publicity was generated by Heinrich Boeters, a medical officer in Saxony, who singlehandedly referred several dozen patients, mostly children, for eugenic sterilization, covering himself by obtaining parental consent and a certificate from a psychiatrist that they suffered from a hereditary malady. The final straw was the depression of 1929–32, which led to the need to cut welfare costs. In 1932 the Prussian government was persuaded by the arguments of eugenicists to introduce several eugenic proposals, including a draft sterilization law, but this law never came into effect because of the political instability at that time.

In the spring of 1933 the Weimar Republic collapsed and the Nazis under Adolf Hitler came into power. Nazism can be regarded as a political religion (Burleigh 2000) whose mission was the preservation and improvement of the German (Aryan) race. Its aims were to reverse the humiliation of the Versailles Treaty of 1918; to create a Greater Germany, by expansion to the East, in which all German-speaking people could be united; to exclude from this Reich, by expulsion or otherwise, all non-Aryans and in particular Jews; and to improve the health of Aryans within the Reich by the promotion of public health and by eugenic measures. Nazi philosophy rejected the liberal view that there might be a conflict between the interest of the individual and that of the state, and maintained that the good of the whole always comes before the good of the individual (*Gemeinnutz geht vor Eigennutz*); it was therefore receptive to eugenic ideas.

The superiority of the Aryan race and the danger of its dilution by the Jewish race was at the heart of Nazism. Hermann Göring stated with chilling clarity: "This war is not the Second World War. This is the great racial war. In the final analysis it is about whether the German and Aryan prevails here, or whether the Jew rules the world, and that is what we are fighting for out there" (Burleigh 2000, 571). Twenty years earlier

Hitler had characterized in *Mein Kampf* the difference between the two races:

> All the human culture, all the results of art, science, and technology that we see before us today, are almost exclusively the creative product of the Aryan.... The Aryan is not greatest in his mental qualities as such, but in the extent of his willingness to put all his abilities in the service of the community. In him the instinct of self-preservation has reached the noblest form, since he willingly subordinates his own ego to the life of the community and, if the hour demands, even sacrifices it.... The mightiest counterpart to the Aryan is represented by the Jew. In hardly any people in the world is the instinct of self-preservation developed more strongly than in the so-called chosen.... If the Jews were alone in this world, they would stifle in filth and offal; they would try to get ahead of one another in hate-filled struggle and exterminate one another, in so far as the absolute absence of all sense of self-sacrifice, expressing itself in their cowardice, did not turn battle into comedy here too. (Hitler 1925, 263–274)

Thus the Aryans were the founders of culture and the Jews were the destroyers of culture since they could only exist as parasites on other races, whom they would eventually bring down. He concluded:

> On the basis of this inner realization [that the Jew was the enemy within], there took form in our new movement the leading principles ... not only of halting the decline of the German people, but of creating the granite foundation upon which some day a state will rest which represents, not an alien mechanism of economic concerns and interests, but a national organism: *A Germanic State of the German Nation.* (Hitler 1925, 299)

After they came to power in 1933, the Nazis introduced two types of eugenic measures: to improve the German (Aryan) race, and to prevent its mixture with inferior races, especially Jews, which are considered in turn.

The Law for the Prevention of Hereditarily Diseased Progeny or Sterilization Law (July 14 1933) mandated compulsory sterilization for those suffering from eight allegedly hereditary illnesses: congenital feeble-mindedness, schizophrenia, manic-depressive illness, epilepsy, Huntington's chorea, hereditary blindness and deafness, and severe physical malformations whose hereditary character had been sufficiently established by research; alcoholism was included in a separate section. It was administered by Hereditary Health Courts, with higher courts to hear appeals, each court consisting of a judge, a public health doctor and another medical expert. The system was praised in American eugenic circles because it contained no elements of race or of punishment. It built on pre-Nazi proposals, with the substitution of compulsory for voluntary sterilization. The Jewish eugenicist Richard Goldschmidt, who was forced by the Nazis to flee to America from his position at the Kaiser Wilhelm Institute for Anthropology, Human Heredity, and Eugenics, complained that the Nazis "took over our entire plan of eugenic measures" (Kühl 1994, 52).

It has been estimated that about 400,000 sterilizations were carried out. Feeblemindedness accounted for over half the cases, followed in frequency by schizophrenia (about one quarter), epilepsy, and then alcoholism (Proctor 1988). There was considerable resistance to the program, which the Nazis tried to counter with propaganda films such as *Erbkrank* (Hereditarily Ill); in particular there were complaints when members of the Nazi Party were brought before the Health Courts for sterilization on the grounds of feeblemindedness for not knowing the answer to questions such as "When was Columbus born?" In 1937 the Reich doctors' leader, Gerhard Wagner, complained that the health courts paid insufficient attention to family history in diagnosing feeblemindedness and schizophrenia, and he characterized the responsible civil servant in the Ministry of the Interior as a shortsighted genetic health fanatic who had fallen victim to medical overenthusiasm. The sterilization law was supplemented by laws allowing the castration of dangerous habitual criminals (1933) and prohibiting marriages between healthy and mentally retarded persons (1935). The sterilization program was wound down at the outbreak of war in September 1939, perhaps because the Nazis had decided to replace it by the "euthanasia" program.

In 1928 Hitler wrote of his admiration for the eugenic system practiced in ancient Sparta, designed to breed a race of first-class fighting men, and in particular for the killing of any defective newborn children. He realized that this would be unacceptable in peacetime, but he told the doctors' leader, Gerhard Wagner, at the Nuremberg rally in 1935 that, if war broke out, he would take up the euthanasia question and implement it (Weindling 1989, 545–547; Burleigh 2000, 383). (In this context, euthanasia meant the compulsory killing of the unfit. It was justified on economic grounds as being more cost-effective than sterilization.) He did this in 1939, but the whole program was kept secret, with false causes of death and death certificates being given to the relatives; it was never legalized but was carried out on the personal authority of the Führer. One part of the program involved the killing of children with congenital deformities such as Down's syndrome by lethal injection or by gassing, though it was made to look as if death had been from natural causes. About 5000 children died in this way. Another program involved the killing of adults, mostly in psychiatric institutions, by gassing. About 70,000 individuals were killed by September 1941, when the centrally organized program was halted due to public unease, since such a large operation could not be completely concealed; but it continued after that date on an individual hospital basis, with death due to lethal injection, poisoning, or starvation, which were easier to conceal than large-scale gassing.

Positive eugenic measures before the war included funding for non-Jewish couples free of mental and physical illness (1933) and for "hereditarily valuable" farmers of "good stock" (1933), and the requirement of a certificate of health for a marriage license (1935). Mention should also be made of Himmler's *Lebensborn* program which encouraged unmarried Aryan girls to "give the Führer a child" through extramarital relations with an SS officer or party member; the program provided maternity homes and child-care institutions, and the children were considered future material for the SS.

The most notorious aspect of Nazi eugenics was their discrimination against Jews, leading ultimately to the Holocaust. Before the war (1933–39), there was extensive harassment of Jews, culminating in a pogrom on November 9 1938 known as *Reichskristallnacht*. The pretext was the death on that day of a legation secretary in the German embassy in Paris as a result of being shot by a Polish Jew; the name *Kristallnacht* comes from the smashed glass windows of Jewish-owned shops. The main discriminatory legislation were the Nuremberg Laws of 1935, comprising the Citizenship Law under which Jews lost German citizenship, becoming "state subjects," and the Law for the Protection of German Blood and Honor, which prohibited marriage and sexual intercourse between Aryans and Jews. There were complicated rules about how these laws were to be interpreted for the existing offspring of mixed marriages. These laws and the anti-Jewish sentiment that underpinned them led to the exclusion of Jews from most professional and business activity and persuaded many of them to emigrate.

After the war began in 1939, the Jews in German-occupied territories added to the problem of the German Jews. Many of the despised eastern Jews in Poland and western Russia and the Baltic states were murdered, but the Germans were unsure how to tackle the general problem of eliminating all Jews from the territories they controlled. At first they toyed with the idea of transporting the Jews to the French-controlled island of Madagascar off the east coast of Africa. But the Madagascar solution depended on obtaining control of the seas following the defeat of Britain. When this did not happen, they turned to the "final solution" of the problem. With Hitler's approval, Reinhard Heydrich, the SS officer in charge of the "Jewish problem," called a meeting at Wannsee in January 1942 to draw up plans for the extermination of all Jews in territories under German control in death camps, of which Auschwitz was the most notorious. Between five and six million Jews died in the Holocaust during the Second World War.

The Rationale of Eugenics

It is interesting to contrast Galton's views on eugenics with those of Charles Darwin, described in *The Descent of Man*. Darwin accepted Galton's argument about the dysgenic effects of civilization: "With savages, the weak in body or mind are soon eliminated; and those that survive commonly exhibit a vigorous state of health. We civilised men, on the other hand, do our utmost to check the process of elimination; we build asylums for the imbecile, the maimed, and the sick; we institute poor-laws; and our medical men exert their utmost skill to save the life of every one to the last moment.... Thus the weak members of civilised societies propagate their kind. No one who has attended to the breeding of domestic animals will doubt that this must be highly injurious to the race of man" (1871a, 1:168). But he did not accept Galton's conclusion: "The surgeon may harden himself whilst performing an operation, for he knows that he is acting for the good of his patient; but if we were intentionally to neglect the weak and helpless, it could only be for a contingent benefit, with a great and certain present evil. Hence we must bear without complaining the undoubtedly bad effects of the weak surviving and propagating their kind" (169).

This is the classical libertarian argument, that it is wrong to sacrifice the rights of the individual to the benefit of society as a whole. It provides a strong argument for opposing state-controlled eugenic measures, however well-intentioned. Galton's enthusiasm for eugenics reveals his desire to find a substitute for the religious convictions that he had rejected and his tough-minded subordination of the good of the individual to that of society. The latter is exemplified by his group selectionist interpretation of Darwinism: "Individuals appear to me as partial detachments from the infinite ocean of Being, and this world as a stage on which Evolution takes place, principally hitherto by means of Natural Selection, which achieves the good of the whole with scant regard to that of the individual" (1908, 323).

Enthusiasm for eugenics is strongly correlated with disregard for individual rights. In Britain, eugenics became popular in the early 1900s, particularly among Fabian and other like-minded socialists (for example, Sydney and Beatrice Webb, George Bernard Shaw, H. G. Wells, Karl Pearson), to whom ideas based on state control appealed; but opposition from libertarians prevented the passage of legislation through Parliament. Kevles (1995) has suggested that it was easier to pass eugenic legislation in America than in Britain, partly because it was the province of state legislatures, but mainly because of the greater willingness to rely on scientific expertise in drafting legislation; in the field of public health this leads to emphasis on the public at the expense of the individual good when there may be a conflict between the two, as seen in debates over compulsory vaccination as well as compulsory sterilization. In Germany, eugenics became popular among public health doctors with their prejudice in favor of the public good, but it made little legislative progress until the Nazis came to power with their extreme philosophy of subordinating the individual to the race.

Galton concluded his autobiography by restating the aim of eugenics:

> Its first object is to check the birth-rate of the Unfit, instead of allowing them to come into being, though doomed in large numbers to perish prematurely. The second object is the improvement of the race by furthering the productivity of the Fit by early marriages and healthful rearing of their children. Natural Selection rests upon excessive production and wholesale destruction; Eugenics on bringing no more individuals into the world than can be properly cared for, and those only of the best stock. (Galton 1908, 323)

The implementation of this program requires a precise definition of what constitutes fitness. Three definitions have been used in practice, based on hereditary disease, class and race.

Under the first definition the unfit were defined as those who were mentally or physically disadvantaged for hereditary reasons. It led to the sterilization program in America and Nazi Germany, which constituted the main non-racial eugenic measure in those countries. Its justification was the public good of the reduction of the numbers of disadvantaged persons in future generations, at the expense of depriving them of the chance of having children. In retrospect, the program was marred by exaggerating the speed with which hereditary defects could be eliminated and the precision with which conditions such as "feeble-mindedness" could be diagnosed and their genetics determined. The history of compulsory sterilization of the unfit confirms the truth of Galton's warning: "I see no impossibility in Eugenics becoming a religious dogma among mankind, but its details must first be worked out sedulously in the study. Over-zeal leading to hasty action would do harm ... and cause the science to be discredited" (1904c, 43).

Positive eugenics in Britain and pre-Nazi Germany was based on defining fitness by the qualities of the professional middle class to which most eugenicists belonged. There was considerable concern at the beginning of the twentieth century about the relative decline in the birthrate among the middle compared with the lower classes, which fueled discussion about reversing this trend, but it led to little action. Apart from its self-serving nature, this concept of eugenics raises the question of whether a society consisting entirely of university professors would be desirable. Galton came to realize this point:

> What is meant by improvement? What by the syllable *Eu* in Eugenics, whose English equivalent is *good*? ... There are a vast number of conflicting ideals of alternative chatacters ... but all are wanted to give fulness and interest to life. Society would be very dull if every man resembled the highly estimable Marcus Aurelius or Adam Bede. The aim of Eugenics is to represent each class or sect by its best specimens; that done, to leave them to work out their common civilisation in their own way. (Galton 1904c, 35–37)

Put more crudely, society needs good plumbers as well as good neurosurgeons; but it is not easy to see how the state can bring this about by selective breeding, except by the methods caricatured in Aldous Huxley's *Brave New World*.

The third definition of fitness was made in terms of race, the fit being those of one's own race, the unfit those of other races. Racial eugenics was unimportant in Britain, which had no serious racial problems until the wave of immigration after the Second World War. In America it led to legislation against mixed marriages and to discrimination against immigrants from southern and eastern Europe. The more respectable American eugenicists, like Charles Davenport, argued along Galtonian lines that selection of immigrants should be on an individual basis, with no national group being prioritized, but prejudice against southern and eastern Europeans on cultural grounds led to legislation on that basis.

Racial eugenics reached its nadir under the Nazi attempt to exterminate the Jews.

In retrospect, the failure of the Galtonian eugenics program can be attributed to the growth of respect for individual rights in the Western world, to the growth in scientific understanding of the complexity of categories such as "mental deficiency" and of the length of time needed to reduce the incidence even of simple genetic disorders, to the self-serving nature of the class-based definition of fitness, and finally to the excesses of Nazi Germany. Galton himself would probably have approved, as did most contemporary eugenicists, of the Nazi law of 1933 for the sterilization of those with hereditary diseases, but he would have been shocked by the substitution in 1939 of "euthanasia" for those with hereditary disease instead of sterilization, and by their attempt to exterminate the Jews. There is little chance that Galtonian, state-sponsored eugenics will find a future place in the Western world (though China already passed a law for sterilization of those with serious genetic diseases in 1995). But the biotechnology revolution has introduced the possibility of a new, individual eugenics in which individual parents can, through embryo selection (adumbrated by Alfred Ploetz in 1895), choose the genetic characteristics of their children (Fukuyama 2002, D. Galton 2001, Stock 2002). The question is no longer whether the state should encourage its "better" citizens to have more children than its "worse" citizens, but whether it should intervene to prevent parents from choosing "designer babies."

4

The Mechanism of Heredity

> We shall therefore take an approximately correct view of the origin of our life, if we consider our own embryos to have sprung immediately from those embryos whence our parents were developed, and those from the embryos of *their* parents, and so on for ever.
>
> Galton, "Hereditary talent and character"

In this chapter I trace the development of Galton's ideas about the mechanism of heredity, from his paper "Hereditary talent and character" in 1865, in which he summarized the facts of heredity but admitted his ignorance of their underlying mechanism, to his mature views expressed rather cryptically in *Natural Inheritance* in 1889.

Galton's thoughts about the mechanism of heredity were stimulated by the publication of Darwin's theory of pangenesis in 1868, which he tried to verify by transfusing blood between different varieties of rabbits; the negative results of these experiments led him in 1871 to reject that part of the theory which required, as he thought, the transportation of hereditary particles in the blood.

He developed his own theory, based largely on the other parts of Darwin's theory, in two papers, published in 1872 and 1875. In the second paper, he coined the name "stirp" for the sum total of hereditary particles or gemmules in the newly fertilized ovum. He supposed that a few of these gemmules became patent and were developed into the cells of the adult person, the residue remaining latent. Having rejected (except as a rare exception) Darwin's idea that cells throw off gemmules that are collected in the germ cells, he supposed that the germ cells contributing to the next generation predominantly comprised the latent residue of gemmules in the stirp that had not developed into the adult person. He did not appreciate the difficulty of accounting for the correlation between parent and offspring under this model.

After 1875 he turned his attention to developing a statistical rather than a physiological theory of heredity (see chapter 7). His statistical work, in particular the law of ancestral heredity (discussed in chapter 8), led him to suppose that latent and patent gemmules were equally frequent and had the same chance of being transmitted to the next generation. He briefly described this idea, under which the correlation between parent and offspring can be explained much more easily, in *Natural Inheritance* (1889), though he did not explicitly acknowledge the change from his previous theory.

Galton's theory of the hereditary mechanism became obsolete after the rediscovery of Mendelism in 1900, but it is still instructive to consider how he, and other brilliant men like Weismann and de Vries, struggled to explain the confusing facts of heredity in the nineteenth century. This chapter is partly based on Bulmer (1999); Olby (1985) provides background reading.

Galton's Knowledge of Heredity in 1865

I begin by discussing Galton's knowledge of the laws of heredity when he published "Hereditary talent and character" in 1865, prior to the publication of Darwin's theory of pangenesis. In this paper, his main aim was to demonstrate the fact of the inheritance of human mental qualities, but he admitted ignorance of the laws which govern the inheritance even of physical features. He mentioned three sources for the information available to him on this subject (Lucas 1847, Lewes 1859, Darwin 1859); and he recognized three principles that would be important in all his future work: biparental inheritance (equal contributions to the offspring from both parents), the rejection of any significant role for the inheritance of acquired characters, and reversion to ancestral characters.

Biparental Inheritance

Opinion in the eighteenth century had been divided between the spermists, who believed that the child was derived from the sperm, the ovists, who believed it was derived from the ovum, and those who held to the two-layer theory, that the outer layer was derived from the father and the inner layer from the mother. The experiments of the plant hybridist Joseph Kölreuter at the end of the eighteenth century showed that hybrids between two species were intermediate between the parental means in many characters, and that offspring from reciprocal crosses were indistinguishable, so that it made no difference whether the pollen of A plants was used to fertilize B plants or vice versa; he concluded that both sexes contribute equally to the offspring. Hybrids are not, of course, intermediate for characters that show dominance, but the evidence from reciprocal crosses is conclusive. Biparental inheritance was generally accepted by the middle of the nineteenth century, though it was not confirmed by cytological evidence until toward its end.

Galton accepted this view in 1865 and suggested that one reason for sons being on average less distinguished than their fathers was that the contribution of the mother was not taken into account. He also applied the principle of equal parental contributions in this passage:

> The share that a man retains in the constitution of his remote descendants is inconceivably small. The father transmits, on an average, one-half of his nature, the grandfather one-fourth, the great-grandfather one-eighth; the share decreasing step by step, in a geometrical ratio, with great rapidity. Thus the man who claims descent from a Norman baron, who accompanied William the Conqueror twenty-six generations ago, has so minute a share of that baron's influence in his constitution, that, if he weighs fourteen stone, the part of him which may be ascribed to the baron (supposing, of course, there have been no additional lines of relationship) is only one fiftieth of a grain in weight—an amount ludicrously disproportioned to the value popularly ascribed to ancient descent. (Galton 1865a, 326–327)

(One-fiftieth of a grain divided by fourteen stone equals $(1/2)^{26}$.)

The argument is straightforward and can be put today like this. If a child inherits half its "genes" from its father (and half from its mother), and if the father inherits half of *his* "genes" from *his* father, then the child must have inherited one quarter of its "genes" from its paternal grandfather, and so on. The argument is valid under any form of biparental inheritance, whatever term is substituted for "gene." It is a restatement of the law of halving known for a long time to breeders in terms of "blood fractions"; this law is a simple consequence of biparental inheritance.

Some confusion has arisen from Karl Pearson's misinterpretation of the above passage as the first enunciation of the law of ancestral heredity (see chapter 8). Quoting the first two sentences in this passage, he wrote:

> And then follows Galton's first enunciation of the Law of Ancestral Heredity.... Galton is clearly on the right track, but the numbers he gives would only be correct if he used parental generation, grandparental generation, great-grandparental generation, instead of father, grandfather, great-grandfather. He has overlooked the mother, and overlooked the multiplicity of the ancestral individuals. The numbers as he gives them in a later publication are one-fourth for the parent, one-sixteenth for the grandparent and one sixty-fourth for the great-grandparent, etc.... Thus in 1865 Galton had already in mind this law of ancestral heredity, although by an obvious oversight he gave the wrong proportions. (Pearson 1924, 84)

This interpretation has been adopted by several authors (Swinburne 1965; Froggatt and Nevin 1971a,b; Cowan 1977), but the passage can be interpreted more simply as a restatement of the law of halving, which is a consequence of biparental inheritance. This interpretation avoids the gratuitous assumption that Galton gave the wrong proportions by an oversight. There is no reason to suppose that he had the ancestral law in mind in 1865.

Galton's argument was later made in a similar way by Weismann, based on cytological observations of fertilization, in an essay on "The continuity of the germ-plasm":

> Each of the two nuclei which unite in fertilization must contain the germ-nucleoplasm of both parents, and this latter nucleoplasm once contained and still contains the germ-nucleoplasm of the grandparents as well as that of all previous generations. It is obvious that the nucleoplasm of each antecedent generation must be represented in any germ-nucleus in an amount which becomes less as the number of intervening generations becomes greater; and the proportion can be calculated after the manner in which breeders, when crossing races, determine the proportion of pure blood which is contained in any of the descendants. Thus while the germ-plasm of the father or mother constitutes half the nucleus of any fertilized ovum, that of a grandparent forms a quarter, and that of the tenth generation backwards only 1/1024, and so on. The latter can, nevertheless, exercise influence over the development of the offspring, for the phenomena of atavism show that the germ-plasm of very remote ancestors can occasionally make itself felt, in the sudden reappearance of long-lost characters. (Weismann 1885, 179)

The Non-Inheritance of Acquired Characters

Galton was ahead of his time in rejecting any significant role for the inheritance of acquired characters:

> Can we hand anything down to our children, that we have fairly won by our own independent exertions? Will our children be born with more virtuous dispositions, if we ourselves have acquired virtuous habits? Or are we no more than passive transmitters of a nature we have received, and which we have no power to modify? There are but a few instances in which habit even seems to be inherited. The chief among these are such as those of dogs being born excellent pointers; of the attachment to man shown by dogs; and of the fear of man, rapidly learnt and established among the birds of newly-discovered islands. But all of these admit of being accounted for on other grounds than the hereditary transmission of habits. Pointing is, in some faint degree, a natural disposition of all dogs. Breeders have gradually improved upon it, and created the race we now possess. There is nothing to show that the reason why dogs are born staunch pointers is that their parents had been broken into acquiring an artificial habit....
>
> If we examine the question from the opposite side, a list of life-long habits in the parents might be adduced which leave no perceptible trace on their descendants. I cannot ascertain that the son of an old soldier learns his drill more quickly than the son of an artizan. I am assured that the sons of fishermen, whose ancestors have pursued the same calling time out of mind, are just as sea-sick as the sons of landsmen when they first go to sea. I cannot discover that the castes of India show signs of being naturally endowed with special aptitudes. If the habits of an individual are transmitted to his descendants, it is, as Darwin says, in a very small degree, and is hardly, if at all, traceable. (Galton 1865a, 321–322)

He concluded with the epigraph at the beginning of this chapter.

The inheritance of acquired characters (the effect of use and disuse) was widely accepted from ancient times through to the nineteenth century (Zirkle 1946). Darwin, though no believer in Lamarckism, nevertheless thought that use and disuse played a definite role in evolution, albeit secondary to that of natural selection. He devoted a whole section of *The Origin of Species* to the effects of use and disuse, and in discussing the habits of the pointer dog he wrote: "It may be doubted whether any one would have thought of training a dog to point, had not some one dog naturally shown a tendency in this line; and this is known occasionally to happen, as I once saw in a pure terrier. When the first tendency was once displayed, methodical selection *and the inherited effects of compulsory training* in each successive generation would soon complete the work (my italics)" (1859, 214). Thus Darwin attributed the development of the pointing instinct partly to selection and partly to the inherited effects of training; Galton disputed the second part of this claim. (Galton exaggerated in attributing to Darwin his own belief that the inheritance of acquired characters is hardly, if at all, traceable. Darwin [1868] used the inheritance of acquired characters to explain the small wings and large legs of domestic ducks that fly less and walk more than wild ducks, the enlarged udders in milking breeds of cows and goats, the presence of drooping ears in domestic animals not alarmed by danger, and the loss of sight in cave-dwelling animals.)

Ruth Cowan has argued that Galton's motivation for rejecting the inheritance of acquired characters was sociopolitical: "Were this doctrine to be proven true Galton's scheme for controlled human breeding would be useless; an individual would be able to improve upon or destroy whatever qualities his parents had handed down to him.... This situation would be inimical in Galton's utopian state, a state in which control of breeding is thought to be the only effective way to improve the human race. Thus it is not surprising that Galton was opposed to the doctrine of the inheritance of acquired characteristics" (1977, 141).

This argument seems to me to be back to front. Galton was undoubtedly interested in improving the human race, but there is no reason to suppose that he first devised a scheme of race improvement by controlled breeding and then looked for a scientific theory to underpin that scheme. It is more natural to suppose that he first looked for the theory in best agreement with the facts known to him, and then considered how it could be applied to race improvement.

Why then did Galton reject the inheritance of acquired habits except as a factor that is "hardly, if at all, traceable"? The passage quoted above is well argued and suggests that he had thought carefully about the evidence for the inheritance of acquired characters, and concluded that it was weak. Given Galton's independence of mind, it is not surprising that he should have reached this conclusion.

The Law of Reversion

The third principle mentioned briefly in "Hereditary talent and character" is the law of reversion or atavism, whereby an individual may resemble a grandparent or more distant ancestor for some character not possessed by either of its parents. Galton suggested this law, which was acknowledged by all three of his sources of information about heredity, as one of the reasons why sons were often less distinguished than their fathers: "Lastly, though the talent and character of both of the parents might, in any particular case, be of a remarkably noble order, and thoroughly congenial, yet they would necessarily have such mongrel antecedents that it would be absurd to expect their children to invariably equal them in their natural endowments. The law of atavism prevents it" (1865a, 319). In other words, another reason for sons being less distinguished than their parents was their tendency to revert to more remote ancestors, who would on average be less distinguished than very distinguished parents. (Atavism is another name for reversion.) The law of reversion was a recurrent theme throughout Galton's work on heredity, particularly in connection with the law of ancestral heredity. It is convenient to discuss reversion more fully in the next section, since it also played an important part in Darwin's thoughts on inheritance.

Darwin's Provisional Hypothesis of Pangenesis

The Origin of Species proposed a dual thesis: that the diversity of life is the product of gradual and branching evolution, a process Darwin called "descent with modification"; and that the mechanism driving this process is natural selection acting on inherited variations. In his final chapter, Darwin described the whole volume as "one long argument" in favor of "the theory of descent with modification through natural selection" (1859, 459); and he concluded: "I have now recapitulated the chief facts which have thoroughly convinced me that species have changed, and are still changing by the preservation and accumulation of successive slight favorable variations" (480).

The first part of this thesis, the fact of evolution, was quickly accepted by most biologists because of the overwhelming evidence documented in *The Origin of Species*. But the second part, the mechanism of evolution, aroused much hostility that led to the "eclipse" of Darwinism around 1900 (Bowler 1983, Gayon 1998). The reasons for this hostility were partly a reluctance to believe that evolution had occurred in such an undirected way (John Herschel called it "the law of higgledy-piggledy"), but more substantively that Darwin could not give a detailed account of how natural selection worked. Natural selection requires the existence of heritable variations, but knowledge of how variation is either produced or inherited was sketchy in the nineteenth century. Darwin invoked "the strong principle of inheritance," summarized as "like produces like," but he could go no farther: "The laws governing inheritance are quite unknown" (1859, 13). Detailed understanding of how natural selection works had to wait until the rediscovery of Mendelism and the development of population genetics in the twentieth century.

Darwin realized this weakness and thought hard about the questions of variability and inheritance. In 1868 he published in two volumes the results of his work in *The Variation of Animals and Plants under Domestication*, in which he summarized all the relevant facts and then proposed a hypothesis to explain them which he called pangenesis. The present account is based on his more mature thoughts in the second edition (1875). In a chapter entitled "Provisional hypothesis of pangenesis," he began by listing the facts to be explained:

> Every one would wish to explain to himself, even in an imperfect manner, how it is possible for a character possessed by some remote ancestor suddenly to reappear in the offspring; how the effects of increased or decreased use of a limb can be transmitted to the child; how the male sexual element can act not solely on the ovules, but occasionally on the mother-form; how a hybrid can be produced by the union of the cellular tissue of two plants independently of the organs of generation; how a limb can be reproduced on the exact line of amputation, with neither too much nor too little added; how the same organism may be produced by such widely different processes, as budding and true seminal generation; and lastly, how of two allied forms, one passes in the course of its development through the most complex metamorphoses, and the other does not do so, though when mature both are alike in every detail of structure. (Darwin 1875, 2:349)

It is usual today to distinguish the rules of inheritance from the study of how genes control development. This distinction was not made in the nineteenth century, and Darwin wanted to find a theory to explain both inheritance and development. This made his task much harder; the success of modern genetics was made possible by decoupling the two disciplines, so that the rules of inheritance could be studied while leaving development as a black box. Of the seven "facts" that Darwin wanted to explain, the first three concern the rules of inheritance (the first is true and the other two false); the fourth embodies the mistaken belief that a graft-hybrid could sometimes behave like a sexually produced hybrid between two species or varieties; and the last three concern development.

The hypothesis of pangenesis, constructed to explain the above "facts" in which Darwin believed, made these assumptions. (1) The cells of the body throw off minute granules, called gemmules, each type of cell producing a different type of gemmule, that can be thought of as a miniature replica of itself. (2) The gemmules can multiply by self-division, and can in the right conditions develop into a cell like those from which they were originally derived. (3) The gemmules "are collected from all parts of the system to constitute the sexual elements, and their development in the next generation forms a new being; but they are likewise capable of transmission in a dormant state to future generations and may then be developed" (374).

Darwin explained the facts of development by assuming that each gemmule has an "elective affinity for that particular cell which precedes it in due order of development" (374); after attachment to that cell, the gemmule develops into its appropriate cell type. To explain how an amputated limb can be regenerated and how an organism can reproduce by budding as well as sexually, he supposed that gemmules of every variety are present in every tissue, as a consequence of their free circulation throughout the body. I pass over the details to discuss more fully his explanation of the three facts of inheritance in which he believed, reversion, the inheritance of acquired characters, and the direct action of the male gamete on the mother-form.

Reversion

Reversion or atavism occurs when a child resembles a grandparent or more distant ancestor for some character not possessed by either of its parents. When two varieties are crossed, Darwin knew that, as a general rule, the "offspring in the first generation are nearly intermediate between their parents, but their grandchildren and succeeding generations continually revert, in a greater or lesser degree, to one or both of their progenitors" (1875, 2:23). Mendelian genetics would attribute this fact to segregation, the separation of allelic differences at meiosis. Naudin proposed a similar idea in 1865, though his hypothesis fell short of Mendel's because he believed that the hereditary elements of a species or variety segregate together without recombination. Naudin's hypothesis of segregation was not fully accepted by Darwin because it did not explain distant reversion, in which the offspring from a cross between two races resemble a very distant ancestor. For example, Darwin showed that all races of domestic pigeon arose from the wild rock pigeon *Columbia livia,* which is slaty blue in color with two black wing bars. The purebred domestic races have lost the wild color and wing bars, but occasionally revert to the ancestral appearance, blue with two black wing bars. When two domestic races are crossed, for example a black with a white-colored race, the progeny or grandprogeny often revert to the ancestral appearance, despite the fact that the races from which they were derived have bred true for several hundred years. Darwin was greatly struck by this phenomenon, which would today be attributed to complementary gene action; the presence of two genes is required to produce the ancestral appearance, one of which is present in the first race and the other in the second. Darwin, however, sought an explanation that would account both for near and for distant reversion.

He explained reversion from the assumption that only some of the gemmules inherited by the offspring develop into cells, the rest remaining dormant or latent. Consider first reversion in the grandchildren following a cross between two varieties. As a hypothetical example, later used by Darwin in correspondence with Galton, suppose that a white and a black variety of plant are crossed to produce gray offspring, the usual case in which hybrid offspring are intermediate between the parents. The offspring possess both white and black gemmules, which produce gray tissue when it develops, and this gray tissue produces gray gemmules; but the plant also possesses unmodified, dormant white and black gemmules. Thus the gonads of the hybrid contain white, gray, and black gemmules, so that when two hybrids pair they can have white, pale gray, gray, dark gray, or black offspring. This is a modified version of Naudin's hypothesis of segregation, which Darwin put as follows:

> Each [cell] in a hybrid must throw off, according to the doctrine of pangenesis, an abundance of hybridized gemmules, for crossed plants can be readily and largely propagated by buds; but by the same hypothesis dormant gemmules derived from both pure parent-forms are likewise present.... Consequently the sexual elements of a hybrid will include both pure and hybridised gemmules; and when two hybrids pair, the combination of pure gemmules derived from the one hybrid with the pure gemmules of the same parts derived from the other, would necessarily lead to complete reversion of character.... Pure gemmules in combination with hybridised gemmules would lead to partial reversion. And lastly, hybridised gemmules derived from both parent-hybrids would simply reproduce the original hybrid form. All these cases and degrees of reversion incessantly occur. (Darwin 1875, 2:395)

Hybrids are not always intermediate between the two parental forms but may resemble one of them in a particular character. Darwin called this form prepotent in the transmission of the character (we should today call the character dominant), and explained it by "assuming that the one form has some advantage over the other in the number, vigour, or affinity of its gemmules" (386). For example, there are two shapes of flower in the snapdragon *Antirrhinum majus*, the normal, irregular shape and the peloric, regular shape. Darwin crossed normal with peloric plants and all the offspring were normal, whether the cross was of normal pollen with a peloric mother plant or the reciprocal. Darwin allowed these hybrid offspring to sow themselves, "and out of a hundred and twenty-seven seedlings, eighty-eight proved to be common snapdragons, two were in an intermediate condition between the peloric and normal state, and thirty-seven were perfectly peloric, having reverted to the structure of their one grandparent" (46). Today this is regarded as a classic example of Mendelian inheritance for a dominant gene; ignoring the two intermediate plants, the proportion of the dominant, normal type in the F_2 generation was 0.70, close to the Mendelian probability of 0.75. Darwin's explanation was different. He supposed that the tendency to produce the normal character prevailed in the first generation because of the advantage of normal over peloric gemmules; but that the tendency to produce pelorism had gained in strength by the intermission of a generation because the dormant peloric gemmules had increased in number relative to the normal gemmules since the latter had been used up in producing normal flowers. He attributed to the same cause the "fact" that certain diseases regularly appear in alternate generations.

Distant reversion in crosses was explained in a rather similar way. When a black domestic pigeon is crossed with a white pigeon, the progeny only have half the number of the black or white gemmules present in the respective parent. This may not be enough to allow the development of either of these characters, leaving it open for the dormant blue gemmules inherited from both parents to develop, causing reversion to the ancestral rock-pigeon color.

The Inheritance of Acquired Characters

The inheritance of acquired characters and of the effects of use or disuse are easily explained under the hypothesis of pangenesis by supposing that when a cell has become structurally modified it throws off similarly modified gemmules which are inherited. Darwin knew that mutilations were not usually inherited: "Dogs and horses formerly had their tails docked during many generations without any inherited effect. . . . Circumcision has been practised by the Jews from a remote period, and in most cases the effects of the operation are not visible in the offspring" (1875, 2:391). Darwin's explanation was that gemmules formerly derived from the part are multiplied and transmitted from generation to generation. He also believed that when removal or mutilation of parts is followed by morbid action, the deficiency is sometimes inherited. "The evidence which admits of no doubt is that given by Brown-Séquard with respect to guinea-pigs, which after their sciatic nerves had been divided, gnawed off their own gangrenous toes, and the toes of their offspring were deficient in at least thirteen instances on the corresponding feet" (392). His explanation was that in this case "the gemmules of the mutilated or amputated part are gradually attracted to the diseased surface during the reparative process, and are there destroyed by the morbid action".

Xenia and Telegony

Darwin believed that the pollen and sperm could have a direct effect on the mother in addition to fertilizing the ovum. In plants the phenomenon has been called "xenia," and occurs when foreign pollen affects the appearance of the endosperm, which contains the reserve materials for nourishing the embryo. Darwin explained xenia by supposing that the gemmules of the pollen "can unite with and modify the partially developed cells of the mother-plant.... According to this view, the cells of the mother-plant may almost literally be said to be fertilized by the gemmules derived from the foreign pollen" (1875, 2:379). This is not far from the modern explanation that the endosperm results from the fertilization of a maternal cell by the second nucleus of the pollen-tube (Dunn 1973).

In animals the phenomenon is called "telegony" and occurs if a male who has been mated to a female affects the appearance of the subsequent offspring of that female by other males. The best known example is that of Lord Morton's quagga, a species of zebra now extinct. Lord Morton had mated a male quagga to an Arab mare, which was subsequently mated to an Arab stallion. The foals from the latter mating showed features resembling the quagga (the "quagga taint"), in particular conspicuous stripes on the legs, which contemporary scientists accepted as proof of telegony. Darwin explained telegony by supposing that gemmules introduced into the female when she mates with the first male can survive and divide in her until subsequent matings. Later control experiments showed that stripes could be found on foals from mothers who had never been mated to a quagga or zebra, and belief in telegony was dead by the end of the nineteenth century (Burkhardt 1979).

Galton's Reaction to Pangenesis

Galton read Darwin's *Variation of Animals and Plants under Domestication* (1868) with great interest. His copy of the book is extensively annotated, particularly the chapters on inheritance and on pangenesis (Cowan 1977). His pencil notes indicate that he reacted favorably to the theory of pangenesis, although he had reservations about the inheritance of acquired characters, and he incorporated a discussion of the theory at the end of *Hereditary Genius*, with the enthusiastic introduction: "This theory ... is—whether it be true or not—of enormous service to those who enquire into heredity. It gives a key that unlocks every one of the hitherto unopened barriers to our comprehension of its nature" (Galton 1869, 364).

In this section I first describe Galton's use of metaphor to illustrate the theory of pangenesis in *Hereditary Genius*, which provides some insight into the way in which his mind worked. I then discuss his attempt to verify the theory experimentally by blood transfusion in rabbits.

Galton's Political Metaphor of Pangenesis

In *Hereditary Genius*, Galton likened the way in which development of the body occurs, through the tendency of gemmules to attach themselves to particular cells, to the development of human assemblages through the free interaction of individual men. He took the development of a "watering place" (seaside resort) as an example. To begin with,

> two or three houses were perhaps built for private use, and becoming accidentally vacant, were seen and rented by holiday folk, who praised the locality, and raised a demand for further accommodation; other houses were built to meet the requirements; this led to an inn, to the daily visit of the baker's and the butcher's cart, the postman, and so forth. Then as the village increased and shops began to be established, young artisans, and other floating gemmules of English population, ... became fixed.... The general result of these purely selfish affinities is, that watering-places are curiously similar, even before the speculative builder has stepped in. (Galton 1869, 365)

He observed that a watering-place would breed true to its kind, either "asexually" by detaching an offshoot or "sexually" by a "mating" between two watering places at some distance apart which might between them afford material to raise another in an intermediate locality. He added that the same remarks might be made about fishing villages, or manufacturing towns, or an encampment of gold diggers, and that each of them would breed true to its kind. However, the result of a "hybrid mating" between different types was less predictable. The union of a watering place and a fishing village would be favorable, because the two elements support each other; the picturesque seaside life is an attraction to visitors, who in turn buy fish from the fishermen. But the union of a manufacturing enterprise with a watering place would be unfavorable, because the two elements are discordant: "Either the place will be forsaken as a watering-place, or the manufacturer will be in some way or other got rid of.... The dirt and noise and rough artisans engaged in the manufactury, are uncongenial to the population of a watering-place.... The moral I have in view will be clear to the reader. I wish to show that because a well-conditioned man marries a well-conditioned woman, each of pure blood as regards any natural gift, it does not in the least follow that the hybrid offspring will succeed" (1869, 366–367).

This passage has considerable period charm, but it reveals more to us about Galton's habit of thought than about the theory of pangenesis. He used the same metaphor to introduce latent gemmules as an explanation of reversion:

> I will continue to employ the same metaphor, to explain the manner in which apparent sports of nature are produced, such as the sudden appearance of a man of great abilities in undistinguished families. Mr. Darwin maintains, in the theory of Pangenesis, that the gemmules of innumerable qualities, derived from ancestral sources, circulate in the blood and propagate themselves, generation after generation, still in the state of gemmules, but fail in developing themselves into cells, because other antagonistic gemmules are prepotent and overmaster them, in the struggle for points of attachment. Hence there is a vastly larger number of capabilities in every living being, than ever find expression, and for every *patent* element there are countless *latent* ones. The character of a man is wholly formed through those gemmules that have succeeded in attaching themselves; the remainder that have been overpowered by their antagonists, count for nothing; just as the policy of a democracy is formed by that of the majority of its citizens, or as the parliamentary voice of any place is determined by the dominant political views of the electors: in both instances, the dissentient minority is powerless....
>
> Suppose that by some alteration in the system of representation, two boroughs, each containing an Irish element in a large minority, the one having always returned a Whig and the other a Conservative, to be combined into a single borough returning one member. It is clear that the Whig and the Conservative party will neutralize one another, and that the union

> of the two Irish minorities will form a strong majority, and that a member professing Irish interests is sure to be returned. This strictly corresponds to the case where the son has marked peculiarities, which neither of his parents possessed in a patent form. (Galton 1869, 367–368)

Galton assumed that the gemmules "circulate in the blood." He also inferred that there are far more latent than patent elements because "there is a vastly larger number of capabilities in every living being, than ever find expression." We shall see that this inference had unfortunate consequences for his theory of heredity, from which he only freed himself in 1889. He then used a political metaphor to explain reversion under the theory of pangenesis; he translated Darwin's explanation of distant reversion in pigeons into human terms by equating Whigs to a black race of pigeon, Conservatives to a white race, and the Irish to the ancestral blue rock-pigeon. He concluded, rather optimistically, that "these similes ... give considerable precision to our views on heredity" (368).

An Experimental Test of Pangenesis

Darwin had written in his original account of the theory that the gemmules "circulate freely throughout the system" (1868, 2:370), and Galton assumed that they must therefore circulate in the blood. He therefore devised a careful series of experiments to test the theory of pangenesis by transfusing blood between different varieties of rabbits. They were carried out at the London Zoo between 1869 and 1871, the prosector to the zoo making the operations, with Galton's assistance. Galton summarized his aims and conclusions:

> It occurred to me ... that the truth of Pangenesis admitted of a direct and certain test. I knew that the operation of transfusion of blood had been frequently practised with success on men as well as animals, and that it was not a cruel operation.... I therefore determined to inject alien blood into the circulation of pure varieties of animals (of course, under the influence of anaesthetics), and to breed from them, and to note whether their offspring did or did not show signs of mongrelism. If Pangenesis were true, according to the interpretation which I have put upon it, the results would be startling in their novelty, and of no small practical use; for it would become possible to modify varieties of animals, by introducing dashes of new blood, in ways important to breeders. Thus, supposing a small infusion of bull-dog blood was wanted in a breed of greyhounds, this, or any more complicated admixture, might be effected (possibly by

> operating through the umbilical cord of a newly born animal) in a single generation.
>
> I have now made experiments of transfusion and cross circulation on a large scale in rabbits, and have arrived at definite results, negativing, in my opinion, beyond all doubt, the truth of the doctrine of Pangenesis. (Galton 1871a, 395)

Two types of experiment were performed. In the first, defibrinized blood from a common lop-eared rabbit was injected into a silver-gray rabbit; common rabbits of different colors (yellow, common gray, and black and white) were used. The transfused silver-gray rabbits were allowed to breed, and out of 36 offspring, 35 were silver-gray and one was silver-gray with a white foot. Galton wrote: "This white leg gave me great hopes that Pangenesis would turn out to be true, though it might easily be accounted for by other causes" (1871a, 402). He also found that some of the does were sterile when wholly rather than partially defibrinized blood was used, and he wondered if the general failure of the experiments to lead to mongrelism was due to the gemmules being removed with the fibrin.

He therefore did a second type of experiment in which a cross-circulation was established between the carotid arteries of a common rabbit and a silver-gray rabbit, the transfused rabbits subsequently being allowed to breed with their own kind. Out of 50 offspring of transfused silver-gray rabbits, 49 were silver-gray and one was Himalayan (sandy with black tips); Galton's stock of silver-gray rabbits was known to throw the occasional Himalayan in the absence of treatment, so that this could not be attributed to the transfusion. Out of 38 offspring of transfused common rabbits, all were like their parents and none were silver-gray.

Galton concluded that the doctrine of pangenesis, pure and simple, as he had interpreted it, was incorrect. He considered two alternative hypotheses: (1) the reproductive elements reside in the gonads, whence they are set free by an ordinary process of growth, the blood merely supplying nutriment to that growth; (2) they reside in the blood itself, being derived from somatic cells and transported to the gonads. He distinguished between two variants of the second hypothesis: (2a) the reproductive elements are independent residents in the blood; (2b) they are only temporary residents in it, being continually renewed by fresh arrivals from the framework of the body. He identified (2a) with Darwinian pangenesis, which his experiments had disproved, but pointed out that they prove nothing against (2b) since "in this latter case, the transfused gemmules would have perished, just like the blood-corpuscles, long before the period had elapsed when the animals had recovered from their operations" (404). He urged that experiments should be done to test this possibility by trying to get the male rabbits to couple immediately, and on successive days after they had been operated on.

Galton asked Darwin's advice about these experiments and kept him informed of their progress. It is clear from the surviving correspondence (Pearson 1924, 156–202), as well as from the quotation above, that both men were hoping for the results to be positive. Thus Mrs. Darwin wrote to her daughter: "F. Galton's experiments about rabbits (viz. injecting black rabbit's blood into gray and *vice versa*) are failing, which is a dreadful disappointment to them both" (158); and Galton wrote to Darwin: "Good rabbit news! One of the litters has a white forefoot" (160). (This was the case mentioned above. Pearson remarks that the appearance of a white foot is a common event, and notes that Galton seized any feature he could that supported mongrelization and hence demonstrated pangenesis.)

It is also clear that Darwin knew the nature of the experiments and did nothing to discourage them. However, when he read Galton's paper he sent a rebuttal to *Nature* (Darwin 1871b), pointing out that his theory of pangenesis did not presuppose the circulation of gemmules in the blood, and adding that when he first heard of Galton's experiments, he did not reflect sufficiently on the subject, and had not seen the difficulty of believing in the presence of gemmules in the blood.

Darwin's rebuttal seems a little unkind in view of his support for the experiments, but Galton took it in good part. In a reply to *Nature*, he stated that he had been misled by the ambiguity of Darwin's language, and concluded his letter with a memorable passage acknowledging Darwin as his leader:

> I do not much complain of having been sent on a false quest by ambiguous language, for I know how conscientious Mr. Darwin is in all he writes, how difficult it is to put thoughts into accurate speech, and, again, how words have conveyed false impressions on the simplest matters from the earliest times. Nay, even in that idyllic scene which Mr. Darwin has sketched of the first invention of language, awkward blunders must of necessity have often occurred. I refer to the passage in which he supposes some unusually wise, ape-like animal to have first thought of imitating the growl of a beast of prey so as to indicate to his fellow monkeys the nature of an expected danger. For my part, I feel as if I had just been assisting at such a scene. As if, having heard my trusted leader utter a cry, not particularly well articulated, but to my ears more like that of a hyena than any

> other animal, and seeing none of my companions stir a step, I had, like a loyal member of the flock, dashed down a path of which I had happily caught sight, into the plain below, followed by the approving nods and kindly grunts of my wise and most-respected chief. And I now feel, after returning from my hard expedition, full of information that the suspected danger was a mistake, for there was no sign of a hyena anywhere in the neighbourhood. I am given to understand for the first time that my leader's cry had no reference to a hyena down in the plain, but to a leopard somewhere up in the trees; his throat had been a little out of order—that was all. Well, my labour has not been in vain; it is something to have established the fact that there are no hyenas in the plain, and I think I see my way to a good position for a look out for leopards among the branches of the trees. In the meantime, *Vive* Pangenesis. (Galton 1871b, 6)

The friendship between the two men remained unbroken, and Darwin even took a more active role in the blood transfusion experiments, which went on for some time, by housing and breeding from some of the rabbits at Down. The results of these further experiments were negative, but they were never published; it is not known whether they were intended to test the theory that the gemmules are temporary residents in the blood under hypothesis (2b) above, but Galton's subsequent theories assumed this process to be of minor importance, if it existed at all.

Galton's Theory of Heredity in the 1870s

Having rejected Darwin's theory of pangenesis, Galton set to work to develop an alternative theory, in which he incorporated those parts of Darwin's theory that had not been experimentally disproven. He published his conclusions in two papers, "On blood relationship" and "A theory of heredity" (Galton 1872b and 1875b, respectively). In this section, I describe this theory, contrasting it with Darwin's theory of pangenesis.

The theory of heredity developed in 1872 is shown in fig. 4.1. The fertilized ovum consists of a very large number of hereditary elements, each with the potential to develop into a particular cell type. Galton argued from the facts of reversion that some of these elements were expressed in the adult while others remained latent: "Each individual may properly be conceived as consisting of two parts, one of which is latent and only known by its effects on his posterity, while the other is patent, and constitutes the person manifest to our senses" (1872b, 394). He supposed, as he did in *Hereditary Genius*, that the latent elements were much more

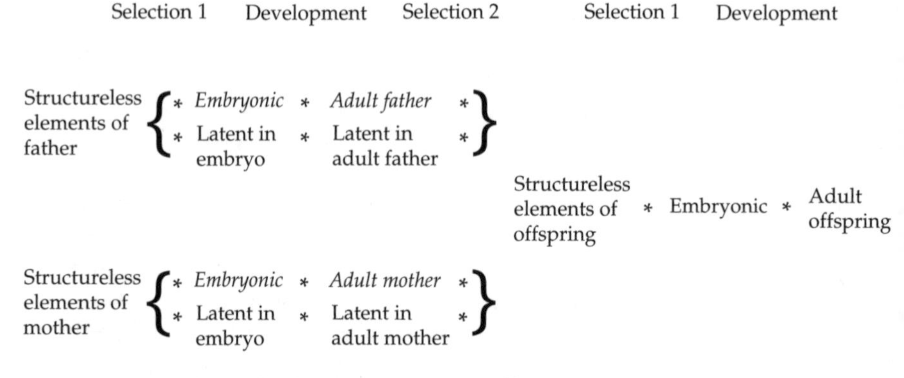

Fig. 4.1. Relationship between parents and offspring according to Galton 1872b

numerous than the patent ones, arguing from the fact that a single strain of impure blood can reassert itself after more than eight generations, which corresponds to a dilution of 1 in 256. The primary elements in the fertilized ovum were structureless (first column in fig. 4.1), and a small number of them were selected (Selection 1) to become the patent embryonic elements which were developed (Development) into the adult person, the residue remaining latent. The patent embryonic elements were selected by a process which Galton called "class representation," by which he meant that the elements in the ovum fall into a number of classes determining, if expressed, different characters, and that a small number was chosen from each class to become patent. He repeated the metaphor of the election of a "representative assembly" used in *Hereditary Genius*, but was deliberately noncommital about the method of election since nothing is known.

Finally, Galton considered how the patent and latent elements in one generation were transmitted via the germ cells to the structureless elements of the next. Not all the elements could be transmitted, otherwise the variety of elements would increase without bound, so that only some of them were selected (Selection 2). He concluded from the fact that acquired characters were rarely inherited that patent elements were transmitted more feebly than latent elements. The combination of the small number of patent elements with their small power of transmission meant that the italicized lines of transmission through patent elements in fig. 4.1 were "a nearly sterile destination" compared with the nonitalicized lines through latent elements.

Galton (1875b) restated, with some change of terminology, the theory of the mechanism of heredity put forward in 1872, and contrasted it with Darwin's theory of pangenesis. He began by coining the word "stirp" to express the sum-total of the hereditary elements, or "germs" as he now called them, which are to be found in the newly fertilized ovum. His theory of heredity in this terminology has four elements. (1) The stirp contains many individuals of many varieties of germs. (2) Each of the enormous number of quasi-independent units (cells) in the body has developed from a separate germ in the stirp. (3) The stirp contains many more germs than there are cells in the body, so that only a small proportion of the germs develop into cells. (4) The residue of undeveloped (latent) germs contributes to the stirp of the next generation, the developed germs (with rare exceptions postulated to allow the possibility of the inheritance of acquired characters) do not and are thus nearly sterile.

In 1872 Galton was noncommital about how the patent elements were selected, but in 1875 he stressed the idea of competition between germs: "We may compare the stirp to a nation and those among its germs that achieve development, to the foremost men of that nation who succeed in becoming its representatives" (1875b, 336). These dominant germs which achieve development are not transmitted, with an unfortunate consequence: "Another result of the best elements of the stirp being rendered sterile, is the strong tendency to deterioration in the transmission of every exceptionally gifted race" (340). In this argument, Galton was referring to competition between germs to become patent (Selection 1 in fig. 4.1), but he has been misled by his metaphor to assume that dominant germs which succeed in the competition to become patent also confer greater fitness on the individuals in whom they are expressed. We see in chapter 5 that in 1896 he used a variant on this argument, which he called sequestration, as an explanation of the advantage of sexual over asexual reproduction. His idea of competition between germs is analogous to Weismann's later theory of germinal selection.

In conclusion, Galton acknowledged his debt to Darwin's theory of pangenesis for his account of the facts to be explained, and for the idea that elements (gemmules) of different types are developed directly into the respective cell types; he also accepted Darwin's explanation of reversion through some elements remaining latent rather than developing into cells. But he rejected, except as a rare exception, the idea that cells throw off elements that travel freely throughout the body and aggregate in the germ cells. His rejection of this transportation hypothesis, except as a very minor process, was based on his failure to demonstrate the existence of these elements in the blood, and on his skepticism about the importance of the inheritance of acquired characters, which the transportation hypothesis was designed to explain. Galton's acceptance of Darwin's idea that some gemmules develop into cells and are no longer available for transmission to the next generation, together with his rejection of the idea that cells can throw off new gemmules, led inevitably to the conclusion that only the latent gemmules are transmitted through inheritance.

The difference between the two theories is encapsulated in two quotations. Darwin concluded his summary of the theory of pangenesis: "Hence, it is not the reproductive organs or buds which generate new organisms, but the [cells] of which each individual is composed" (1875, 2:370). On the other hand, Galton concluded: "We cannot now fail to be impressed with the fallacy of reckoning inheritance in the usual way, from parents to offspring, using those words in their popular sense of visible personalities. The span of the true hereditary link connects, as I have already insisted upon, not the parent with the offspring, but the primary elements of the two, such as they existed in the newly impregnated ova, whence they were respectively developed" (1872b, 400).

The Danish biologist Wilhelm Johannsen contrasted these two ideas as the *transmission-conception* and the *genotype-conception* of heredity. In the transmission conception, an individual's *personal* qualities are transmitted to his or her offspring: "The view of natural inheritance as realized by an act of transmission, viz., the transmission of the parent's (or ancestor's) *personal qualities* to the progeny, is the most naive and oldest conception of heredity. We find it clearly developed by Hippocrates.... Darwin's hypothesis of pangenesis is in this point very consistent with the Hippocratic view, the *personal* qualities of the parent or the ancestor in question being the heritage" (1911, 129). (Latent gemmules are the exception to this rule in Darwin's theory.) In the genotype conception, on the other hand, "The *personal qualities* of any individual organism do not at all cause the qualities of its offspring; but the qualities of both ancestor and descendant are in quite the same manner determined by the nature of the 'sexual substances'—*i.e.,* the gametes—from which they have developed. Personal qualities are then *the reactions of the gametes* joining to form a zygote; but the nature of the gametes is not determined by the personal qualities of the parents or ancestors in question" (130). The genotype conception had, he thought, been initiated by Galton and Weismann, but had been "revised as an expression of the insight won by pure line breeding and Mendelism" (132).

Galton's rejection of the transportation hypothesis affected his theory of both sexual and asexual reproduction. In sexual reproduction, he thought, following Darwin, that the patent elements had been used to develop into somatic (nonreproductive) cells, but he did not think that developed cells throw off new elements that can accumulate in the germ cells; he was therefore led to believe that the germ cells are directly descended from the residue of latent elements that have not developed into cells. However, he allowed a very moderate transgression of this general rule to allow for the occasional possibility of the inheritance of acquired characters: "Each cell may be supposed to throw off a few germs that find their way into the circulation, and thereby to acquire a chance of occasionally finding their way to the sexual elements, and of becoming naturalised among them" (1875b, 346).

Galton also needed to find a physiological explanation for asexual reproduction. Darwin thought that sexual and asexual reproduction were similar. The buds of a tree can reproduce the parent, as can a small piece of a freshwater worm. He concluded that all the different varieties of gemmule must be present in every tissue, which he explained as a result of their free travel throughout the body. Galton accepted that all varieties of gemmules must be present in all tissues, but Darwin's explanation was not open to him. He supposed that the fertilized ovum contains elements (gemmules) of every variety, that the different varieties are segregated by mutual attractions into different cell lines during the first few cell divisions of the embryo, but that the segregation is inexact so that the different cell lines that give rise to different tissues all contain a few latent elements of every variety and are thus totipotent.

Similarities between Relatives

The weakness of Galton's theory was that it gave him little insight into the reasons for the similarities between relatives, the question in which he was most interested. In 1872, he discussed the parent-offspring relationship in fig. 4.1:

> We see that parents are very indirectly and only partially related to their own children, and that there are two lines of connexion between them, the one of large and the other of small relative importance. The former is a collateral kinship and very distant, the parent being descended through two stages (two asterisks) from a structureless source, and the child (so far as that parent is concerned) through five totally distinct stages from the same

> source; the other, but unimportant line of connexion, is direct and connects the child with the parent through two stages. (Galton 1872b, 400)

In interpreting this passage I assume that the collateral kinship is (a) {Structureless elements in parent * *Embryonic* * *Adult parent*}, and (b) {Structureless elements in parent * Latent in embryo * Latent in adult parent * Structureless elements of offspring * Embryonic * Adult offspring}; this has two asterisks in (a) and five asterisks in (b). I also assume that the unimportant, direct line of connection is {*Adult parent* * Structureless elements of offspring * Embryonic * Adult offspring}; but I cannot reconcile this with Galton's statement that it has two, rather than three, asterisks. In any case, it seems arbitrary to describe the closeness of the relationship simply by the number of stages separating parent and child without discussing the amount of variability generated at each stage.

There is also a serious problem in accounting for the correlation arising from the more important, collateral line of connection. The parent-child correlation is a correlation between patent elements. Since only latent elements are transmitted in the collateral line, this is dependent on a correlation between latent and patent elements within parents; but under the simplest model of random mating and random selection of patent elements within each class, there would be no correlation between latent and patent elements within an individual, and therefore no correlation between parent and child!

How did Galton explain the resemblance between parent and child in view of this problem? He wrote: "I maintain that the personal manifestation is, on the average, though it need not be so in every case, a certain proof of the existence of some latent elements" (1872b, 399). In effect, he assumed that there was, in general, a positive correlation between the latent and the patent elements, though he had no explanation how this correlation was maintained. In fact, it requires nonrandom selection of the patent elements so as to make them more typical of all the elements of the same class than they would be by random choice.

It may be that Galton did not appreciate the distinction between sampling with and without replacement. His model assumes that the patent elements are selected *without* replacement from the pool of structureless elements in the embryo, leaving a residue of latent elements from which the germ cells contributing to the next generation are largely drawn. But he wrote:

> An approximate notion of the nearest conceivable relationship between a parent and his child may be gained by supposing an urn containing a great number of balls, marked in various ways, and a handful of them to be drawn out of them as a sample: this sample would represent the person of a parent. Let us next suppose the sample to be examined, and a few handfuls of new balls to be marked according to the patterns of those found in the sample, and to be thrown *along with them* [my italics] back into the urn. Now let the contents of another urn, representing the influences of the other parent, to be mixed with the first. Lastly, suppose a second sample to be drawn out of the combined contents of the two urns, to represent the offspring. (Galton 1872b, 400)

This passage is rather unclear, but the words italicized do seem literally to envisage selection *with* replacement, which would allow a small correlation between parent and child. (I am grateful to David Burbridge for drawing this passage to my attention.)

In 1875, Galton suggested, as we have seen, that some germs are dominant over others and are therefore more likely to be selected for development. He also supposed that these dominant germs would be unlikely to be transmitted to the next generation, which was predominantly recruited from the residue of undeveloped germs, and he concluded that this might give rise to a negative correlation between parent and offspring. He wrote:

> The hypothesis that the developed germ is less fecund ... than the germ that continues latent, agrees singularly well with many classes of fact. Thus it explains why, although hereditary resemblance is the general rule, the offspring is frequently deficient in the very peculiarity for which the parent was exceptionally remarkable. We can easily understand that the dominant characters in the stirp will, on the whole, be faithfully represented in the structure of the person who is developed out of it; but if the personal structure be a faithful representative of the dominant germs, it must be an over-favourable representative of the germs generally, and therefore, *a fortiori*, of the undeveloped residue; nay in extreme cases, the personal elements may be absolutely unrepresentative of the residual elements, the accidental richness of the sterile sample in some particularly valuable variety of germ, having drained the fertile residue of every germ of that variety.... Experience testifies to the fact that children of men of extraordinary genius have not unfrequently been singularly deficient in ability. (Galton 1875b, 339)

Despite this reason for expecting a negative correlation between parent and child, Galton accepted without comment that "hereditary resemblance is the general rule."

Galton discussed the relationship between siblings in this way:

> The great dissimilarity between brothers and sisters is to be accounted for and easily illustrated by a political metaphor. We have to recognise, on the one hand, that the stirps of the brothers and sisters must have been nearly alike, because the germs are simple organisms, and all such organisms breed true to their kind, and on the other hand, that very different structures have been developed out of these stirps. A strict analogy and explanation of all this is afforded by the well-known conditions and uncertainties of political elections. We have abundant experience that when a constituency is very varied, trifling circumstances are sufficient to change the balance of parties, and therefore, although there may be little real variation in the electoral body, the change in the character of its political choice at successive elections may be abrupt. (Galton 1875b, 336)

In this passage he assumed that selection of elements for the germ cells (Selection 2 in fig. 4.1) gave rise to little or no variability, so that the stirps of brothers and sisters were almost identical. Thus differences between brothers and sisters must be due to variability in the choice of patent elements from the stirp (Selection 1)—in other words, to developmental variability.

Galton's views on the reasons for similarities and dissimilarities between twins, which were discussed in chapter 2, are very instructive. He was familiar with the existence of two types of twins: "The word 'twins' . . . covers two very dissimilar events—the one corresponding to the progeny of animals that have usually more than one young at a birth, each of which is derived from a separate ovum, while the other is due to the development of two germinal spots in the same ovum. In the latter case, they are enveloped in the same membrane, and all such twins are found invariably to be of the same sex" (1875a, 392). They are today called dizygotic and monozygotic twins, respectively, the former being derived from two zygotes or fertilized ova, the latter from a single zygote. Galton believed that all twins, dizygotic as well as monozygotic, had identical stirps, just as siblings did, and he attributed the fact that most monozygotic twins are almost identical to the similarity of their developmental environment: "As regards the similarity of true twins, there can be little difficulty; we should expect, on statistical grounds, that the two halves of any assemblage of germs would be much alike. The secondary stirps of the twins being alike, *and the circumstances under which the bodily structure is developed out of them being almost identical*, the results must be closely similar" (1875b, 337, my italics).

His remark "we should expect, on statistical grounds, that the two halves of any assemblage of germs would be much alike" suggests why he thought that selection of elements for the germ cells gives rise to little or no variability, since this selection amounts to taking half an assemblage of germs. The italicized clause suggests that the near identity of most monozygotic twins compared with the differences between brothers and sisters is due to the absence of developmental variability in the former. As discussed in chapter 2, he also believed that late division of the zygote generated a small group of markedly dissimilar monozygotic twins; it was thought until quite recently that division of the embryo after its right and left sides had been established led to twins who were "mirror images" of one another, but this belief is unfounded (Bulmer 1970, 67).

In conclusion, Galton's ideas about the relationship between relatives at this time were confused. His theory of heredity provided no reason to explain the resemblance between parent and child, except for the unwarranted expectation of a correlation between patent and latent elements. In considering the resemblance between siblings, he supposed that they had identical genotypes (stirps) and attributed their differences to developmental variability; with one exception, to be discussed later, he discounted variability due to segregation.

Galton's Ideas on Heredity in 1889

After 1875 Galton turned his attention to developing a statistical theory of heredity, summarized in *Natural Inheritance* (1889). His statistical work, including his law of ancestral heredity, led him to modify his theory of the mechanism of heredity, in particular the role of latent and patent elements; but he never explicitly recognized his change of position, nor did he adhere to it consistently.

In the Introduction to *Natural Inheritance*, Galton wrote: "Though one half of every child may be said to be derived from either parent, yet he may receive a heritage from a distant progenitor that neither of his parents possessed as *personal* characteristics. Therefore the child does not on the average receive so much as one half of his *personal* qualities from each parent, but something less than a half" (1889, 2). This passage distinguishes clearly between the phenotype (personal features) and the genotype (of which every child receives half from each parent). He believed that reversion, which gave rise to this distinction, was due to the fact that only some of the hereditary elements (the patent or personal elements) were expressed, the rest being latent or dormant, unexpressed but capable of transmission. But he has changed his viewpoint from what he believed in 1875 about the transmission of latent and patent elements. He was clearly thinking of the fraction (something less than a half) of the personal qualities that a child received from each parent as being patent elements in the child which were patent in and transmitted by that parent. His estimate of this fraction, which led to his law of ancestral heredity (see chapter 8), was that one quarter of the personal qualities of a child were received directly from each parent, that is to say from hereditary elements patent in the parent.

This interpretation is confirmed by the discussion of latent elements:

> *Latent Elements not very numerous*.—It is not possible that more than one half of the varieties and number of each of the parental elements, latent or personal, can on the average subsist in the offspring. For if every variety contributed its representative, each child would on the average contain actually or potentially twice the variety and twice the number of the elements (whatever they may be) that were possessed at the same stage of its life by either of its parents, four times that of any one of its grandparents, 1024 times as many as any one of its ancestors in the 10th degree, and so on, which is absurd. Therefore as regards any variety of the entire inheritance, whether it be dormant or personal, the chance of its dropping out must on the whole be equal to that of its being retained, and only one half of the varieties can on the average be passed on by inheritance. Now we have seen that the *personal* heritage from either Parent is one quarter, it follows that the Latent Elements must follow the same law of inheritance as the Personal ones. In other words, either Parent must contribute on the average only one quarter of the Latent elements, the remainder of them dropping out and their breed becoming absolutely extinguished. (Galton 1889, 187–188)

In this passage, Galton implicitly assumed that patent and latent elements were equally likely to be transmitted, reversing his previous view that patent elements were unlikely to be transmitted since they had been developed into cells. He then argued that only half the elements of either kind could be transmitted under biparental inheritance if the total number of elements was to remain constant. Since he had estimated that a child received one quarter of its personal (patent) elements from each parent, that is to say as elements patent in the parent, it followed that there must be equal numbers of patent and latent elements. (The hidden assumption is that an element has the same chance, one half, of becoming patent in the child whether it was patent or latent in the parent.)

Galton had previously argued that latent elements must be more numerous than patent ones because a single strain of impure blood can reassert itself after more than eight generations. He now refuted this argument on the grounds that it ignored the stochastic nature of transmission:

> There seems to be much confusion in current ideas about the extent to which ancestral qualities are transmitted, supposing that what occurs occasionally must occur invariably. If a maternal grandparent be found to contribute some particular quality in one case, and a paternal grandparent in another, it seems to be argued that both contribute elements in every case. This is not a fair inference, as will be seen by the following illustration. A pack of playing cards consists, as we know, of 13 cards of each sort—hearts, diamonds, spades, and clubs. Let these be shuffled together and a batch of 13 cards dealt out from them, forming the deal, No. 1. There is not a single card in the entire pack that may not appear in these 13, but assuredly they do not all appear. Again, let the 13 cards derived from the above pack, which we shall suppose to have green backs, be shuffled with another 13 similarly obtained from a pack with blue backs, and that a deal, No. 2, of 13 cards be made from the combined batches. The result will be of the same kind as before. Any card of either of the two original packs may be found in the deal, No. 2, but certainly not all of them. So I conceive it to be with hereditary transmission. No given pair can possibly transmit the whole of their ancestral qualities; on the other hand, there is probably no description of ancestor whose qualities have not been in some cases transmitted to a descendant....
>
> If the Personal and Latent Elements are transmitted on the average in equal numbers, it is difficult to suppose that there can be much difference in their variety. (Galton 1889, 188–189)

If he had followed through the card-dealing analogy, he might have been led to attribute the differences between siblings to segregation, since they are the result of independent deals, but he did not do so.

Galton explained above his reason for changing his mind about the relative frequencies of latent and patent elements. One can understand how reflection on the statistical theory of heredity might have led him to abandon the view that patent elements were unlikely to be transmitted, since this made it much easier to explain the resemblance between parent and child. But this would also require him to abandon the view that the patent elements were transformed into cells and were therefore not available for transmission if the transportation hypothesis were rejected. Galton did not discuss this point explicitly, but there was a hint of a new viewpoint in a passage on the relationship between parent and child:

> A conviction that inheritance is mainly particulate and much influenced by chance, greatly affects our idea of kinship and makes us consider the parental and filial relation to be curiously circuitous. It appears that there is no direct hereditary relation between the personal parents and the personal child, except perhaps through little-known channels of secondary importance, but that the main line of hereditary connection unites the sets of elements out of which the personal parents had been evolved with the set out of which the personal child was evolved. The main line may be rudely likened to the chain of a necklace, and the personalities to pendants attached to its links. We are unable to see the particles and watch their grouping, and we know nothing directly about them, but we may gain some idea of the various possible results by noting the differences between the brothers in any large fraternity (as will be done further on with much minuteness), whose total heritages must have been much alike, but whose personal structures are often very dissimilar. (Galton 1889, 19–20)

Three points may be noted in this passage. First, he repeated his earlier argument that the main line of hereditary connection between parent and child is between the structureless elements of the parent and those of the child, but he refrained from stating that this is a collateral connection because elements that have developed into the person of the parent cannot be transmitted to the child. Second, he likened the main line of hereditary connection between parent and child to the chain of a necklace, and their expressed personalities to pendants attached to its links. This suggests that a link is not used up in making a pendant but is still available for transmission, providing a mechanism for patent elements in the parent to be transmitted to the child. Third, he repeated his earlier belief that siblings are very similar in their heritages, and that differences between them are due to developmental variability rather than to segregation.

Galton's theory of heredity in 1889 may be summarized as follows. Inheritance is mediated through particulate elements in the germ plasm. In bisexual inheritance each parent transmits half of his or her elements to the offspring, thus maintaining the total number of elements in successive generations. Elements may be latent or patent, only the patent ones being expressed, but a latent element may become patent in a subsequent generation. Latent and patent elements are equally numerous, they are equally likely to be transmitted, and an element has the same chance of one half of becoming patent in the child regardless of its status in the parent. Galton's views on the role of latent elements in explaining reversion had changed between 1875 and 1889. In particular, he held in 1875 that latent elements were much more numerous and much more likely to be transmitted than the patent elements expressed in the phenotype. Both these views were abandoned without comment in 1889. But he was not consistent in thinking that patent elements were equally likely to be transmitted; in 1896, for example, he used the idea of sequestration, which assumed that patent elements were not transmitted, to explain the advantage of sexual over asexual reproduction (chapter 5).

Discussion

Galton's theory of heredity was based on Darwin's hypothesis of pangenesis, with modifications to allow for his rejection of the transportation hypothesis. In particular, he adopted Darwin's explanation of reversion, that some gemmules develop into cells while others remain latent and capable of expression in subsequent generations; but he rejected the idea that cells produce new gemmules that could be incorporated into the germ cells. This idea is dependent on the transportation hypothesis, and was in any case not needed by Galton since he did not accept the inheritance of acquired characters. He was thus forced in the 1870s into supposing that, with rare exceptions, only the undeveloped, latent gemmules are capable of transmission to the next generation. To explain the similarity between parent and child under this model requires a correlation between the latent and patent elements within individuals, possibly due to a high degree of assortative mating. Galton probably came to realize the problem of this requirement during his subsequent statistical work on heredity. He quietly dropped the idea that patent elements are not transmitted and that they are much less frequent than latent elements in *Natural Inheritance* (1889), though he did not discuss why he had changed his view or what was the role of patent elements in development.

To conclude, I compare Galton's theory of heredity with the theories of Weismann and de Vries; I discuss his flirtation with segregation; and I consider the concept of blending inheritance and its relationship to the objections of Fleeming Jenkin to Darwinian theory.

Weismann and the Continuity of the Germ-Plasm

It has been suggested that Galton foreshadowed Weismann's theory of the continuity of the germ-plasm, but this is only true in a weak sense. Weismann assumed "the existence in the germ-cell of a reproductive substance, the *germ-plasm,* which cannot be formed spontaneously, but is always passed on from the germ-cell in which an organism originates in direct *continuity* to the germ-cells of the succeeding generations.... The germ-cells alone transmit the reproductive substance or germ-plasm in uninterrupted succession from one generation to the next, while the body (soma) which bears and nourishes the germ-cells, is, in a certain sense, only an outgrowth from one of them." (1892, 9). One consequence of this theory of "hard" heredity is that it makes impossible the inheritance of characters acquired by somatic cells, since they are a dead end, making no contribution to the next generation.

Galton concluded his discussion of the inheritance of acquired characters with the sentence: "We shall therefore take an approximately correct view of the origin of our life, if we consider our own embryos to have sprung immediately from those embryos whence our parents were developed, and those from the embryos of *their* parents, and so on for ever" (1865a, 322). Karl Pearson regarded this statement as amounting to Weismann's theory of the continuity of the germ-plasm, and quoted a letter from Weismann to Galton in 1889 in which Weismann wrote: "You have exposed in your paper an idea which is in one essential point nearly allied to the main idea contained in my theory of the continuity of the germ-plasm" (Pearson 1930a, 341). But there is an important distinction. Galton expressed an idea similar to Weismann's in less precise language (with "embryos" instead of "germ-plasm"), but with the qualifying word "approximately" and with the argument reversed. To Weismann, the continuity of the germ-plasm is an empirical fact which makes the inheritance of somatically acquired characters impossible; to Galton, the empirical fact is that the inheritance of acquired characters is of little, if any, importance, so that inheritance can be regarded as taking place from embryo to embryo.

With regard to Galton's theory of the 1870s, Romanes wrote that "there is not merely resemblance, but virtual identity, between the theories of stirp and germ-plasm.... Galton anticipated by some ten years all the main features of Weismann's theory of heredity" (1893, 59). However, Weismann thought that there was a fundamental difference between the two theories: "According to my idea, the active and the reserve germ-plasm contain precisely similar primary constituents, gemmules, or determinants; and on this the resemblance of a child to its parent depends. The theory of the continuity of the germ-plasm, as I understand it, is not based on the fact that each 'gemmule' necessary for the construction of the soma is present many times over, so that a residue remains from which the germ-cells of the next generation may be formed: it is founded on the view of a special adaptation, which is inevitable in the case of multicellular organisms, and which consists in the germ-plasm of the fertilized egg-cell becoming doubled primarily, one of the resulting portions being reserved for the formation of germ-cells" (1892, 200).

Thus there is considerable resemblance between the views of Galton and Weismann. Both believed in what Johannsen (1911) called the genotype conception of heredity, and they believed that the inheritance of acquired characters was rare (Galton) or impossible (Weismann). But Galton did not anticipate the essential part of Weismann's theory, the partially mistaken idea that the germ-plasm of the zygote is doubled, with one part being reserved for the formation of the germ cells. (Weismann's hard and fast distinction between the germ line and the soma is valid for higher animals but not for plants.) The irreversibility of translation from DNA into protein is today thought to be the real obstacle to the inheritance of acquired characters, rather than the separation of germ cells from somatic cells, which does not hold in plants or lower animals. Weismann himself later stressed the importance of the problem of translation as a barrier to the inheritance of acquired characters: "But, as these primary constituents [the hereditary determinants] are quite different from the parts themselves [the adult organs], they would require to vary in quite a different way from that in which the finished parts had varied: which is very like supposing that an English telegram to China is there received in the Chinese language" (1904, 2:63).

De Vries's Theory of Intracellular Pangenesis

In 1890 the Dutch botanist Hugo de Vries (1848–1935) published a book (translated as *Intracellular Pangenesis* in 1910) in which he developed a theory of heredity that had considerable similarities with Galton's theory. He began by observing that Darwin's hypothesis of pangenesis was based on two assumptions:

1. In every germ-cell (egg-cell, pollen-grain, bud, etc.) the individual hereditary qualities of the whole organism are represented by definite

material particles. These multiply by division and are transmitted during cell-division from the mother-cell to the daughter-cells.

2. In addition, all the cells of the body, at different stages of their development, throw off such particles; these flow into the germ-cells, and transmit to them the qualities of the organism, which they are possibly lacking. (Transportation-hypothesis). (De Vries 1890, 5)

The second assumption, the transportation hypothesis, was needed by Darwin to explain the inheritance of acquired characters and the more obscure topics of xenia and the behavior of graft-hybrids. But the work of Weismann had shattered the doctrine of the inheritance of acquired characters, and both xenia and the behavior of graft-hybrids were open to serious doubt; hence the transportation hypothesis, involving the intercellular transportation of hereditary particles, was no longer needed. Most authors, according to de Vries, had considered unthinkingly that, by refuting the transportation hypothesis, they had refuted Darwin's theory of pangenesis. His intention was to construct a theory of heredity based solely on Darwin's first assumption. This is of course exactly what Galton had tried to do, though de Vries did not mention his theory.

De Vries assumed, in accordance with current knowledge, that the hereditary particles were contained in the nucleus. Since he wanted to construct a theory of development as well as inheritance, he had to consider how these particles determined development. There were at the time two alternative theories of development, which Weismann called the dissection theory and the activation theory (Mayr 1982). The dissection theory proposed that during development there was a progressive dissection or subdivision of the total genetic material into ever smaller groups to be segregated into different cells. Under this theory, as de Vries put it, "every somatic cell receives, at the time of its origination, only those hereditary elements which will be needed by itself and its descendants" (1890, 56). The activation theory proposed that the determinants of all characters remained together in the nuclei of all the cells of the developing organisms, but that only some of them are activated by appropriate stimuli, depending on the cell type.

Influenced by his theory of a hard distinction between germ line and somatic cells, Weismann chose the dissection theory as being more probable. But botanists knew that in plants it was usually possible to regenerate the whole organism from a tiny fragment of somatic cells, so that nearly all plant cells must contain all the hereditary elements. Hence botanists rejected Weismann's distinction between germ line and somatic cells, and they rejected the dissection theory of development in favor of the activation theory.

De Vries therefore proposed that all the hereditary elements, which he called "pangenes," were present in the nuclei of all cells, both germinal and somatic. The pangenes were inactive in the nucleus, except for a small number of them needed for nuclear function. They became active when they were transported from the nucleus into the cytoplasm. He called this hypothesis "intracellular pangenesis" to emphasize the fact that it is derived from Darwin's hypothesis of pangenesis, but that the pangenes can be transported only within the cell and not, as Darwin supposed, between cells. He summarized his theory:

> Every hereditary character ... has its special kind of pangene.... In the nucleus every kind of pangene of the given individual is represented; the remaining protoplasm in every cell contains chiefly only those that are to become active in it.... With the exception of those kinds of pangenes that become directly active in the nucleus, as for example those that dominate nuclear division, all the others have to leave the nucleus in order to become active. But most of the pangenes of every sort remain in the nuclei, where they multiply, partly for the purpose of nuclear division, partly in order to pass on to the protoplasm. This delivery always involves only the kinds of pangenes that have to begin to function. During this passage they can be transported by the currents of the protoplasm and carried into the various organs of the protoplasts. (De Vries 1890, 215–216)

As an explanation of development, this theory is superior to the ideas of Darwin and Galton, since it rejects the idea that the hereditary particles develop directly into cells, which was at variance with the cell theory that cells can only arise from cells. Indeed, it is not far from the current theory that DNA in the nucleus is copied into messenger RNA, which is transported into the cytoplasm and translated into protein. Its main weakness as a theory of inheritance was the assumption, common to all non-Mendelian theories, including those of Galton and Weismann, that there are multiple copies of each kind of pangene. Based on this assumption, de Vries explained the existence of two kinds of variability: fluctuating variability, due to different numbers of the individual kinds of pangenes; and "species-forming" variability, due to mutation of pangenes in their division. This was the basis of de Vries's mutation theory, which underlay his belief in the discontinuity of evolution. The mutation theory is discussed in chapter 9, together with the contradiction between his belief that pangenes exist in multiple copies and his status as one of the rediscoverers of Mendelism.

Segregation

Galton adopted from the start the principle of biparental inheritance, which was generally accepted in the middle of the nineteenth century, and he drew two conclusions from it. The first, which he stated in 1865, is that, if an individual inherits one-half of its hereditary particles from each parent, it must inherit one-fourth from each grandparent, one-eighth from each grandparent, and so on.

The second consequence of biparental inheritance is that, if the number of hereditary particles is to remain constant from one generation to the next, an individual can transmit only half of its particles to each offspring. Galton wrote: "As regards the large variety of adult elements, they cannot all be transmitted, for the following very obvious reason—the corresponding qualities of no two parents can be considered exactly alike; therefore the accumulation of subvarieties, if they were all preserved as the generations rolled onwards, would exceed in multitude the wildest flights of rational theory.... The contributions from the latent adult elements are therefore no more than *Representative*" (1872b, 397). Three years later he formulated the concept more precisely in terms of the number rather than the varieties of elements at the end of a discussion of the advantage of biparental over uniparental reproduction (see chapter 5): "There is yet another advantage in double parentage, namely, that as the stirp whence the child sprang, can be only half the size of the combined stirps of his two parents, it follows that one half of his possible heritage must have been suppressed" (1875b, 334). This is a clear statement of the necessity for the halving of the number of hereditary elements, which was again restated in *Natural Inheritance*: "It is not possible that more than one half of the varieties and number of each of the parental elements, latent or personal, can on the average subsist in the offspring" (1889, 187).

The halving of the number of hereditary elements suggests at least a weak form of the law of segregation. Galton used it in this way in *Natural Inheritance* in discussing the correlation between parent and child. It is therefore surprising that he never used it as an explanation of genetic variability between siblings, which was first clearly stated by Weismann. In an essay, "On the number of polar bodies and their significance in heredity," Weismann concluded that there must be a "reducing division" during gametogenesis in which the number of hereditary elements in the nucleus is halved, and he suggested that this reducing division would inevitably give rise to variability in the germ-cells: "It is quite impossible for the 'reducing division' of the nucleus to take place in an identical manner in all the germ-cells of a single ovary, so that the same ancestral germ cells would always be removed in the polar bodies. But if one group of ancestral germ-plasms is expelled from one egg, and a different group from another egg, it follows that no two eggs can be exactly alike as regards their contained hereditary tendencies; they must all differ" (1887, 379). He attributed the differences between brothers and sisters, and between dizygotic twins, to this fact and also used it to account for the similarity of monozygotic twins. Galton had recognized the need for halving of the number of hereditary elements in biparental inheritance, but he did not connect it with the origin of genetic variability in siblings. Weldon (1890a) drew Galton's attention to Weismann's "delightful hypothesis, which people seem to have gone quite mad about," but Galton did not take up the hint.

The empirical evidence known in the nineteenth century which might have led investigators to infer segregation was the variability observed in the second and subsequent generations after a hybrid cross. Galton did in fact consider this evidence in an isolated passage in his paper "On blood-relationship," and came to the correct conclusion: "Lastly, it is often remarked (1) that the immediate offspring of different races or even varieties resemble their parents equally, but (2) that great diversities appear in the next and in succeeding generations. In which stage does the variability occur? It cannot be in the first [Selection 1 in fig. 4.1] nor in the second (Development), else (1) could not have been true; therefore it must be in the third stage [Selection 2]. A white parent necessarily contributes white elements to the structureless stage of his offspring, and a black, black; but it does not in the least follow that the contributions from a true mulatto must be truly mulatto" (1872b, 402).

Galton clearly had in mind in this passage that variability in the offspring of mulattos resulted from the process of selection of a representative sample of latent elements in the germ cells of mulatto (F_1) individuals, but he did not pursue the question further and never referred to it again. In fact, it seems likely that Galton added this passage to the paper as an afterthought in response to a question put by Darwin to whom Galton had sent the manuscript for comment before publication. Galton wrote to Darwin the week before he read the paper at the Royal Society:

> Your criticisms on my paper are very gratifying to me, the more so that the question you put is one to which I can at once reply. You ask, why hybrids of the first generation are nearly uniform in character while great diversity appears in the grandchildren and succeeding generations? I answer, that the diagram shows (see next page) that only 4 stages separate the children from the parents, but 20 from their grandparents and therefore, judging from these limited data alone, (ignoring for the moment all considerations of unequal variability in the different stages and of prepotence of particular qualities etc.,) the increase of the mean deviation of the several grandchildren (from the average hybrid) over that of the several children is as $\sqrt{20}:\sqrt{4}$, or more than twice as great. The omitted considerations would make the deviation (as I am prepared to argue) still greater. (Pearson 1924, 169)

In the paper published in the *Proceedings of the Royal Society*, Galton (1872b) substituted for the above rather obscure argument the suggestion about variability arising at the stage of the second selection, though this passage is absent from the version published in *Nature* (Galton 1872c).

Blending Inheritance

R. A. Fisher began his influential book *The Genetical Theory of Natural Selection* (1930) with a chapter on "The nature of inheritance." The opening sentence reads: "That Charles Darwin accepted the fusion or blending theory of inheritance, just as all men accept many of the undisputed beliefs of their time, is universally admitted." He went on to show that this led to a serious problem because, under this type of inheritance, the hereditary variance is halved in each generation under random mating; to maintain a stationary variance, fresh mutations must be available in each generation to supply the half of the variance that is lost. This is a reformulation of Jenkin's swamping argument (see below). Fisher implied that this problem in evolutionary theory was not solved until the acceptance of Mendel's particulate theory of inheritance in the years following 1900.

Because of Fisher's great authority, his statement that Darwin, and by implication most other nineteenth-century biologists, accepted "the fusion or blending theory of inheritance," has been uncritically accepted by many evolutionary biologists (see for example Wright 1968), but it is misleading. Much of the confusion has arisen from failure to distinguish between physical blending of the hereditary particles during fertilization, which is what Fisher meant, and the blending of phenotypes in the sense that offspring are often intermediate in appearance between their parents. To avoid this confusion, I refer to physical blending of hereditary particles as fusion, and I reserve the term blending for phenotypic blending, which was its usual meaning in the nineteenth century. Mayr (1982) has shown that Nägeli, and possibly Hertwig, were the only nineteenth-century biologists to adopt a pure fusion theory, and that the theory of particulate inheritance, in which the genetic factors derived from the parents do not fuse after fertilization, was widely accepted in the nineteenth century. There was, however, some confusion of thought due to the failure to distinguish between physical blending (fusion) and phenotypic blending.

Darwin's theory of pangenesis involved partial fusion, since the patent elements fuse but the latent elements do not. Suppose that a cross between a white and a black plant produces gray offspring. Under Darwin's theory, the patent white and black gemmules are transformed into gray tissue which produces gray gemmules, but the latent gemmules retain their identity; all three types of gemmule find their way to the gonads and are transmitted to the next generation.

Galton distinguished between blended inheritance and exclusive or alternative inheritance:

> As regards heritages that blend in the offspring, let us take the case of human skin colour. The children of the white and the negro are of a blended tint; they are neither wholly white nor wholly black, neither are they piebald, but of a fairly uniform mulatto brown. The quadroon child of the mulatto and the white has a quarter tint; some of the children may be altogether darker or lighter than the rest, but they are not piebald. Skin-colour is therefore a good example of what I call blended inheritance. It need be none the less "particulate" in its origin, but the result may be regarded as a fine mosaic too minute for its elements to be distinguished in a general view.
>
> Next as regards heritages that come altogether from one progenitor to the exclusion of the rest. Eye-colour is a fairly good illustration of this, the children of a light-eyed and of a dark-eyed parent being much more apt to take their eye-colours after the one or the other than to have intermediate and blended tints. (Galton 1889, 12)

By blended inheritance Galton clearly meant phenotypic blending, and he pointed out that it did not imply fusion of the hereditary particles. He rejected Darwin's idea, resulting from the transportation hypothesis, that gray tissue produces gray gemmules which find their way to the gonads, together with the idea that the existence of gray tissue indicates the fusion of white and black gemmules. Darwin wrote to Galton in 1875, asking him how, on the theory put forward in "A theory of heredity," he would explain the fact that if two varieties of plant (say black and white) are crossed, the hybrid is often intermediate in character (say gray); and that this hybrid could produce millions of buds all exactly reproducing the intermediate, gray character (Olby 1985, 55–57). Darwin thought that this could occur only if the hybrid cells produced gray gemmules which give rise to gray buds. Galton replied that the facts could be explained as well on his hypothesis: all somatic tissue contains surplus latent gemmules from the original stirp (in this case white and black gemmules), and these surplus gemmules can give rise to gray buds, in the same way that the original gray tissue was formed.

But the most interesting part of Galton's reply was his discussion of the possible structure of the gray tissue in the hybrid. It may take only one gemmule to form a cell, or it may take several. If it takes only one, the tissue would consist of equal numbers of white and black cells, which could be distinguished under high magnification but would look gray when less highly magnified. He continued: "If there were two gemmules only, each of which might be white or black, then in a large number of cases one quarter would be always quite white, one quarter quite black, and one half would be gray. If there were 3 [gemmules], we should have 4 grades of color (1 quite white, 3 light gray, 3 dark gray, 1 quite black), and so on according to the successive lines of 'Pascal's triangle'. This way of looking at the matter would perhaps show (a) whether the number in each given [cell] was constant, and (b), if so, what those numbers were" (Pearson 1924, 189–190).

Olby (1985) remarks that the proposal of a 1:2:1 ratio is not an independent discovery of Mendelism, since it concerns only the constitution of the somatic cells, but that it does suggest that Galton was well equipped to appreciate Mendel's discovery. Galton did not know of Mendel's work until after 1900, when he was too old to appreciate it properly. Bateson (1900a) wrote to Galton suggesting that he look up Mendel's paper. In 1905 Galton wrote to Weldon: "I don't believe anybody would have appreciated your work more than Mendel himself had he been alive. Dear old man; my heart always warms at the thought of him, so painstaking, so unappreciated, so scientifically isolated in his monastery. And his face is so nice—I can't give you any useful hints. I wish I could. I am just a learner, and bad at that now" (Pearson 1930b, 542). Around the same time he tried unsuccessfully to develop a Mendelian model for the advantage of sex (see chapter 5). In his autobiography he wrote: "I must stop for a moment to pay a tribute to the memory of Mendel, with whom I sentimentally feel myself connected owing to our having been born in the same year 1822.... Mendel clearly showed that there were such things as alternative atomic characters of equal potency in descent. How far characters generally may be due to simple, or to molecular characters more or less correlated together, has yet to be discovered" (1908, 308).

Galton's theory of heredity changed through his lifetime, but it was throughout a particulate theory in which the hereditary particles derived from the parents do not fuse after fertilization; this type of theory was widely accepted in the nineteenth century. But he did not anticipate the other parts of Mendel's discovery, that the hereditary particles exist in pairs, and that they undergo segregation and independent assortment. His comprehension of segregation was tenuous. Furthermore, he had no way of inferring from study of the polygenic, continuous characters in which he was interested that there were only two hereditary particles for each unit character, one inherited from each parent; and even if he had considered more carefully data on unit characters, such as Darwin's data on flower shape in snapdragons, he would have been prevented from giving them a Mendelian interpretation by his commitment to the idea of reversion. There is no justification for the view that he might easily have discovered Mendelism independently in the 1880s.

Fleeming Jenkin and the Problem of Swamping

Fleeming (pronounced "Flemming") Jenkin (1833–85) was a professor of engineering at Edinburgh University and a member of the North British school of energy physics, which was hostile to Darwinian evolution (Smith 1998). In 1867, he published an anonymous critical review of *The Origin of Species*, which Darwin found very perceptive and which he discussed in the fifth and sixth editions (see Vorzimmer 1963; Gould 1991; Morris 1994; Cookson and Hempstead 2000). His son Francis wrote: "It is not a little remarkable that the criticisms, which my father, as I believe, felt to be the most valuable ever made on his views should have come, not from a professed naturalist but from a Professor of Engineering" (F. Darwin 1887, 107).

Jenkin produced several objections to Darwin's theory. He thought that variability within species was confined within strict limits, so that transformation of one species into another was impossible (see chapter 9). Following his mentor William Thomson (Lord Kelvin), he argued that the age of the earth was too short to have allowed evolution to occur (see Burchfield 1990). But his most influential argument was that natural selection would be ineffective because of the swamping of new variants by backcrossing to the original population. He distinguished between two kinds of variability: "*First,* that kind of common variation which must be conceived as not only possible, but inevitable, in each individual of the species, such as longer and shorter legs, better or worse hearing, etc.; and, *secondly,* that kind of variation which only occurs rarely, and may be called a sport of nature, or more briefly a 'sport,' as when a child is born with six fingers on each hand" (Jenkin 1867, 286–287). Darwin called these two kinds of variation "individual differences" and "single variations," respectively. Individual differences were universal and were admitted by Jenkin to be subject to natural selection: "If we could admit the principle of a gradual accumulation of improvements, natural selection would gradually improve the breed of everything, making the hare of the present generation run faster, hear better, digest better, than his ancestors; his enemies, the weasels, greyhounds, etc., would have improved likewise, so that perhaps the hare would not be really better off; but at any rate the direction of the change would be from a war of pigmies to a war of Titans" (287). But this would only lead to the improvement of hares as hares and of weasels as weasels. It could not lead to the evolution of a new species, such as hares with prehensile tails or with burrowing habits like rabbits. The latter would require the occurrence of a sport, which because of its rarity would be subject to swamping, according to Jenkin, for the following reason.

Consider a population of a million newborn individuals, of whom ten thousand survive to produce offspring. Suppose that one of the newborns is a "sport" with a survival rate of 2 percent instead of 1 percent. This individual has a negligible effect for two reasons. First, it has a 98 percent chance of immediate elimination. Second, if it does survive, it will soon be swamped by the normal individuals for this reason. It will mate with a normal individual and have, say, 100 offspring, who will be intermediate between the two parents and have a survival rate of 1.5 percent. In the next generation, the 1.5 surviving sports will almost certainly mate with normal individuals and have 150 offspring with a survival rate of 1.25 percent; the 1.875 surviving sports in this generation will almost certainly mate with normal individuals and have 187.5 offspring with a survival rate of 1.125 percent, and so on. The point is that the selective advantage of the sports will be halved in each generation since they nearly always mate with non-sports, while the total number of sports only increases very slowly (1, 1.5, 1.875, ...), asymptotically reaching 2.384 sports with a negligible selective advantage. The single original sport with a twofold advantage has been swamped by the much larger normal population. Jenkin illustrated his argument by an example based on contemporary racial prejudice:

> Suppose a white man to have been wrecked on an island inhabited by negroes.... Suppose him to possess the physical strength, energy, and ability of a dominant ... race.... Our shipwrecked hero would probably become king; he would kill a great many blacks in the struggle for existence; he would have a great many wives and children, while many of his subjects would live and die as bachelors.... In the first generation there will be some dozens of intelligent young mulattoes, much superior in average intelligence to the negroes. We might expect the throne for some generations to be occupied by a more or less yellow king; but can anyone believe that the whole island will gradually acquire a white, or even a yellow population, or that the islanders would acquire the energy, courage, ingenuity, patience, self-control, endurance, in virtue of which qualities our hero killed so many of their ancestors, and begot so many children? (Jenkin 1867, 289–290)

Jenkin implicitly assumed a fusion theory of inheritance, so that all the offspring of a sport and a non-sport are intermediate in survival rate between the two parents. He also assumed that each mated pair had 100 offspring, but to maintain constant population size he should have assumed that in the original population each pair produced 200 offspring. (Each adult pair in a stationary population must have two surviving offspring. This point was made by Davis 1871.) With this correction, the number of surviving sports increases rapidly (1, 3, 7.5, ...), though the selective advantage of a sport is halved in each generation, as before. Thus Jenkin's analysis exaggerated the effect of swamping. It is no longer adequate to assume that a sport always mates with a normal individual, but it is clear that after a long enough time the population will reach a stable state in which all individuals have the same survival rate, which is very slightly greater than 1 percent. Selection on survival rate has ceased because the population has become homogeneous as a result of the halving of the genetic variance in each generation under fusion. Davis also pointed out that recurrent occurrence of a favorable sport could effect substantial change: "Though any favourable sport occurring once, and never again, except by inheritance, will effect scarcely any change in a race, yet that sport, arising independently in different generations, though never more than once in any one generation, may effect a very considerable change. These conclusions are opposed to those which the writer of the article is endeavouring to establish" (1871, 161).

Despite some confusion in Jenkin's subsequent discussion, Darwin was strongly influenced by his review to place even more importance on individual differences as opposed to single variations (sports) than he had done previously. In the fifth and sixth editions of *The Origin of Species*, he added a passage in which he stated: "Until reading an able and valuable article in the 'North British Review' (1867), I did not appreciate how rarely single variations, whether slight or strongly-marked, could be perpetuated" (Darwin 1872, 178). He then summarized Jenkin's original swamping argument, though unfortunately he got it wrong. Like Jenkin, he postulated a sport with twice the survival rate of other individuals, but he went on (my italics): "Supposing it to survive and to breed, and that *half its young inherited the favourable variation*" (178). Swamping does not occur under this model, since there is no fusion.

Under Jenkin's model, *all* the young inherited *half* the favorable variation. This model clearly does not apply to sports that do not show blending inheritance; the children of a man with six fingers on each hand who is married to a normal woman do not have five and a half fingers on their hands. To counter this objection Jenkin set up a model in which all the offspring of a sport mated to a normal individual inherited in full vigor the peculiarity of the sport. "Let an animal be born with some useful peculiarity, and let all his descendants retain his peculiarity in an eminent degree, however little of the first ancestor's blood be in them, then it follows, from mere mathematics, that the descendants of our gifted beast will probably exterminate the descendants of his inferior brethren" (1867, 291). But he argued, correctly, that "this theory of the origin of species is surely not the Darwinian theory" (292). Under this theory the sport would increase in frequency merely from its prepotency in transmission even though it survived slightly less well than the other type. Jenkin was not just tilting at windmills in analyzing this model since it was seriously entertained by T. H. Huxley: "Indeed, there seems to be, in many instances, a prepotent influence about a newly-arisen variety which gives it what one may call an unfair advantage over the normal descendants from the same stock" (1860, 37). Jenkin never considered the model envisaged by Darwin in which half the offspring of the sport inherited its character.

Despite the weaknesses of Jenkin's analysis and Darwin's failure to understand it, the swamping argument in its original form, with a rare sport with blending inheritance involving fusion, became very influential. It was discussed briefly by Galton (1887c) in his presidential address to the Anthropological Institute. Like Jenkin, he supposed that a single white individual was introduced into an effectively infinite black population, so that intermarriage with the black parent stock was the rule. He also supposed for simplicity that all individuals had the same fitness, and that each pair left two children to succeed them.

To illustrate blending (really fusion) inheritance, he imagined a large number of glasses each representing an individual, one of them filled with white and the rest with black liquid. The result of mating between two individuals is represented by mixing the contents of the two glasses and pouring the mixture into two new glasses representing the children. In the first generation there will be two glasses of mulatto tint, then four of quadroon tint, then eight of octoroon tint, and so on. This is similar to the paintpot metaphor later popularized by Hardin (1959).

To illustrate mutually exclusive inheritance he imagined that each glass contained a colored cylinder representing the tint of the individual. To start with there were a large number of glasses containing black cylinders and one glass containing a white cylinder. The result of mating between two individuals is represented by mixing the contents of the two glasses, that is, throwing and shaking together the two cylinders in a separate jar and filling two new glasses from out of the jar. Thus the result of mating the white individual to a black one is not two mulatto children but one black and one white child. In the next generation there are four grandchildren, one white and three black, and then eight great-grandchildren, one white and seven black, and so on. Galton continued:

> It would be tedious and of little profit to endeavour to modify this rude but distinct illustration so as to apply to families of varying numbers of children. In some cases the offspring would fail and the race of the white cylinder would come altogether to an end, in others it would be prolific and increase. In all cases the broad fact remains conspicuous that when heritages are mutually exclusive a rare variety may have numerous chances of establishing itself, one in each of many successive generations. Until it is wholly established, it will present itself again and again for competitive examination without diminution of vigour, and if it has natural advantages over the general population it has a corresponding number of chances of profiting by them. The conditions are very different with heritages that blend. (Galton 1887c, 402)

He concluded his account with a tantalizing passage: "It is between these two extreme conditions that the facts of inheritance really lie. They might be roughly illustrated by supposing each of the glasses to contain neither a volume of fluid nor yet a single cylinder, but a moderate number of large beads partly strung together as on a broken necklace, from which some fall off each time it is handled; but I will not pursue this illustration further" (1887c, 402). He concluded that the difficulty of a rare variant spreading suggested by the swamping argument had been exaggerated.

Galton realized the distinction between a fusion model of inheritance, represented by glasses full of liquid, in which the swamping argument was a real problem, and exclusive inheritance, which necessitated a particulate model without fusion, represented by glasses each containing a single cylinder. In the latter case, he correctly concluded that the result of mating a white with a black cylinder would be an equal number of black and white cylinders, so that the swamping problem vanished, and he discussed the stochastic nature of the outcome. It also seems that he had some insight into the distinction between a blending model with fusion, represented by glasses full of liquid, and a blending model without fusion, represented by glasses containing beads on a necklace (foreshadowing his metaphor in *Natural Inheritance* and reminding one of beanbag genetics), in which the swamping problem disappears; but his insight became rather opaque at this point.

5

Four Evolutionary Problems

> In considering the Origin of Species, it is quite conceivable that a naturalist ... might come to the conclusion that each species had not been independently created, but had descended, like varieties, from other species. Nevertheless, such a conclusion, even if well founded, would be unsatisfactory, until it could be shown how the innumerable species inhabiting this world have been modified, so as to acquire that perfection of structure and coadaptation which most justly excites our admiration.... I am convinced that Natural Selection has been the main but not exclusive means of modification.
>
> Darwin, *The Origin of Species*

Galton was not a naturalist, but he was stimulated by *The Origin of Species* to apply evolutionary thinking to several problems related to his primary interest in human biology and heredity. In this chapter we discuss four specific evolutionary questions that attracted his attention. The first two, the domestication of animals and the evolution of gregariousness, arose from his observations in Southern Africa; the third, the inheritance of human fertility and the consequent danger of marrying heiresses, arose out of data on the extinction of peerages which he gathered during his work on *Hereditary Genius*; the fourth, the evolution of sex, is discussed in his 1875 paper "A theory of heredity" and in later unpublished manuscripts.

The Domestication of Animals

The neolithic revolution about ten thousand years ago saw the change from a hunter-gatherer to an agricultural way of life, involving the domestication of animals and plants. Galton read a paper on "The first steps towards the domestication of animals" to the British Association in

1864; it was published in full next year (Galton 1865b), and reprinted with minor changes in *Inquiries into Human Faculty* in 1883. With some changes of wording, this paper still provides a comprehensive summary of what is known about the subject today (Clutton-Brock 1999).

He started from the observation that nearly all domestic animals were first reclaimed from wildness in prehistoric times; men of modern times had only been able to improve the races of animals that they had received from their forefathers in an already domesticated condition. He put forward the explanation that every animal, of any pretensions, has had numerous opportunities of becoming domesticated, since savages are fond of keeping them as pets, or as sacred animals, or in menageries. But these opportunities have only rarely led to any result since no animal is fitted for domestication unless it fulfills certain stringent conditions. Thus only a few species are capable of domestication, all of which were domesticated long ago.

He listed six conditions needed for the domestication of a wild species of animal:

1. It should be hardy, and able to shift for itself and to thrive, although neglected; if it wanted much care, it would not be worth its keep. As evidence of the hardiness of domestic animals, he cited the rapidity with which they establish themselves in new lands: "The goats and hogs left on islands by the earlier navigators throve excellently on the whole" (1865b, 131).

2. It should have an inborn liking of man. He noted that attachments and aversions between different species occurred in nature, and that animals were only likely to be domesticated which had a mutual attachment with man: "Two herds of animals would hardly intermingle, unless their respective languages of action and of voice were mutually intelligible. The animal which above all others is a companion to man is the dog, and we observe how readily their proceedings are intelligible to each other.... A man irritates a dog by an ordinary laugh, he frightens him by an angry look, or he calms him by a kindly bearing; but he has less spontaneous hold over an ox or a sheep. He must study their ways and tutor his behaviour before he can either understand the feelings of those animals or make his own intelligible to them. He has no natural power at all over many other creatures. Who, for instance, ever succeeded in frowning away a mosquito, or in pacifying an angry wasp by a smile?" (133).

3. It should be comfort loving and attracted to human habitations. For example, antelope were not suitable for domestication because they were

adapted to flee from fast-moving predators: "From my own recollection, I believe that every antelope in South Africa has to run for its life every one or two days upon an average, and that he starts or gallops under the influence of a false alarm many times a day.... Now this hourly life-and-death excitement is a keen delight to most wild creatures, but must be peculiarly distracting to the comfort-loving temperament of others. The latter are alone suited to endure the crass habits and dull routine of domesticated life" (133).

4. It should be useful to man, either as a source of food or clothing or through the pleasure of possessing them. As an example he mentioned seals, of which he had heard this story when he visited Shetland as a young man: "A fisherman caught a young seal; it was very affectionate, and frequented his hut, fishing for itself in the sea. At length it grew self-willed and unwieldy; it used to push the children and snap at strangers, and it was voted a nuisance, but the people could not bear to kill it on account of its human ways. One day the fisherman took it with him in his boat, and dropped it in a stormy sea, far from home; the stratagem was unsuccessful; in a day or two the well-known scuffling sound of the seal, as it floundered up to the hut, was again heard; the animal had found its way home. Some days after the poor creature was shot by a sporting stranger, who saw it basking and did not know it was tame. Now had the seal been a useful animal and not troublesome, the fisherman would doubtless have caught others, and set a watch over them to protect them; and then, if they bred freely and were easy to tend, it is likely enough he would have produced a domestic breed" (134–135).

5. It should breed freely under confinement. He noted that this was one of the most important of all the conditions that have to be satisfied, as has been confirmed by the difficulty of keeping breeding colonies of many species in zoos.

6. Finally, it should be easy to tend. He observed that the instinct of gregariousness made it easy for large numbers of animals to be controlled by a few herdsmen, and he remarked that "the cat is the only non-gregarious domestic animal. It is retained by its extraordinary adhesion to the comforts of the house in which it is reared" (136).

He concluded that animals satisfying these conditions were domesticated, not by a preconceived intention, followed by elaborate trials, nor by one successful effort made by an individual, but that "a vast number of half-unconscious attempts have been made throughout the course of ages" (138). Once the process of domestication began, it would be reinforced by unconscious selection: "The irreclaimably wildest members of every flock would escape; the wilder of those that remained would assuredly be selected for slaughter, whenever it was necessary that one of the flock should be killed. The tamest cattle—those that kept the flock together, and led them homewards—would be preserved alive longer than any of the others. It is therefore these that chiefly become the parents of stock, and bequeath their domestic aptitudes to the future herd. I have constantly witnessed this process of selection among the pastoral savages of South Africa" (137).

The Evolution of Gregariousness

The sixth condition making animals suitable for domestication was that they should be gregarious. Galton (1871c) discussed the evolution of this character in an article which was reprinted with minor changes in *Inquiries into Human Faculty* in 1883. He introduced some of the ideas about the advantages and disadvantages of living in groups which are of current interest in behavioral ecology (Krebs and Davies 1993, chap. 6), but he linked his discussion to some ideas about the evolution of "slavish instincts" in man, which today seem rather naive. Perhaps for this reason, the paper has been largely ignored by biologists, with the notable exception of Hamilton (1971).

Galton pointed out that, earlier in his life, he had gained an intimate knowledge of gregarious animals, first by observing during his North African travels the urgent need of the camel for the close companionship of his fellows, and then by studying at greater leisure the habits of the half-wild cattle of the Damaran people of South West Africa. He remarked that the blind gregarious instincts of these cattle were conspicuously distinct from the ordinary social desires. "In the latter they are deficient; thus they are not amiable to one another, but show, on the whole, more expressions of spite and disgust than of forbearance or fondness.... Yet, although the ox has so little affection for, or individual interest in, his fellows, he cannot endure even a momentary severance from his herd. If he be separated from it ... he strives with all his might to get back again, and when he succeeds, he plunges into its middle to bathe his whole body with the comfort of closest companionship" (1871c, 354).

He suggested that the instinct to live in herds had evolved through natural selection to avoid predation by lions and other carnivores. Cattle on their own are not defenseless, and their horns are feared by lions. For this reason a cow who has calved by the wayside, and has been temporarily abandoned by the caravan, is never seized by lions because she is so eager for the safety of her young that no beast of prey can approach her unawares. However cattle must normally spend most of the day eating grass or chewing the cud, when they are not on the alert and can be easily surprised by a predator if alone. "But a herd of such animals, when considered as a whole, is always on the alert; at almost every moment some eyes, ears, and noses will command all approaches, and the start or cry of alarm of a single beast is a signal to all his companions.... The protective senses of each individual who chooses to live in companionship are multiplied by a large factor, and he thereby receives a maximum of security at a minimum cost of restlessness" (356). He concluded that "it follows from the law of natural selection, that the development of gregarious, and therefore of slavish, instincts must be favoured in such cattle" (356).

He had also observed that there was considerable variability in the degree of independence of the Damaran cattle. It was difficult to procure animals capable of acting as fore-oxen to the team of oxen, since the majority of the wild herd were unfitted to move in such an isolated position. But a few animals who showed a more independent (less gregarious) nature, by grazing apart or ahead of the rest of the herd, could be broken in for fore-oxen, and even more exceptionally it was possible to find an ox who could be ridden apart from the companionship of others; an example was his ox Ceylon (see fig. 1.4). At the other end of the scale, he had a general impression of oxen showing a deficiency from the average ox standard of self-reliance about equal to the excess of that quality found in ordinary fore-oxen (exemplified by running more madly than the rest into the middle of the herd when they were frightened). He concluded that the law of deviation from an average (the normal distribution) was likely to be applicable to independence of character among cattle.

He then considered the question: "Why is the range of deviation from the average such that we find about one ox out of fifty to possess sufficient independence of character to serve as a pretty good fore-ox? Why is it not one in five, or one in five hundred? The reason undoubtedly is, that natural selection tends to give but one leader to each suitably-sized herd, and to repress super-abundant leaders. There is a certain size of herd most suitable to the geographical and other conditions of the country; it must not be too large, or the scattered puddles which form their only watering-places for a great part of the year would not suffice.... It must not be too small, or it would be comparatively inefficient" (356). If this optimal herd had several independent members, one of them would become the leader, and the rest would graze on the outside of the herd. "The leaders are safe enough from lions, because their flanks and rear are guarded by their followers; but each of those who graze apart, and who represent the superabundant supply of self-reliant animals, have one flank and the rear exposed, and it is precisely those whom the lions take. Looking at the matter in a broad way, we may justly assert that wild beasts trim and prune every herd into compactness, and tend to reduce it into a closely-united body with a single, well-protected leader. The development of independence of character in cattle is thus suppressed far below its healthy natural standard by the influence of wild beasts, as is shown by the greater display of self-reliance among cattle whose ancestry, for some generations, have not been exposed to such danger" (356–357).

In this paper, Galton foreshadowed some of the ideas about the evolution of group living which are under active investigation today. He expressed clearly the advantage of increased alertness to avoid predation which could be obtained by living in a group. He proposed the idea of marginal predation, so that animals at the center of a group are safer than those at its edge. In discussing the effect of selection on the degree of gregariousness he appealed to forces of individual rather than group selection. Thus, in discussing the advantage of increased alertness through belonging to a group, he wrote: "It also follows from the same law [of natural selection], that the degree in which those instincts are developed is, on the whole, the most conducive to their safety. If they were more gregarious, they would crowd so closely as to interfere with each other, when grazing the scattered pasture of Damara land; if less gregarious, they would be too widely scattered to keep a sufficient watch against the wild beasts" (356). In other words, the degree of gregariousness is selected to optimize the spacing between animals.

He also put forward the idea that there was an optimal group size determined by the balance between costs and benefits, but he used a group selection argument to explain how this optimal group size is maintained. In discussing how it has come about that there is only one good fore-ox in fifty, he supposed that there has been selection to ensure that there is only one leader in each group. Thus he proposed two conflicting selection pressures on the degree of gregariousness, to optimize the spacing between animals and to ensure that there is only one leader in each group, without suggesting how the conflict between them has been resolved. These ideas may not convince those who distrust the adaptationist program because they think it promotes Kiplingesque Just-So stories or Darwinian fairytales (Gould and Lewontin 1979, Stove 1995). It is true that Galton provided little empirical evidence, but it is remarkable how close his ideas are to current theory which has ample empirical support (Krebs and Davies 1993, chap. 6).

Galton linked his percipient discussion of the evolution of gregariousness in cattle to a rather naive analogy with the evolution of "slavish" aptitudes in man, "from which the leaders of men, and the heroes and the prophets, are exempt, but which are irrepressible elements in the disposition of average men. I refer to the natural tendency of the vast majority of our race to shrink from the responsibility of standing and acting alone, ... to their willing servitude to tradition, authority and custom" (353). He argued that the inhabitants of South West Africa were divided into a large number of tribes, all more or less at war with one another, that tribes of intermediate size were more stable than very small or very large ones, and that in consequence selection would favor a race that supplied an appropriate proportion of self-reliant individuals. "The law of selection ... must discourage every race of barbarians which supplies self-reliant individuals in such large numbers as to cause their tribe to lose its blind desire of aggregation. It must equally discourage a breed that is incompetent to supply such men, in a sufficiently abundant ratio to the rest of the population, to ensure the existence of tribes of not too large a size" (357).

He concluded: "What I wish to prove in the present essay is the steady influence of social conditions, all through primaeval periods, down, in some degree, to the present day, in destroying the self-reliant, and therefore the nobler races, of men. I hold that the blind instincts evolved under those long-continued conditions have been deeply engrained into our breed.... The hereditary taint due to the primaeval barbarism of our race, and maintained by later influences, will have to be bred out of it before our descendants can rise to the position of free members of a free and intelligent society" (357).

The Fertility of Heiresses

A chapter of *Hereditary Genius* was devoted to "English peerages, their influence upon race." It had been frequently remarked that the families of great men are apt to die out, from which it was argued that men of ability were unprolific. If this were the case, any attempt to produce a highly gifted race of men by breeding from gifted individuals would be doomed to failure.

To investigate this question, Galton began by examining the descendants of English judges between 1660 and 1840 who had gained peerages. There were 31 such peerages, of which 19 remained and 12 had become extinct. He summarized his conclusion:

> I … tabulated [the results]; when, to my astonishment, I found a very simple, adequate, and novel explanation, of the common cause of extinction of peerages, stare me in the face. It appeared, in the first instance, that a considerable proportion of the new peers and of their sons married heiresses. Their motives for doing so are intelligible enough, and not to be condemned.… But my statistical lists showed, with unmistakeable emphasis, that these marriages are peculiarly unprolific. We might, indeed, have expected that an heiress, who is the sole issue of a marriage, would not be so fertile as a woman who has many brothers and sisters. Comparative infertility must be hereditary in the same way as other physical attributes, and I am assured it is so in the case of the domestic animals. Consequently, the issue of a peer's marriage with an heiress frequently fails, and his title is brought to an end. (Galton 1869, 131–132)

He based this conclusion on the observation that, out of the 12 peerages that had failed in the direct male line, no less than 8 failures were accounted for by heiress marriages. But his most convincing evidence came from a direct comparison of the fertility of 50 heiresses with 50 non-heiresses shown in table 5.1. One-fifth of the heiresses leave no male children, compared with only 2 percent of the non-heiresses (though Galton suspected that the latter figure was too small). He concluded that "although many men of eminent ability … have not left descendants behind them, it is not because they are sterile, but because they are apt to marry sterile women, in order to obtain wealth to support the peerages with which their merits have been rewarded. I look upon the peerage as a disastrous institution, owing to its destructive effects on our valuable races" (1869, 139–140).

Galton thought that the effect of marrying an heiress was mediated directly through the inheritance of physiological infertility. A similar idea had occurred to his grandfather, Erasmus Darwin, who wrote in his notes to *The Temple of Nature:* "As many families become gradually extinct by hereditary diseases, as by scrofula, consumption, epilepsy, mania, it is often hazardous to marry an heiress, as she is frequently the

Table 5.1. Fertility of 50 Heiresses and 50 Non-Heiresses

Number of sons to each marriage	Number of cases in which mother was an heiress	Number of cases in which mother was not an heiress
0	11	1
1	8	5
2	11	7
3	11	17
4	5	10
5	3	4
6	1	4
7	0	2
>7	0	0
Total number of sons	104	168
Total number of daughters	103	142

Source: Galton 1869

last of a diseased family." (This passage was cited by Galton on the interleaf of his copy of *Hereditary Genius* [Pearson 1924, 95], though it was apparently not known to him when he wrote the book.)

Pearson and Lee (1899) later studied the inheritance of human fertility, obtaining data from reference books on the Peerage and the Landed Gentry. The significance of these sources is that entries of women in the Landed Gentry are very often entries of heiresses, while the entries of women in the Peerage are entries because of class. They concluded that there was evidence of inheritance of fertility, estimated from the correlation between mothers and daughters. But they found that the average number of children of the heiresses (from the Landed Gentry) was as large as that of women of whom most were not heiresses (from the Peerage). They concluded that "heiresses are not on the whole the children of sterile mothers; ... [they] are rather the daughters of mothers whose apparent fertility is fictitious. They have, owing to the sterility or early death of their husband, to their own marriage late in life, or to some physical disability, or other restraint, never reached their true fertility" (284).

If this conclusion is correct, the marked reduction in fertility among heiresses in table 5.1 is not a direct effect of the inheritance of physiological fertility but must be due to an indirect effect of the marriage between a peer and an heiress. Perhaps such marriages occur late in life, or perhaps a marriage for money on the one side and status on the other does not lead to conjugal harmony. But Galton's main conclusion remains valid. There is no reason to suppose that men of ability are unprolific, as long as they refrain from marrying heiresses.

The Extinction of Surnames

This study led Galton to develop a mathematical theory of the extinction of surnames, in collaboration with a mathematical friend, the Rev. Henry Watson, which began the mathematical study of branching processes. As we saw in chapter 2, Alphonse de Candolle had read *Hereditary Genius* (1869) just before completing his own work on science and scientists in 1873. Motivated by Galton's discussion of the extinction of peerages, de Candolle pointed out that a large proportion of family names are continually dying out simply by chance, so that one should know what that proportion is before postulating an additional factor of reduced fertility. He wrote: "Among the accurate data and the very sensible views of Benoiston de Châteauneuf, Galton, and other statisticians, I have not found any discussion of an important point that they should have made, the *inevitable* extinction of family names. It is obvious that all surnames must eventually become extinct. . . . A mathematician could calculate how the diminution of surnames or titles should occur, based on the probability distribution of family size and the probability of childless marriages" (De Candolle 1873, quoted by Watson and Galton 1874, 138 [my translation]).

Galton states that he had some years previously tried to obtain some numerical results for this very problem, "but the computation became intolerably tedious after a few steps, and I had to abandon it" (Watson and Galton 1874, 138). He now applied to several mathematicians for a solution, without success, and then proposed it in 1873 as a problem in a mathematical periodical, the *Educational Times*:

> PROBLEM 4001: A large nation, of whom we will only concern ourselves with the adult males, N in number, and who each bear separate surnames, colonise a district. Their law of population is such that, in each generation, a_0 per cent of the adult males have no male children who reach adult life; a_1 have one such child; a_2 have two; and so on up to a_5 who have five.

> Find (1) what proportion of the surnames will have become extinct after r generations; and (2) how many instances there will be of the same surname being held by m persons. (Quoted in Kendall 1966, 386)

The only answer it received was totally erroneous, and Galton finally appealed to Watson for help. The latter published a solution in the *Educational Times* in 1873, and extended his results next year in a joint paper with Galton; the problem became known as the Galton-Watson process. Since Watson also helped Galton in his study of correlation, described in chapter 6, it may be of interest to cite some biographical details recorded by Galton (1908, 305–307), beginning with an anecdote that provides an early example of Anglo-German rivalry on vacation.

Henry William Watson (1827–1903) was a keen alpinist and helped to found the Alpine Club in 1857. On one occasion Watson and a friend "set off at a good pace to vanquish some new but not difficult peak, and passed on their way a somewhat plodding party of German philosophers bound on the same errand. One of Watson's shoes had shown previous signs of damage, but he thought he could manage to get on for a day or two longer if he now and then covered it with an indiarubber galosh.... He and his friend reached the top long before the Germans, whom they thought no more about. However, shortly after, a Swiss-German newspaper gave a somewhat grandiose account of the ascent of the mountain by Professors This and That, in which it was remarked that the Professors would have been the very first to reach its summit had not two jealous Englishmen provided themselves with 'Gummi Schuhe' and so were able to outstrip them" (306).

Watson studied mathematics at Trinity College, Cambridge, and was second Wrangler in 1850. He was elected to a Fellowship at Trinity in 1851, which he had to resign on his marriage. He then earned a living by teaching mathematics at the City of London School, King's College, London, and finally Harrow School, before moving to a valuable living (clerical benefice) in 1865, where he continued his mathematical interests, chiefly in mathematical physics. Galton records how he obtained this living: "He was a Master at Harrow when some scrape had occurred, and a boy in whom he was interested was judged guilty and sent up to be flogged. The boy protested his innocence so vehemently, that although appearances were sadly against him, Watson was ready to believe what he said, and took unusual pains to investigate the matter. The result was that the boy was completely exculpated. A few years after, the boy's father bought the property at Berkswell in which the gift of the living was included. It happened to be then vacant, and the new proprietor found he must either nominate some one at once, or the nomination would lapse, and fall (I think) to the Bishop. He knew of no suitable clergyman. Then the boy called out, 'Give it to Mr. Watson,' which the father, knowing the story, did" (306–307). It is not known how Galton and Watson became friends. They may have met climbing in the Alps, or at Harrow, where Galton's brother-in-law was headmaster.

We now return to Watson's solution of Problem 4001 (Watson and Galton 1874). To simplify the problem, he neglected the overlapping of generations and assumed that the percentages a_i were constant from one generation to the next. He worked with probabilities $p_i = a_i/100$ and removed the artificial restriction that $i \le 5$. He then defined the generating function

$$f(s) = p_0 + p_1 s + p_2 s^2 + p_3 s^3 + \cdots \qquad (0 \le s \le 1), \qquad [1]$$

which is a function of the dummy variable s increasing from p_0 when $s = 0$ to 1 when $s = 1$. Finally he defined the series of generating functions

$$f_1(s) = f(s),\ f_{n+1}(s) = f(f_n(s)), \quad n = 1,2,3,\cdots \qquad [2]$$

and showed that the coefficient of s^r in the power series expansion of f_n is the probability that there are r males in the nth generation derived from a single male in generation 0. He also observed that the probability q_n of extinction by the nth generation satisfies the equations

$$q_1 = p_0,\ q_{n+1} = f(q_n). \qquad [3]$$

Watson concluded that the problem had been reduced to the mechanical but generally laborious process of successive substitution (functional iteration), given the probabilities p_0, p_1, p_2, etc., and that no further progress could be made until these probabilities had been determined. To simplify the calculations he assumed that these probabilities might be approximated by the binomial probabilities

$$p_i = \frac{n!}{i!(n-i)!} P^i Q^{n-i}, \quad 0 \le i \le n \qquad [4]$$

for appropriate values of n, P and $Q = 1 - P$. In this case, the generating function takes the simple form

$$f(s) = (Q + Ps)^n. \qquad [5]$$

He considered the example $n = 5$, $P = 0.25$, giving values of the probabilities $p_0, \ldots, p_5$ as .237, .396, .264, .088, .015, .001, which he thought plausible. The probabilities of extinction in the first ten generations can be calculated from eq.[3] as

.237, .347, .410, .450, .478, .497, .511, .521, .528, .534.

(I have corrected some minor discrepancies in the third decimal place. These calculations must have been quite time consuming in Watson's time, but can today be performed in seconds using a computer program.) The disappearances are much more rapid in the earlier than in the later generations; while 237 names out of a thousand disappear in the first step, and an additional 110 names in the second step, there are only 28 disappearances in the fifth step, and only six in the tenth step.

It is natural to ask what is the ultimate probability of extinction after a very long time. Watson correctly observed from eq.[3] that this is the solution of the equation

$$s = f(s). \qquad [6]$$

But he now fell into a trap. He observed that eq.[6] has the solution $s = 1$ since $f(1) = 1$. He concluded that all surnames are ultimately doomed to extinction: "All the surnames, therefore, tend to extinction in an indefinite time, and this result might have been anticipated generally, for a surname once lost can never be recovered, and there is an additional chance of loss in every successive generation" (Watson and Galton 1874, 143). He failed to observe that, in the example he considered, eq.[3] has another solution, $s = .553$, which turns out to be the relevant solution. Under this model, only 55 percent of the surnames will ever go extinct, starting from a unique representation of each name at time zero. The reason is that the total male population is expanding, since each male leaves on average $nP = 1.25$ males in the next generation. Thus the names that have not gone extinct early on will each have a large number of representatives in later generations and will be protected from extinction. Suppose in general that each male contributes on average M surviving adult males to the next generation. It turns out quite generally that, when $M \leq 1$, $s = 1$ is the relevant solution of eq.[6], so that any surname is ultimately doomed to extinction; this is obviously true when $M < 1$ because the whole population is contracting and will eventually become extinct. But when $M > 1$, so that the population is expanding, eq.[6] has a second solution less than 1, which is the relevant solution, so that a non-zero proportion of the original surnames will ultimately survive. Which of the names survive and which of them go extinct is determined by stochastic events in the first few generations. This important result, the criticality theorem, was not discovered until the period 1920–30. The branching process was reinvented at this time in a genetic context by J. B. S. Haldane and R. A. Fisher, apparently independently of the work of Galton and Watson. In particular, Fisher (1930) used the model, in conjunction with the assumption of a Poisson distribution of offspring and stationary population size, to show that the ultimate probability of survival of a single mutant with a small selective advantage s over the wild type is approximately $2s$. The first complete proof of the criticality theorem was published in 1930 in Danish by J. F. Steffensen.

The story so far is that the idea of a branching process was invented by Galton and Watson in 1873 to study the extinction of surnames, but that they did not discover its most important property, the criticality theorem. The process was reinvented and the theorem discovered 50 or 60 years later, by Haldane, Fisher, and Steffensen, among others. It was shown in 1972 that this story is incomplete. In 1845 the mathematician I. J. Bienaymé published a brief note in an obscure French journal on "The law of multiplication and duration of families." He stated that he intended to publish a full account of his work in a special memoir, which has never been discovered, but it is clear from the note that he was interested in the same problem as Galton, that he approached it in the same way as Watson did, and that furthermore he was aware of the criticality theorem. Bienaymé's work went unrecognized until 1972 and played no role in the development of the theory of branching processes. (This account is largely based on Kendall 1966 and 1975, and on Guttorp 1995. See also Heyde and Seneta 1977.)

The Evolution of Sex

Most plants and animals reproduce sexually; that is to say, each new individual contains genetic material from two parents. This fact presents a challenge to evolutionary biologists today, and a bewildering number of theories have been suggested to explain it (Maynard Smith 1978; Bell 1982 and 1988; Bulmer 1994, chap. 12). Galton proposed a remarkably prescient theory to account for the advantage of sexual over asexual reproduction in 1875; his later thoughts are embodied in three unpublished essays (1890c, 1896, and c. 1905).

"A Theory of Heredity" (1875)

In a long paragraph in "A theory of heredity," Galton put forward two theories to account for the advantage of sexual (biparental) over asexual (uniparental) reproduction; the first of them is remarkably prescient. He began by stating the problem:

> Much wonder is expressed by physiologists at the apparent fact that none, at least of the higher races, admits of being long maintained through any system of unisexual parentage; but that a deterioration, which we may reasonably ascribe to a deficiency of some of the structural elements, is always observed to set in and gradually to increase, the race ultimately perishing from that cause. A system of double parentage is therefore a very important requirement, some think an essential one, to secure the indefinite maintenance of any race whose organisation is complex.... In many of the lowest forms of organised life, double parentage exists, but sex apparently does not, because any two cells seem able to conjugate and to combine their contents within a single cell; these forms are also capable of easy unisexual multiplication by self-division or by budding. Proceeding higher in the scale of life, the sexual differentiation becomes increasingly marked, and unisexual propagation is of rarer occurrence. At length we reach a stage where the differentiation of sex is complete, and the power of unisexual propagation is wholly lost. Now the necessity of a system of double parentage in complex organisations, is the immediate consequence of a theory of organic units and germs, as we shall see if we fix our attention upon any one definite series of unisexual descents, and follow out its history. (Galton 1875b, 332–333)

He then developed his theory in two stages. First, he considered a clonal line of descent in which a single bud was chosen from a plant, cut off, and grown to maturity; in the next generation, a single bud was chosen from this mature plant, cut off, and grown to maturity, as before; and so on, indefinitely. Under this system of clonal reproduction, he argues, there is a chance in each generation "of some one or more of the various species of germs in the stirp dying out, or being omitted; and of course when they are gone they are lost forever" (1875b, 333). The loss of a particular species of germ is sometimes unfavorable, and this process leads to a gradual deterioration of the race and the eventual extinction of the line. At this point he refers in a footnote to the Galton-Watson theory of the extinction of surnames, which is analogous to the extinction of species of germs.

Second, he recognized that the situation is complicated by natural selection between lines in a state of nature "where the weakly plants are supplanted by those that remain sound." There is thus a balance between the growing chance of deterioration in a single line and the growing number of all possible lines of descent. This would lead to extinction of the population when the former exceeds the latter, and he suggested that this would be the case in complex organisms, since the chance of deterioration increases while the fecundity diminishes with increasing complexity. But this would not happen in biparental reproduction since the chance deficiency of a particular species of germ inherited from one of them would be supplied by the other:

> On the other hand, when there are two parents, and therefore a double supply of the material, the chance deficiency in the contribution from either of them, of any particular species of germ, tends to be supplied by the other. No doubt, cases must still occur, though much more rarely than before, in which the same species of germ is absent from the contribution of both, and a very small proportion of the families will thereby perish. But what if they do become extinct? The remaining families are perfectly sound, or tend to become so in each succeeding generation, and they fill up, only too easily, the gap. Thus we see that in any specified course of unisexual generation, every line of descent is doomed to extinction, sooner or later; but that in bisexual, only a very small proportion of families become extinct, or even temporarily suffer, from the cause we are considering, while the great majority do not suffer a whit, and those few who do, tend to become rehabilitated. (Galton 1875b, 334)

This completes his first theory to explain why higher plants and animals nearly always reproduce sexually. It contains the germs of some modern theories, but it was not appreciated at the time. Charles Darwin, to whom Galton had sent the paper for comment, made the objection: "If gemmules (to use your own term) were often deficient in buds I could but think the bud-variations would be commoner than they are in a state of nature; nor does it seem that bud-variations often exhibit deficiencies which might be accounted for by absence of the proper gemmules. I take a very different view of the meaning or cause of sexuality" (Pearson 1924, 187).

Galton's theory has some similarity to Muller's ratchet, one of the theories for the maintenance of sex current today. Muller argued that "an asexual population incorporates a kind of ratchet mechanism, such that it can never get to contain, in any of its lines, a load of mutations smaller than that already existing in its at present least-loaded lines. However, the latter lines can ... become more heavily loaded by mutation" (1964, 8). The ratchet does not operate in a sexual population because a deleterious mutant in one line can be substituted by a good copy from another line. Mathematical analysis of Muller's ratchet shows that it operates very weakly in organisms with very large populations or very small genomes, that is to say in simple organisms, so that it would only be expected to give a significant advantage to sexual reproduction in complex organisms. (The "two-fold cost of sex" adds another twist to modern accounts.)

Galton then rather spoiled his case by adding a secondary theory based on the idea of competition between germs, which carries little conviction today:

> There is yet another advantage in double parentage, namely, that as the stirp whence the child sprang, can be only half the size of the combined stirps of his two parents, it follows that one half of his possible heritage must have been suppressed. This implies a sharp struggle for place among the competing germs, and the success, as we may infer, of the fitter half of their numerous varieties. (Galton 1875b, 334)

In this passage he is referring to competition during the process of selection of the germs to be transmitted to the next generation ("Second selection" in fig. 4.1). The problem with this explanation, even under Galton's model of heredity, is that there is no reason why a germ that is "fitter" in the sense of having a greater chance of transmission to the next generation should confer greater "fitness" on the individual to whom it is transmitted.

Three Unpublished Essays

Galton returned to the subject in three unpublished essays preserved in the Galton papers. The first essay was entitled "Sexual generation and cross fertilisation" (Galton 1890c); it can be dated to 1890 from a letter in response to it from Weldon. Galton began: "I propose to discuss more fully, in the light of more recent knowledge, a cause that I pointed out many years ago (space for reference, presumably 1875) of the advantage both of sexual generation and of cross-fertilisation to plants and animals of complex structure, and to give some numerical notion of the enormous magnitude of the advantage." He supposed that n different types of germinal particles were needed in the new embryo (which he rather confusingly called a "somatic cell") to form the different tissues of the soma. In asexual generation he supposed that the chance that one of these species was completely absent was $1/r$; thus the chance that a particular species was present in one or more copies was $(r - 1)/r$, and the chance of the presence of one or more germinal particles of every one of the n species was $[(r - 1)/r]^n$. In sexual generation, the chance that neither the father nor the mother would contribute a particle of a particular species was $1/r^2$, so that the chance that either or both of them would contribute a representative of every one of the n species was $[(r^2 - 1)/r^2]^n$. This would give an enormous advantage to sexual generation in complex organisms in which n was much larger than r; for example, if $n = 500$, $r = 100$, the chance of completeness would be .007 for an asexual and .95 for a sexual embryo. The same reasoning would account for the benefit of cross-fertilization, since it relies on independent contributions from the two parents.

In his comments on this essay, Weldon discussed recent histological work, and in particular urged Galton to add a discussion of the significance of polar bodies. In his second essay, "The service of sex" (Galton 1896), which can be dated to 1896 from a letter in response to it by Weldon, Galton emphasized the importance of what he called sequestration: "During the very earliest stages in a reproductive cell the strong organisation of the incipient soma enables it to sequestrate for its own use, the fittest among each class of the available germinal elements, leaving the residue of them in that cell so much the poorer. This act of *sequestration* is held to be a factor of very great importance in the processes of heredity and by its supposed existence the advantage to a complex organism of bisexual production will be explained and to some degree formulated." He thought that the germ cells were imperfect, either in quantity or in quality, in respect to a few out of very many different classes of elements, owing to sequestration of the best elements for the soma. In sexual reproduction between unrelated individuals, the chance of the same class of elements being imperfect in both parental cells was small, so that a perfect set of the classes may usually be made up out of their joint contents. In sexual reproduction between related individuals, the same class of element was often imperfect in both germ cells, so that a perfect set of the classes could not usually be made up out of their joint contents; the same argument would apply with greater force to asexual reproduction.

Galton thought that the formation of polar bodies was an adaptation to circumvent the consequences of sequestration:

> The very complicated process of karyokinesis, which consists in the duplication of the elements in the cell and the subsequent extrusion and waste

> of about one half of them in the form of a polar body, becomes intelligible if it may be assumed that the fitter half is retained and the residue is extruded.... The same idea also applies to bisexual generation, where a second halving takes place in each of the two parental cells, reducing the bulk of their joint contents to that of the original contents of one of them. Here also it is to be presumed that the fitter is kept and the other part extruded. (Galton 1896)

In his comments, Weldon pointed out that Galton had misunderstood the cytology of cell division, and that in any case there was no evidence that the separation of chromosomes was selective.

The most interesting part of the essay is this analogy:

> Suppose then it is desired to transport a complicated machine to a distant island, for immediate service. Also that the only means of transport is by one of the small and similar boats in use at the port, and that one of these is just large enough to carry the machine after it has been taken to pieces and packed into n different boxes each labelled with a distinctive number. There are supposed to be means on arriving at the island, for putting the machine together, but none for supplying any part of it that may be missing or for repairing it if it be much damaged. This corresponds to unisexual generation. If however, to lessen the risk of mischance, a duplicate machine, similarly packed and labelled with similar numbers, be sent in a second boat, there is good hope that such parts of the machine as may receive damage in one of the boats will be uninjured [in] the other, so that all the parts of one complete machine may be obtained out of the joint cargoes. This corresponds to bisexual generation. Let us now consider the difference of risk in the two cases. Let the risk of any one box being damaged be $1/r$ and let the risk to each box be *independent* of that to every other box. (We must begin with the simplest case.) Then the chance of a damage occurring among the n boxes in one boat is n/r. [This approximation is only accurate when $n \ll r$.] So again the chance of any given box being damaged in both boats is $1/r^2$, and that of a damage to a similar box among the n boxes in either boat is n/r^2. Therefore in this case the two risks would be as 1 to $1/r$, or as r to 1, and it is independent of n. If for example the risk to a single element was as 1 to 50, or such that one out of fifty perished then bisexual generation would be fifty times as safe as unisexual. (Galton 1896)

The model depends on the assumption that the mechanic who puts the machine together can distinguish a damaged from an undamaged box. A modern analogy is the construction of a complete car from two broken-down cars of the same make.

The third essay (Galton c. 1905) is called "On the large advantage of biparental over uniparental generation." This short note appears to be in a hand other than Galton's, with corrections by him; the original draft was probably dictated by him. It is undated, but the reference to Mendelians shows that it must postdate 1900, and it may be tentatively dated as c. 1905. It is of little intrinsic merit, but it is interesting because it shows Galton trying to come to grips with Mendelism.

He first outlined the argument:

> An organism presumably grows out of many units all of which are necessary to its development. Any one of them may be absent from a single germ, but it is less likely to be absent from both of the two germs whose combination forms a zygote. The resulting gain in safety will now be shown to be enormous. (Galton c. 1905)

He began by considering a single unit:

> Let M be one of these necessary units & let it be absent on the average in one out of every r germs, the chance of its absence in any one germ will therefore be $1/r$. Let a zygote be represented by a couple of letters, m being used to signify the presence of M, and μ its absence. Then if the first letter in the couple refers to the male germ, and the second to the female germ, the four varieties of zygote will take the form so familiar to Mendelians of mm, mμ, μm, μμ, and they will occur with equal frequency. In other words, the absence of M from a particular germ will be three times as frequent as from a zygote, whatever the value of r may be. (Galton c. 1905)

He then supposed that there were n units, the absence of any one of which was lethal, and he concluded that

> the chance of all these units being present in the zygote will be 3^n times greater than in a single germ. This increase of safety becomes enormous when n is only moderate in value. Even if n were = 10, 3^n would exceed fifty-nine thousand. (Galton c. 1905)

There are several problems with this analysis. Galton assumed that the four types of zygote would be equally frequent under Mendelism, a common mistake before the discovery of the Hardy-Weinberg law in 1908 (see chapter 10); even under this assumption it is not obvious how he obtained the factor of three. He should have argued that if the frequency of μ in the germs was $1/r$, then the frequency of the lethal recessive μμ in zygotes would be $1/r^2$ under random mating, so that the absence of M from a particular germ would be r times as frequent as from a zygote. Even with moderate values of r this would give a much greater advantage than he found. Like Galton's 1890 model, this model makes the incorrect assumption that r would be the same in haploid and diploid organisms. If recessive lethal genes are maintained in the population by a balance between selection and mutation, then a lethal gene is maintained at a much higher frequency in diploid than in haploid organisms because it is sheltered in heterozygotes.

Galton struggled to understand the advantage of sexual reproduction for thirty years, but he was hampered by his inadequate understanding of cytology and of Mendelian genetics. He never achieved a better insight into the problem than in his first thoughts of 1875.

6

The Charms of Statistics

> *The Charms of Statistics.*—It is difficult to understand why statisticians commonly limit their inquiries to Averages, and do not revel in more comprehensive views. Their souls seem as dull to the charm of variety as that of the native of one of our flat English counties, whose retrospect of Switzerland was that, if its mountains could be thrown into its lakes, two nuisances would be got rid of at once. An Average is but a solitary fact, whereas if a single other fact be added to it, an entire Normal Scheme, which nearly corresponds to the observed one, starts potentially into existence.
>
> Some people hate the very name of statistics, but I find them full of beauty and interest.... They are the only tools by which an opening can be cut through the formidable thicket of difficulties that bars the path of those who pursue the Science of man.
>
> Galton, *Natural Inheritance*

Historians of science have recognized that there was a "probability revolution" in the period 1830–1930, which led to the application of probability theory and statistical models to a wide range of problems in the natural and social sciences (Porter 1986; Krüger, Daston, and Heidelberger 1987; Gigerenzer et al. 1989; Hacking 1990). The revolution began with the demonstration of statistical regularities in social data by Quetelet, on which he built a science of "social physics"; Quetelet's work inspired both Galton's statistical theory of heredity and Maxwell's statistical interpretation of the kinetic theory of gases. This growing range of applications reflects the increasing acceptance of statistical laws as valid scientific explanations. It was accompanied by a reinterpretation of the meaning of probability, from rational degree of belief to relative frequency in the long run.

These statistical applications were accompanied by the development of the corresponding statistical theory. Of particular importance were the invention of the concepts of regression and correlation by Francis Galton, and their development on a sound mathematical basis by Karl Pearson, which are described in this chapter; their attempt to construct a statistical theory of heredity based on these ideas is discussed in the next two chapters.

The normal distribution played a key role in these developments. It was usually represented in the nineteenth century by the function

$$f(x) = \frac{1}{c\sqrt{\pi}} \exp\left[-(x-\mu)/c\right]^2. \qquad [1]$$

This formula defines a family of distributions with the same bell shape but depending on two parameters, μ and c, the mean and the modulus, which respectively determine its location and dispersion. (The square of the modulus is twice the variance, σ^2.)

The importance of the normal distribution is due to the central limit theorem, which states that the sum of a large number of independent random variables of individually small effect follows a normal distribution, almost regardless of their individual distributions. This theorem was formulated as a general result by Laplace about 1810 and was applied by him and others to explain why errors of measurement, particularly in astronomy, were approximately normal; for this reason, it was often called the "law of error." It was applied to the human sciences by Quetelet, with whose work Galton became familiar from his book *Letters on Probabilities*, published in 1846 and translated into English in 1849. (It was written in the form of letters to the Grand Duke of Saxe Coburg and Gotha, the father of Prince Albert.) It was not called the "normal distribution" until the 1870s, when this term was independently used by three men, Charles S. Peirce, Francis Galton, and Wilhelm Lexis; since the same date it has also been called the "Gaussian distribution" after the great mathematician Carl Friedrich Gauss, who associated it with the method of least squares in 1809 (Stigler 1999).

Quetelet and the Average Man

Adolphe Quetelet (1796–1874) was a Belgian mathematician who made his career as an astronomer and meteorologist at the Royal Observatory in Brussels. On a visit to Paris to learn the practical side of these subjects, he found out about probability and its applications and was struck by the statistical regularity of observations on large numbers of individuals. For example, he found that the stillbirth rate in Belgium was consistently higher in towns than in the country; about 6 per cent of the urban babies and only 3 percent of the rural babies were stillborn. He went on to show that the stillbirth rate was higher in illegitimate than in legitimate births (about 6 percent versus 4 percent), which might explain in part the higher rate in towns, since illegitimate births were much more numerous than in the country, and he concluded "that immorality and misery are destructive causes which affect man even before he has seen the light" (1849, 153). This analysis is quite modern in tone. Quetelet did not give a simultaneous breakdown of the ratio by legitimacy and place of birth (town versus country), which would have resolved the relative importance of these factors, but the data to do this were probably not available to him.

Quetelet is best known today for two related ideas, his concept of the average man, and the importance he attached to the normal distribution. He explained these ideas in his *Letters on Probabilities* (1849). He first observed that "the mean of a series of observations is obtained by dividing the sum of the values observed by the number of observations" (39) but he pointed out that this could be interpreted in two ways, which he called the "true mean" and the "arithmetic mean." If the height of a building is measured twenty times, a different value may be found on each occasion due to errors of measurement, but the building has a determinate height, of which the mean value of the twenty measurements is the best estimate; this is a "true mean," since it is the best estimate of a real value, whose measurement is subject to error. On the other hand, if the heights of the houses in a certain street have been measured, their mean value assists in showing their heights in general, but it is only an "arithmetic mean" since it does not represent the height of any particular house.

Quetelet observed that it was not always obvious whether a mean was a "true mean," which measured a real underlying value, or whether it was only an "arithmetic mean" with only descriptive significance. He proposed that the distribution of the observations provided a test: if they followed what is now called a normal distribution, their mean was a "true mean"; otherwise, it was only an "arithmetic mean."

He imagined as an example that a thousand sculptors had been employed to copy the Greek statue "The Gladiator," and that a bodily measurement was made on each copy. He asserted that these measurements would follow a normal distribution, with mean value determined by that of the original statue and with errors caused by both copying and measurement errors. He continued:

Your Highness smiles. You will doubtless tell me that such assertions will not compromise me, since no one will be disposed to make the required experiment. And why not? I shall perhaps astonish you very much by stating that the experiment has already been made. Yes, surely, more than a thousand copies have been made of a statue, which I do not assert to be that of the Gladiator, but which in all cases differs but little from it. These copies were even living ones, so that the measurements have been taken with all possible chances of error: I will add more, that the copies have been subject to deformity by a host of accidental causes. We ought then to expect here a very considerable probable error. (Quetelet 1849, 92)

He then gave the distribution of the chest measurements of 5,732 Scotch soldiers and showed that it gave a good fit to a normal distribution. Unfortunately, he made several arithmetical errors in compiling the data (Stigler 1986). The correct data are shown in table 6.1, together with numbers predicted by using the normal density function in eq.[1]. There is clearly good agreement between the observed and the predicted numbers. The data are shown graphically in fig. 6.1, with the normal curve superimposed. Quetelet concluded that "the example ... shows us that the results really occur, as though the chests which have been measured had been modelled from the same type from the same individual,—an ideal one if you will, but whose proportions we ascertain by a sufficiently long trial. If such were not the law of nature, the measurements would not (spite of their imperfections) group themselves with the astonishing symmetry which the law of possibility [the normal distribution] assigns them" (93).

Table 6.1. Chest Circumference of 5,732 Scotch Soldiers in Inches

Chest size	Number of men		Chest size	Number of men	
	Observed	Predicted		Observed	Predicted
33	3	5	41	935	945
34	19	21	42	646	644
35	81	72	43	313	347
36	189	197	44	168	149
37	409	429	45	50	50
38	753	741	46	18	14
39	1062	1014	47	3	3
40	1082	1100	48	1	1

Source: Stigler 1986

The constants μ and c in the fitted normal distribution were found by first calculating the sample mean and variance:

$$m = \sum_{i=1}^{n} x_i / n = 39.85 \tag{2a}$$

$$s^2 = \sum_{i=1}^{n} (x_i - m)^2 / (n-1) = 4.30 \tag{2b}$$

and then estimating $\mu = m = 39.85$, $c = \sqrt{(2s^2)} = 2.93$. This was the standard procedure in the nineteenth century. (See the texts of Airy 1861 or Merriman 1877, which were both known to Galton. The divisor $(n-1)$ was used in calculating s^2 to allow for the error in estimating the true mean by the sample mean m. The usual measure of dispersion used in practice in the nineteenth century was the probable error, defined as the deviation from the mean as likely as not to be exceeded in absolute value; it was estimated for a normal distribution as $0.6745\,s$.)

Thus Quetelet interpreted the normal distribution as evidence that departures from the mean were like errors of measurement, so that the mean value was a "true mean" which represented a real underlying value or type. The importance that Quetelet and his followers gave to the

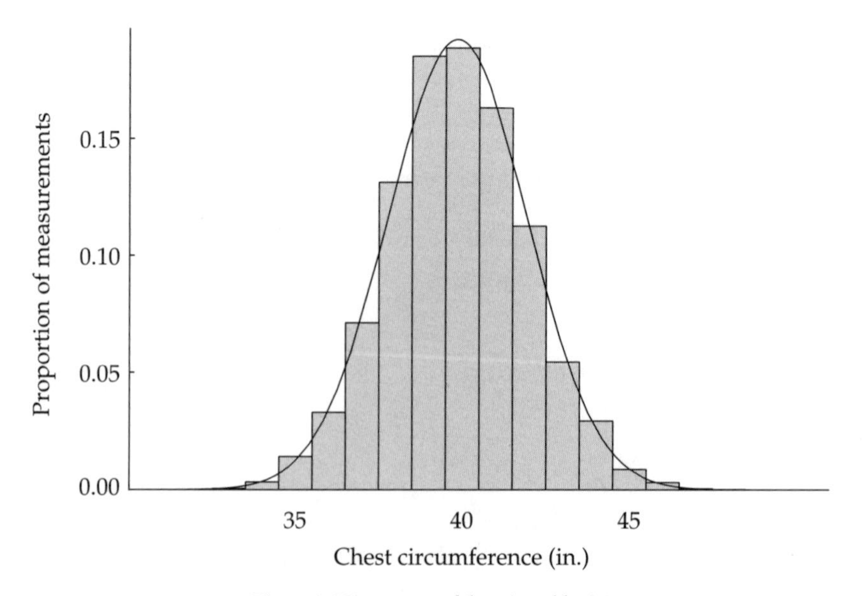

Fig. 6.1. Histogram of data in table 6.1

normal distribution led to an exaggerated idea of its prevalence, which was nicknamed "Queteletismus."

Galton and the Normal Distribution

Galton tells us in his autobiography how he first heard of the normal distribution:

> My first serious interest in the Gaussian Law of Errors was due to the inspiration of William Spottiswoode, who had used it long ago in a Geographical memoir for discussing the probability of the elevations of certain mountain chains being due to a common cause. He explained to me the far-reaching application of that extraordinarily beautiful law, which I fully apprehended. I had also the pleasure of making the acquaintance of Quetelet, who was the first to apply it to human measurements, in its elementary binomial form, which I used in my *Hereditary Genius*. (Galton 1908, 304)

Spottiswoode was a close friend of Galton and had supported his efforts to reform the Royal Geographical Society in the 1860s. He found time to pursue a wide range of scientific interests, later becoming president of the Royal Society, as well as heading the family printing firm. In 1860, he read a paper to the Royal Geographical Society, "On typical mountain ranges: An application of the calculus of probabilities," in which he used Quetelet's method to fit a normal distribution to the directions of twelve mountain ranges in a group of mountains in Central Asia. He found that there was a satisfactory fit, though the sample size was small, with a mean direction of 3° 49′ 36″ from the parallel of latitude passing through the range; and he concluded that the minor deviations could be regarded as "errors" from this true typical direction. Thus Galton was introduced to the normal distribution about 1860 in the context of its use to validate the mean as the typical value. He probably met Quetelet at the International Statistical Congress in London in 1860, and received Belgian meteorological data from him for his weather maps of Europe during December 1861 (see chapter 1).

Hereditary Genius (1869)

Galton discussed the normal distribution in *Hereditary Genius*, calling it "the very curious theoretical law of deviation from an average" (1869, 26), and referring the reader to the English translation of Quetelet's

Letters on Probabilities for further details. He argued that it would be expected to hold whenever there was a large number of similar events, each the result of the same variable conditions. For example, he considered the height of men in a large island inhabited by a single race, who intermarried freely under constant physical conditions. The average height would be constant from one generation to the next and its distribution would follow the normal curve. However, it would not be proper to combine the heights of men belonging to two or more dissimilar races, because the conditions would not be the same. Normality could therefore be used as a test of the homogeneity of the population from which the measurements were drawn. "The law may, therefore, be used as a most trustworthy criterion, whether or no the events of which an average has been taken, are due to the same or to dissimilar classes of conditions" (29).

The main use that Galton made of the normal distribution in *Hereditary Genius* is shown in table 6.2. He argued from analogy with physical characters that mental characters were also likely to be normally distributed: "Analogy clearly shows there must be a fairly constant average mental capacity in the inhabitants of the British Isles, and that the deviations from that average—upwards towards genius, and downwards

Table 6.2. Classification of Men According to Their Natural Gift

Grades of natural ability, separated by equal intervals		Number of men per million in each grade
Below average	Above average	
a	A	256,791
b	B	161,279
c	C	63,563
d	D	15,696
e	E	2,423
f	F	233
g	G	14
x (all grades below g)	X (all grades above G)	1
On either side of average		500,000
Total, both sides		1,000,000

Source: Galton 1869

towards stupidity—must follow the law that governs deviations from all true averages" (32). He used this assumption to divide the population into sixteen grades or classes of natural ability, separated by equal intervals, by using tables of the cumulative normal distribution. Class A contains men between the mean and k probable errors above the mean, class B those between k and $2k$ probable errors above the mean, class G those between $6k$ and $7k$ probable errors above the mean, and class X all those more than $7k$ probable errors above the mean, where $k = 1.03$; the classes below the mean are defined in a similar way. He chose this particular way of dividing ability into grades because he had estimated that 250 men per million become eminent, and he wanted to contrive the classes so that the two highest, F and G, together with X, amounted to about that number. (He had found from the 1865 edition of *Dictionary of Men of the Time* that there were about 500 men who were eminent in the sense of being "decidedly well known to persons familiar with literary and scientific society," out of a population of about two million adult males in the British Isles above fifty years of age, giving a frequency of about 250 eminent men per million.)

Galton attached some importance to this classification, but his use of it in *Hereditary Genius* reveals more about contemporary opinion than about the real world. The six mediocre classes a, b, c, A, B, C, represent "the standard of intellectual power found in most provincial gatherings, because the attractions of a more stirring life in the metropolis and elsewhere, are apt to draw away the abler classes of men, and the silly and the imbecile do not take a part in the gatherings" (35). The class F of dogs is nearly commensurate with the f of the human race, in respect to memory and powers of reason. Toward the end of the book, he discussed the comparative worth of different races, and concluded, on very slight evidence, that the average intellectual standard of the Negro race was about two grades below, while that of the ancient Athenians was nearly two grades higher, than "our own" race.

Natural Inheritance (1889)

Galton had no way of measuring human mental abilities, or even of placing them in rank order, except very crudely. When he later tried to construct a statistical theory of heredity, he wisely concentrated on physical characters, such as height, which could be measured directly. His anthropometric work convinced him that such characters were to a good approximation normal, so that he could rely on the well-known properties of the normal distribution in constructing his statistical theory. He published a review of this work in 1889 in his most influential book, *Natural Inheritance*, which inspired the early biometricians to tackle evolutionary questions by statistical methods (see chapter 10). I now discuss the fourth and fifth chapters of this book, on "Schemes of distribution and of frequency" and "Normal variability," in which he outlined his statistical methodology.

Galton equipped and maintained an Anthropometric Laboratory at the International Health Exhibition of 1884, at which measurements were made of many physical and physiological characters. In chap. 4, he used the data on the strength of pull of 519 young men, shown in table 6.3, to illustrate the idea of a frequency distribution. (I have converted percentages to proportions in the third and fourth columns.)

The third column of the table, headed "Proportion of cases," can be represented graphically in the histogram shown in fig. 6.2, in which the proportion of cases per pound is plotted against strength of pull. The smooth curve is the normal curve with the same measures of location and dispersion.

The histogram is a useful way of showing the main features of the distribution, but Galton placed greater emphasis on the cumulative plot shown in fig. 6.3. In this figure, the cumulative proportion has been plotted against strength of pull, and the points have been joined by straight lines. (In fact, he reversed the axes, plotting strength of pull on the vertical axis against cumulative percentage on the horizontal axis, but the

Table 6.3. Frequency Distribution of Strength of Pull of 519 Young Men

Strength of pull (lb.)	No. of cases	Proportion of cases	Cumulative proportion
< 50	10	0.02	0.02
50–60	42	0.08	0.10
60–70	140	0.27	0.37
70–80	168	0.32	0.69
80–90	113	0.22	0.91
90–100	22	0.04	0.95
>100	24	0.05	1.00
Total	519	1.00	

Source: Galton 1889b

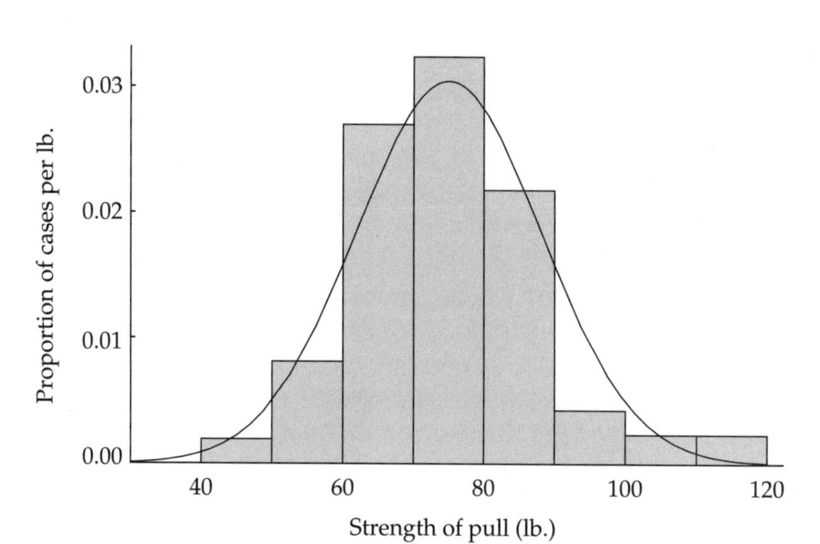

Fig. 6.2. Histogram of data in table 6.3

presentation in fig. 6.3 is more familiar today. The orientation of the axes changes the appearance of the graph but makes no difference to its interpretation.)

Galton liked the cumulative graph because the percentiles of the distribution could be determined directly from it. He then calculated appropriate measures of location and dispersion from the percentiles. The nth percentile is the value such that n percent, a proportion $n/100$, of the observations are below it. For example, it can be seen from table 6.3 that 70 pounds is the 37th percentile of the distribution of strength of pull; that is to say, 37 percent of the young men had strengths below, and 63 percent above, this value. Other percentiles can be determined from fig. 6.3 as shown. The 50th percentile, or median, is $M = 74.0$, the 25th percentile, or lower quartile, is $L = 65.6$, and the 75th percentile, or upper quartile, is $U = 82.6$. (Galton called them centiles and centesimal grades, as well as percentiles, but the latter word is in common usage today.)

Appropriate measures of location and dispersion must be estimated from the data before a normal distribution can be fitted. As we have seen, the standard method started by calculating the arithmetic mean and the mean square error (variance) from the data (eqs.[2] and [3] above). Galton disliked the arithmetic calculations needed to find these estimates, preferring estimates based on the percentiles, which, with his strong geometrical intuition, he found easier both to calculate and to

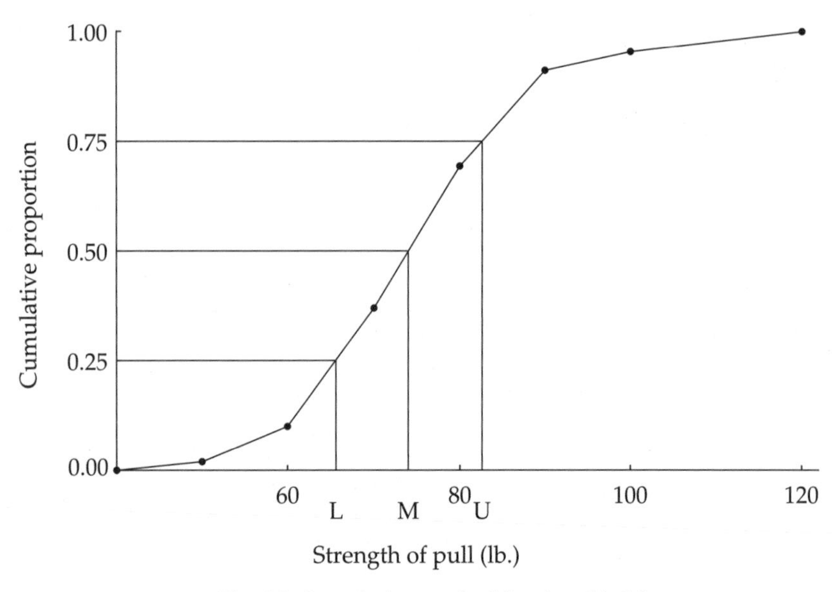

Fig. 6.3. Cumulative graph of data in table 6.3

understand. Assuming a symmetrical distribution, he used the median M instead of the arithmetic mean m, and he estimated the probable error either from the difference between the upper quartile and the median, $U - M$, or preferably from the half-interquartile range, $Q = (U - L)/2$. In advocating this method he wrote: "Suppose that I want to get the average height and 'probable error' of a crowd of savages. Measuring them individually is out of the question; but it is not difficult to range them—roughly for the most part, but more carefully near the middle and one of the quarter points of the series. Then I pick out two men, and two only—one as near the middle as may be, and the other near the quarter point, and I measure them at leisure. The height of the first man is the average of the whole series, and the difference between him and the other man gives the probable error" (1874b, 343). Even when individual measurements had been made, he preferred to calculate the median and the quartiles from the grouped frequency distribution than to find m and s^2. He used this graphical methodology based on percentiles in making statistical calculations on height, weight, and strength for the Anthropometric Committee of the British Association in 1881; more traditional methods were preferred by other members of the committee, such as

Charles Roberts, who did much of its statistical work, and William Farr, whom Galton succeeded as chairman in 1881.

For the strength of pull data in table 6.3, $Q = (82.6 - 65.6)/2 = 8.5$; for a symmetrical distribution, Q is equal to the probable error. On the assumption of normality, the mean μ can be estimated by the median and the modulus as $\sqrt{2}Q/0.6745$, which give $\mu = 74.0$, $c = 17.85$ for the strength of pull data. The smooth curve fitted to the histogram in fig. 6.2 is the normal curve defined in eq.[1] with these parameter values.

Galton was aware that the empirical frequency distribution shown in figs. 6.2 and 6.3 would tend more and more closely to a theoretical probability distribution as the sample size increased. He wrote:

> When Schemes are drawn from different samples of the same large group of measurements, though the number in the several samples may differ greatly, ... the shapes of the Schemes drawn from different samples will be little affected by the number of observations used in each, supposing of course that the numbers are never too small for ordinary statistical purposes. The only recognisable differences between the Schemes will be, that, if the number of observations in the sample is very large, the upper margin of the Scheme will fall into a more regular curve, specially towards either of its limits. Some irregularity will be found in the above curve of the Strength of Pull; but if the observations had been ten times more numerous, it is probable, judging from much experience of such curves, that the irregularity would have been less conspicuous, and perhaps would have disappeared altogether. (Galton 1889, 44)

This statement foreshadows the modern frequentist interpretation, that as the sample size increases and the class interval at the same time decreases, the histogram tends to a smooth limiting curve, the probability density function, $f(x)$, and the cumulative graph tends to a smooth curve, the cumulative probability function, $F(x)$. (The probability density function is so called because, for a small increment dx in x, $f(x)dx$ defines the probability that the random variable lies between x and $x + dx$. The cumulative probability function represents the probability that the random variable is less than x; it is the area under the density function up to x.)

Galton summarized the data on eighteen schemes of distribution of characters measured at his Anthropometric Laboratory. (They were sitting and standing height, arm span, weight, breathing capacity, strength of pull and of squeeze, swiftness of blow, and keenness of sight, recorded separately for men and women.) He found that they could all be adequately approximated by the normal distribution (see the normal curve fitted to the histogram of strength of pull in fig. 6.2). He wrote:

> It has been objected to some of my former work, especially in *Hereditary Genius*, that I pushed the applications of the Law of Frequency of Error somewhat too far. I may have done so, rather by incautious phrases than in reality; but I am sure that, with the evidence now before us, the applicability of that law is more than justified within the reasonable limits asked for in the present book. I am satisfied to claim that the Normal Curve is a fair average representation of the Observed Curves during nine-tenths of their course; that is, for so much of them as lies between the [fifth and ninety-fifth percentiles.] In particular, the agreement of the Curve of Stature with the Normal Curve is very fair, and forms a mainstay of my inquiry into the laws of Natural Inheritance. (Galton 1889, 56–57)

The Importance of the Normal Distribution to Galton

Three reasons underpinned Galton's enthusiasm for the normal distribution. The first was his scientific delight, expressed in this quotation from *Natural Inheritance*, that such a simple law was so widely exemplified:

> *Order in Apparent Chaos.*—I know of scarcely anything so apt to impress the imagination as the wonderful form of cosmic order expressed by the "Law of Frequency of Error." The law would have been personified by the Greeks and deified, if they had known of it. It reigns with serenity and in complete self-effacement amidst the wildest confusion. The huger the mob, and the greater the apparent anarchy, the more perfect is its sway. It is the supreme law of Unreason. (Galton 1889, 66)

But he regarded the law as an approximation, and his psychological investigations led him to think that physiological and psychological variables obeying the Weber-Fechner law should follow a lognormal distribution; that is to say, they should be normally distributed after a logarithmic transformation (Galton 1879, McAlister 1879). He recognized that it would not apply to weight, which was, he thought, approximately proportional to the square (not the cube) of height (Galton 1874b).

Galton's second reason for emphasizing the normal distribution was his mistaken adherence to Quetelet's idea that it had a privileged position, and that its mean had a special importance as an indicator of the racial type. Quetelet thought that only the mean value of an approximately normal biological variable, such as chest circumference, is important, and that deviations from the mean are meaningless errors, similar to errors of measurement, to be eliminated from the analysis as far as possible. Mayr (1982) has characterized this style of thinking in biology as essentialist or typological, as opposed to population thinking. In population thinking it is recognized that differences between individuals are real and often reveal the hereditary variability on which evolution depends, while mean values are only man-made constructs. It was necessary to progress from typological to population thinking to understand how natural selection worked. Mayr continued: "Francis Galton was perhaps the first to realize fully that the mean value of variable populations is a construct. Differences in height among a group of people are real and not the result of inaccuracies of measurement. The most interesting parameter in the statistics of natural populations is the actual variation, its amount, and its nature" (1982, 47).

This assessment exaggerates Galton's progress from typological to population thinking. He certainly understood the importance of biological variability and rejected the idea that it was just "error." In his autobiography he wrote of his difficulty in collaborating with mathematicians because "the primary objects of the Gaussian Law of Error were exactly opposed, in one sense, to those to which I applied them. They were to get rid of, or to provide a just allowance for errors. But these errors or deviations were the very things I wanted to preserve and to know about" (1908, 305). But at the same time he retained Quetelet's idea that the mean of a normally distributed variable represented a racial "type" which was maintained from one generation to the next. In *Hereditary Genius*, he concluded (see above) that normality could be used as a test of the homogeneity of a set of observations. Much later, in his study of fingerprints, he concluded that the three main types (arches, loops, and whorls) represented distinct "genera" from the fact that variability within each of them was approximately normal (Galton 1891 and 1892b). His belief in the stability of type underlaid his theory of discontinuity in evolution (chapter 9).

The third reason for Galton's dependence on the normal distribution was the technical reason, that its theoretical properties were well understood. In particular, he made use of a well-known theorem about the sum of independent normal variates:

THEOREM *If x and y are independently and normally distributed with means μ_x and μ_y and with moduli c_x and c_y, respectively, their sum $z = x + y$, is normally distributed with mean equal to the sum of the individual means, $\mu_z = \mu_x + \mu_y$, and with squared modulus equal to the sums of the squares of the moduli,*

$$c_z^2 = c_x^2 + c_y^2; \qquad [3a]$$

the same relationship holds for the probable error,

$$Q_z^2 = Q_x^2 + Q_y^2. \qquad [3b]$$

Galton referred to this theorem as the "well-known law of the sum of two fallible measures" (1877, 533), quoting Airy's *Theory of Errors* as his source, and he relied quite heavily upon it. He did not know, as we do today, that the mean of the sum of any two random variables is equal to the sum of their means, regardless of their distribution or of their independence, and that the variance of the sum of any two independent random variables is equal to the sum of their variances.

Galton's Quincunx

With his strong mechanical bent, Galton devised an instrument to demonstrate some of the properties of the normal distribution and to convince himself of their reality. In the simplest form of the apparatus (fig. 6.4a), there is a funnel at the top, a succession of rows of pins below it, and below these are a series of vertical compartments. Lead shot is introduced into the funnel and

> scampers deviously down through the pins in a curious and interesting way; each of them darting a step to the right or left, as the case may be, every time it strikes a pin. The pins are disposed in a quincunx fashion, so that every descending shot strikes against a pin in each successive row ... and, at length, every shot finds itself caught in a compartment immediately after freeing itself from the last row of pins. The outline of the columns of shot that accumulate in the successive compartments approximates to the [Normal] Curve of Frequency, and is closely of the same shape however often the experiment is repeated. The outline of the columns would become more nearly identical with the Normal Curve of Frequency, if the rows of pins were much more numerous, the shot smaller, and the compartments narrower; also if a larger quantity of shot were used. (Galton 1889, 64)

The device, which is called a quincunx after the way in which the pins are arranged, demonstrates the approximation to a normal distribution by a binomial distribution with an equal probability of being deflected to the left or right in a large number of trials (the number of rows of pins). Galton had a quincunx made for him by an instrument maker in 1873, which can be seen today at the Galton Laboratory at University College

London (Stigler 1986). He used it to illustrate the normal distribution at lectures at the Royal Institution in 1874 and 1877 (Galton 1877).

The double quincunx (fig. 6.4b) illustrates the theorem that the sum of two independent normal variates is normally distributed. The apparatus is cut horizontally in two parts through the rows of pins, and a row of vertical compartments AA is interposed between them:

> Now close the bottoms of all the AA compartments; then the shot that falls from the funnel will be retained in them, and will be comparatively little dispersed.... Next, open the bottom of any one of the AA compartments; then the shot it contains will cascade downwards and disperse themselves among the BB compartments on either side of the perpendicular line drawn from its starting point.... Do this for all the AA compartments in turn ... and the final result must be to reproduce the identically same system in the BB compartments that was shown in [fig. 6.4b].... The dispersion of the shot at BB may therefore be looked upon as compounded of two superimposed and independent systems of dispersion.... It is a corollary of the foregoing that a system Z, in which each element is the Sum of a couple of independent Errors, of which one has been taken at random from a Normal system A and the other from a normal system B, will itself be Normal. (Galton 1889, 67–68)

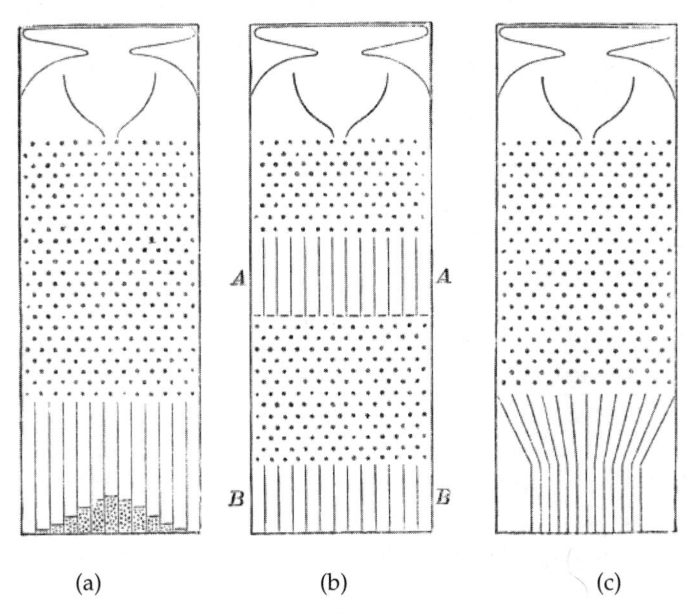

Fig. 6.4. (a) The quincunx, (b) the double quincunx, (c) the convergent quincunx (from Galton 1889)

This was probably a thought experiment to help Galton and his less mathematical readers understand the theorem, since there is no evidence that he had the apparatus made.

The convergent quincunx (fig. 6.4c), in which the compartments at the bottom converge before allowing the shot to drop vertically, is a similar thought experiment to illustrate the theorem that if x is normally distributed with probable error Q, then rx $(0 < r < 1)$ is normal with reduced probable error rQ.

Another example of Galton's practical ingenuity is provided by his letter to *Nature* (1890a) on "Dice for statistical experiments." In this paper he showed how dice could be used to simulate a sample of observations from a normal distribution (see Stigler 1999).

Regression and the Bivariate Normal Distribution

Galton invented the important idea of regression and discovered the properties of the bivariate normal distribution in the course of developing his statistical theory of heredity. (Earlier error theorists had, unknown to Galton, derived the formula for the bivariate normal distribution, but they did not use it to develop the idea of regression, nor did they apply it to observed bivariate frequency distributions. See MacKenzie 1981, Seal 1967, and Stigler 1986.) He first observed reversion toward the mean in 1877 in his experiments on seed size in successive generations of sweet peas (see chapter 7). He carried his analysis much farther and changed the name of the phenomenon to regression in the mid-1880s after obtaining human data on the heights of parents and their offspring. I here describe his work on the statistical theory of regression based on the latter data, consolidating the accounts in Galton 1885c, 1886a, and 1889. His application of these theoretical results to heredity and the underlying cause of regression to the mean is left for discussion in the next chapter.

Table 6.4 shows Galton's data on the joint distribution of the heights of mid-parents (x) and their adult children (y). He had collected data on 205 families, correcting female heights to their male equivalents by multiplying them by 1.08 and averaging the heights of the two parents to compare the mid-parental height with the heights of their children.

Galton first considered the marginal distributions of the mid-parents (shown in the last column) and of their children (shown in the penultimate row). Both of them are approximately normal. He estimated the central value and the probable error of these distributions by the median and the semi-interquartile range, Q, and found that the medians had the same value of about 68.25 in. but that the mid-parents had a lower probable error than their children, with $Q_x = 1.2$ in., $Q_y = 1.7$ in.; their ratio is $Q_x/Q_y = 0.7 = 1/\sqrt{2}$. He had already predicted that this should be the case because he had found no tendency for the heights of husband and wife to be correlated. The heights of men and women had been adjusted to have the same probable error, Q_y, so that the sum of the heights of husband and wife should have probable error $\sqrt{2}Q_y$, from the law of the sum of two independent normal variates, and their average height, which is half this sum, should have probable error $Q_x = Q_y/\sqrt{2}$.

Galton then considered the relationship between the heights of children and their parents. For each row in table 6.4 (corresponding to mid-

Table 6.4. Joint Distribution of Height (in inches) of 928 Adult Children Born of 205 Mid-parents

Height of mid-parents	Heights of the adult children														Total adult children	Median	Total mid-parents
	<62	62–	63–	64–	65–	66–	67–	68–	69–	70–	71–	72–	73–	>74			
>73	0	0	0	0	0	0	0	0	0	0	0	1	3	0	4		5[a]
72–	0	0	0	0	0	0	0	1	2	1	2	7	2	4	19		6
71–	0	0	0	0	1	3	4	3	5	10	4	9	2	2	43	72.2	11
70–	1	0	1	0	1	1	3	12	18	14	7	4	3	3	68	69.9	22
69–	0	0	1	16	4	17	27	20	33	25	20	11	4	5	183	69.5	41
68–	1	0	7	11	16	25	31	34	48	21	18	4	3	0	219	68.9	49
67–	0	3	5	14	15	36	38	28	38	19	11	4	0	0	211	68.2	33
66–	0	3	3	5	2	17	17	14	13	4	0	0	0	0	78	67.6	20
65–	1	0	9	5	7	11	11	7	7	5	2	1	0	0	66	67.2	12
64–	1	1	4	4	1	5	5	0	2	0	0	0	0	0	23	66.7	5
<64	1	0	2	4	1	2	2	1	1	0	0	0	0	0	14	65.8	1
Total	5	7	32	59	48	117	138	120	167	99	64	41	17	14	928		205
Median			66.3	67.8	67.9	67.7	67.9	68.3	68.5	69.0	69.0	70.0					

Source: Galton 1885c

Note: A class like 68– includes all heights greater than or equal to 68 inches but less than 69 inches. Galton found that the mean height of this class for mid-parents was 68.5 inches, but that it was about 68.2 inches for the children because of the tendency to record height to the nearest integer.

[a] Galton noted that there must be a mistake in this row because 4 children cannot have 5 mid-parents.

parents of given height), he calculated the median height of their children (shown in the penultimate column) and plotted these median values against the mid-parental height. He fitted a straight line through the points by eye, and found that its slope was 2/3 (see fig. 6.5). If M is the common mean value of x and y, and $m_{y|x}$ is the mean (or median) value of offspring of mid-parents of height x, the line is

$$(m_{y|x} - M) = \tfrac{2}{3}(x - M), \text{ or}$$
$$m_{y|x} = \tfrac{1}{3}M + \tfrac{2}{3}x. \quad [4]$$

He expressed this in a chart showing the rate of regression in hereditary stature (fig. 6.5). If children were on average as tall as their parents, one would expect the slope to be unity. That the slope was only 2/3 showed that there was, in Galton's phrase, regression toward mediocrity.

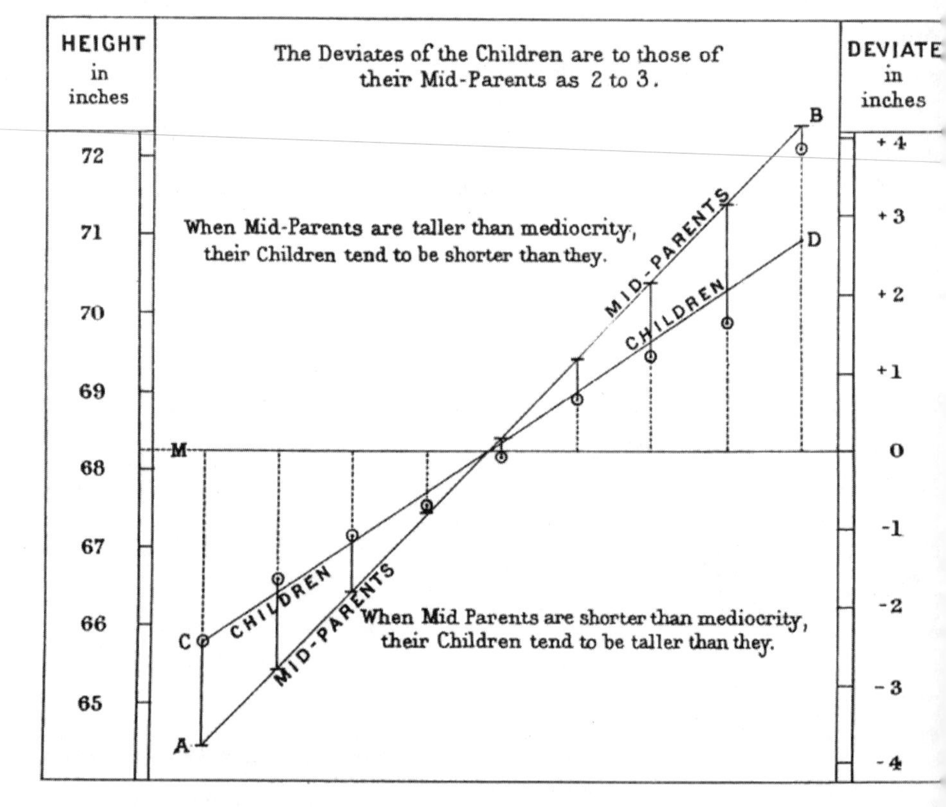

Fig. 6.5. Rate of regression in hereditary stature (from Galton 1885c)

Galton called the children in the same row in table 6.4 a co-fraternity. They form a number of fraternities which have in common that their mid-parents have the same height. He next considered the variability within co-fraternities:

> I concluded after carefully studying the chart upon which each of the individual observations from which [table 6.4] was constructed, had been entered separately in their appropriate places, and not clubbed into groups as in the Tables, that the value of Q in each Co-Fraternal group was roughly the same, whatever their Mid-Parental value might have been. It was not quite the same, being a trifle larger when the Mid-Parents were tall than when they were short. This justifies what will be said in Appendix E about the Geometric Mean; it also justifies neglect in the present inquiry of the method founded upon it, because the improvement in the results to which it might lead, would be insignificant.... The value that I adopt for Q in every Co-Fraternal group, is 1.5 inch. (Galton 1889, 95)

Thus Galton determined that the conditional distribution of the height of the child (y) given that of the mid-parent (x) is approximately normal with mean $0.3M + 0.67x$ and with dispersion about the mean almost the same for all values of x. He also studied the converse problem of determining the conditional distribution of the mid-parent given that of the child by looking vertically down the columns of the table rather than horizontally along its rows. By plotting the median values of mid-parental height in the last row in table 6.4 against the height of their children, he found that the regression of mid-parental height on offspring height was a straight line with a slope of 1/3:

$$(m_{x|y} - M) = \tfrac{1}{3}(y - M), \text{ or}$$
$$m_{x|y} = \tfrac{2}{3}M + \tfrac{1}{3}y. \qquad [5]$$

He remarked that this regression coefficient of 1/3 was very different from the converse of regression of child on mid-parent: "Though the Son deviates on the average from [M] only 2/3 as widely as his Mid-parent, it does not in the least follow that the Mid-parent should deviate on the average from [M], 3/2 or $1\frac{1}{2}$ as widely as the Son" (1889, 99).

Galton was puzzled by the relationship between the two regression lines, and to understand it better, he considered the joint distribution of the two variables by a graphical method. He took a large sheet of squared paper, on which he entered the frequencies in the body of table 6.4. He wrote:

> I found it hard at first to catch the full significance of the entries in the table, which had curious relations that were very interesting to investigate. They came out distinctly when I "smoothed" the entries by writing at each intersection of a horizontal column with a vertical one, the sum of the entries in the four adjacent squares, and using these to work upon. I then noticed (see [fig. 6.6]) that lines drawn through entries of the same value formed a series of concentric and similar ellipses. Their common centre lay at the intersection of the vertical and horizontal lines, that corresponded to 68.25 inches. Their axes were similarly inclined. The points where each ellipse in succession was touched by a horizontal tangent, lay in a straight line inclined to the vertical in the ratio of 2/3; those where they were touched by a vertical tangent lay in a straight line inclined to the horizontal in the ratio of 1/3. These ratios confirm the values of average regression already obtained by a different method, of 2/3 from mid-parent to offspring, and of 1/3 from offspring to mid-parent, because it will be obvious on studying [fig. 6.6] that the point where each horizontal line in succession is touched by an ellipse, the greatest value in that line must appear at the point of contact. The same is true in respect to the vertical lines. These and other relations were evidently a subject for mathematical analysis and verification. (Galton 1885c, 254–255)

(The reader may wonder exactly how he obtained the numbers in fig. 6.6. If the entries in table 6.4 are smoothed by the method suggested, and the numbers obtained are divided by 10 and rounded to make them more manageable, and the rather sparse rows and columns at the edges are discarded, the resulting numbers are similar to but not identical with those in fig. 6.6. Galton may have used a revised version of table 6.4 with observations grouped so that the mean heights in the groups differed from 68.25 by an integral number of inches; a revised grouping was possible because the original measurements were made to the nearest tenth of an inch. The anomalous number of 2 at [midparental deviate = 0.5 in., offspring deviate = 1.5 in.] may be a typographical error.)

The above account reveals the power of Galton's geometrical intuition, which had been honed by his work on interpreting meteorological contour maps, but he did not have the analytic skill to exploit his intuition without mathematical help. He tells the story in his autobiography:

> I had given much time and thought to Tables of Correlations, to display the frequency of cases in which the various deviations say in stature, of an adult person, measured along the top, were associated with the various deviations of stature in his mid-parent, measured along the side.... But I could not see my way to express the results of the complete table in a single formula. At length, one morning, while waiting at a roadside station

near Ramsgate for a train, and poring over the diagram in my notebook, it struck me that the lines of equal frequency ran in concentric ellipses. The cases were too few for certainty, but my eye, being accustomed to such things, satisfied me that I was approaching the solution.

All the formulae of Conic Sections having long since gone out of my head, I went on my return to London to the Royal Institution to read them up. Professor, now Sir James, Dewar, came in, and probably noticing signs of despair in my face, asked me what I was about; then said, "Why do you bother over this? My brother-in-law, J. Hamilton Dickson of Peterhouse, loves problems and wants new ones. Send it to him." I did so, under the form of a problem in mechanics, and he most cordially helped me by working it out, as proposed, on the basis of the usually accepted and generally justifiable Gaussian Law of Error. So I begged him to allow his solution to be given as an appendix to my paper, where it will be found. (Galton 1908, 302–303)

The problem that Galton sent to Hamilton Dickson was this. Suppose that x and y are expressed as deviations from the mean, that x is normal

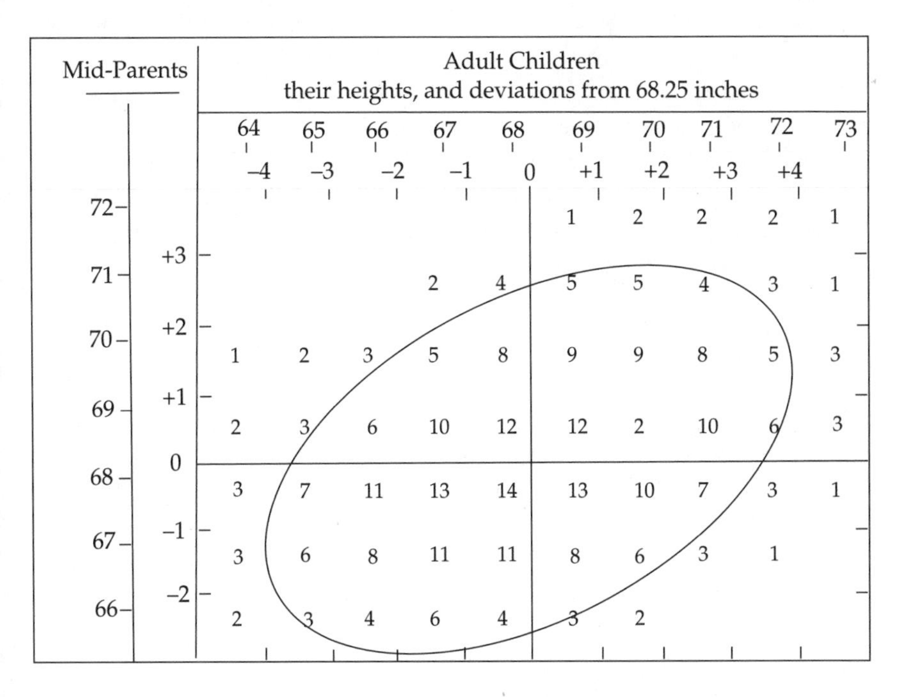

Fig. 6.6. Bivariate distribution of mid-parental and offspring height (redrawn from Galton 1885c)

with zero mean and probable error Q_x, and that conditional on a fixed value of x, y is normal with mean $\beta_{y|x} x$ and with constant probable error $Q_{y|x}$. (1) What is the joint density function of x and y, and in particular are contours of equal probability density elliptical? (2) What is the conditional distribution of x given y, and in particular what is the relation between the two regression coefficients? (Galton 1886a)

Hamilton Dickson's answers to these questions, paraphrased in modern terminology, were as follows. Write f as a generic symbol for a probability density function, so that $f(x)$ is the density function of x, $f(y|x)$ is the conditional density function of y given x, and $f(x, y)$ is the joint density function of x and y. Then the joint density of x and y is the product of the density of x and the conditional density of y given x:

$$f(x, y) = f(x) \times f(y|x)$$

$$= \text{const} \times \exp - 0.2275\left[\frac{x^2}{Q_x^2} + \frac{(y - \beta_{y|x}x)^2}{Q_{y|x}^2}\right]. \qquad [6]$$

(The number 0.2275 arises from the conversion from modulus to probable error.) Contours of equal probability density are obtained by equating the quadratic form in square brackets to a constant, which determines an ellipse, as Galton found empirically.

The joint density function can be factored differently to represent the marginal distribution of y multiplied by the conditional distribution of x given y. Hence it can be shown that y is normal with zero mean, and that for fixed y, x is normal with mean value $\beta_{x|y}\, y$ where

$$\beta_{x|y} = \frac{Q_x^2}{Q_y^2}\beta_{y|x}. \qquad [7]$$

The ratio of the two slopes is determined by the ratio of the squares of the probable errors. (A simpler proof of this result is given below.) For the heights of mid-parents (x) and their children (y), $Q_x = Q_y/\sqrt{2}$ so that $\beta_{x|y} = \beta_{y|x}/2$, as Galton had found empirically.

Galton was already familiar with another relationship between the distributions of x and y. If they are deviations from the mean, then

$$y = \beta_{y|x}x + e, \qquad [8]$$

where e is the residual error about the line, which is normal with zero mean and probable error $Q_{y|x}$. It follows from the law of the sum of two fallible measures in eq.[3] that

$$Q_y^2 = \beta_{y|x}^2 Q_x^2 + Q_{y|x}^2. \qquad [9]$$

Galton noted that for his data on human height, $Q_y = 1.7$, $\beta_{y|x} = 2/3$, $Q_x = Q_y/\sqrt{2}$, $Q_{y|x} = 1.5$. The right-hand side of eq.[9] evaluates to 2.89, exactly equal to 1.7^2, "satisfactorily cross-testing the various independent estimates" (Galton 1886a, 56).

Galton also asked Hamilton Dickson to solve a second problem which had arisen when he was trying to determine the regression between brothers. In *Natural Inheritance* Galton posed the problem by means of a different illustration "for the sake of presenting the same general problem under more than one of its applications" (1889, 69). A man aims a bullet at the center of a target, the mark made being painted red; the lateral deviation, x, between the target center and the red mark is normal with zero mean and probable error Q_x. Another man takes aim, not at the center of the target but at the red mark, and his shot is marked green; the lateral distance, y, between the red mark and the green mark is normal with zero mean and probable error Q_y. If it is only known that the lateral distance of the green mark from the center of the target is z, what is the most probable deviation of the red mark at which it was aimed? What is the probable error of this estimate?

In modern terminology, suppose that

$$z = x + y, \qquad [10]$$

where x is the deviation of the red mark from the center and y is the deviation of the green mark from the red mark. The problem is to find the best estimate of x given z and the probable error of this estimate. The regression of x on z is linear with slope β, which can most easily be calculated as the ratio of the covariance of z and x to the variance of z (see eq.[18] below):

$$\beta = \frac{Q_x^2}{Q_x^2 + Q_y^2}. \qquad [11]$$

The best estimate of x is βz; the probable error of this estimate can be calculated, *mutatis mutandis*, from eq.[9].

Correlation

After revising the proofs of *Natural Inheritance*, Galton began a line of research that led him to widen the application of regression to a broad class of problems and to invent the idea of correlation. He published his conclusions in a paper on "Co-relations and their measurement" in 1888, and in 1890 he wrote a popular account of the invention of correlation called "Kinship and correlation." It began: "Few intellectual pleasures are more keen than those enjoyed by a person who, while he is occupied in some special inquiry, suddenly perceives that it admits of a wide generalization, and that his results hold good in a previously unsuspected direction. The generalization of which I am about to speak arose in this way" (Galton 1890b, 81). He had been busily at work on a new inquiry that had been suggested to him by two circumstances. The first was a renewed discussion among anthropologists about how to predict the height of an unknown man from the length of a particular bone, say a solitary thigh bone dug out of an ancient grave. The second arose out of the interest excited by Bertillon's method of identifying criminals from anthropometric records, which gave rise to the question of its accuracy.

> An additional *datum* was no doubt obtained through the measurement of each additional limb or bodily dimension; but what was the corresponding increase of accuracy in the means of identification? The sizes of the various parts of the body of the same person are in some degree related together. A large glove or shoe suggests that the person to whom it belongs is a large man. But the knowledge that a man has a large glove *and* a large shoe does not give us very much more information than if our knowledge had been confined to only one of the two facts.... The lengths of the various limbs and bodily dimensions of the same person do not vary independently; so that the addition of each new measure adds to the security of the identification in a constantly lessening degree. (Galton 1890b, 81)

He continued that the two problems, that of estimating the stature of an unknown man from the length of one of his bones, and that of the relation between the various bodily dimensions of the same person, were clearly identical. He was able to attack them at once from the anthropometric data he had collected at the International Health Exhibition in 1884. For example, he had obtained data on the stature and the left cubit (the length of the forearm from the elbow to the middle finger) of 348 adult males. He began by plotting pairs of observations against each other, and it suddenly struck him that the form of their distribution was closely similar to that with which he had become familiar when investigating the stature of children and their parents. "Reflection soon made it clear to me that not only were the two new problems identical in principle with the old one of kinship which I had already solved, but that all

Table 6.5. Joint Distribution of Stature and Length of Left Cubit (inches) in 348 Adult Males

Stature	Length of left cubit								
	< 16.5	16.5–16.9	17.0–17.4	17.5–17.9	18.0–18.4	18.5–18.9	19.0–19.4	≥ 19.5	Total
> 70	0	0	0	1	3	4	15	7	30
70	0	0	0	1	5	13	11	0	30
69	0	1	1	2	25	15	6	0	50
68	0	1	3	7	14	7	4	2	38
67	0	1	7	15	28	8	2	0	61
66	0	1	7	18	15	6	0	0	47
65	0	4	10	12	8	2	0	0	36
64	0	5	11	2	3	0	0	0	21
< 64	9	12	10	3	1	0	0	0	35
Total	9	25	49	61	102	55	38	9	348

Source: After Galton 1888

three of them were no more than special cases of a much more general problem—namely, that of Correlation" (82). He continued: "Fearing that this idea, which had become so evident to myself, would strike many others as soon as *Natural Inheritance* was published, and that I should be justly reproached for having overlooked it, I made all haste to prepare a paper for the Royal Society with the title of 'Correlation.' It was read some time before the book was published, and it even made its appearance in print a few days the earlier of the two" (82).

In this paper (Galton 1888), he began by saying that "co-relation or correlation of structure" was a common idea in biology (for example, the length of the arm was said to be co-related with that of the leg, because a person with a long arm usually has a long leg, and conversely), but that no attempt had previously been made to measure its degree. As an example he gave the data shown in table 6.5 on the joint distribution of stature (y) and cubit (x) in 348 men. He plotted the regression of stature on cubit and vice versa in the same way as for the height of mid-parent and child, and found that both regressions were linear. The slope of the regression of stature on cubit, obtained by fitting a straight line by eye, was $b_{y|x} = 2.5$, (i.e. the mean stature increased by 2.5 inches for each increase of an inch in the cubit), while the slope of the regression of cubit on stature was $b_{x|y} = 0.26$ (the mean cubit increased by 0.26 inches for each increase of an inch in stature). These relations show that there is a positive correlation between the two variables, but they do not measure the strength of the correlation because the variables differ in their scale of variability. Galton proposed that a measure of correlation could be obtained by standardizing both variables to have the same variability. The probable error of stature was $Q_y = 1.75$ inch, while that of cubit was $Q_x = 0.56$ inch. If each variable is divided by its probable error, so as to bring it to a common scale of variability, the regression coefficients for the standardized variables $x^* = x/Q_x$ and $y^* = y/Q_y$ become

$$b_{y^*|x^*} = 2.5 \times \frac{0.56}{1.75} = 0.80$$
$$b_{x^*|y^*} = 0.26 \times \frac{1.75}{0.56} = 0.81. \qquad [12]$$

It follows from eq.[7] that the two standardized regression coefficients must be the same, apart from errors of estimation, since $Q_{y^*} = Q_{x^*} = 1$. Their common value provides a measure of correlation, which may be denoted r. Table 6.6 shows some numerical values.

It also follows from eq.[9] that the probable error of the residual error about the standardized regression line of y^* on x^* (or vice versa) is

$$Q_{y^*|x^*} = \sqrt{1 - r^2}\,. \qquad [13]$$

Thus the probable error about the regression line of standardized height on standardized cubit is 0.60. The probable error about the unstandardized regression of height on cubit is $0.60Q_y = 1.05$. Galton attached some

Table 6.6. Regression and Correlation Coefficients

x	y	$b_{y\|x}$	$b_{x\|y}$	$b_{y^*\|x^*}$	$b_{x^*\|y^*}$	$\sqrt{(1-r^2)}$
Left cubit	Stature	2.5	0.26	0.80	0.81	0.60
Head length	Stature	3.2	0.04	0.35	0.35	0.94
Middle finger	Stature	8.2	0.06	0.70	0.70	0.71
Middle finger	Cubit	3.1	0.21	0.84	0.78	0.59
Head length	Head breadth	0.4	0.48	0.45	0.45	0.89
Height of knee	Stature	2.0	0.41	0.90	0.90	0.44
Cubit	Height of knee	1.1	0.56	0.80	0.80	0.60

Source: After Galton 1888

importance to this probable error, remarking that "this value ... has much anthropological interest of its own, especially in connexion with M. Bertillon's system of anthropometric identification" (1888, 144) but he did not elaborate on this brief comment.

Galton rejected the idea of proportional scaling that had been suggested by anthropologists for predicting the height of an unknown man from the length of a particular bone, such as a thigh bone: "The fact that the average ratio between the stature and the cubit is as 100 to 37, or thereabouts, does not give the slightest information about the nearness with which they vary together. It would be an altogether erroneous inference to suppose their average proportion to be maintained so that when the cubit was, say, one-twentieth longer than the average cubit, the stature might be expected to be one-twentieth greater than the average stature, and conversely. Such a supposition is easily shown to be contradicted both by fact and theory" (136). His solution was to use the regression of stature on cubit. If an individual's cubit exceeds its average value by 0.9 in. (which is one-twentieth of the average cubit), his stature is predicted to be 2.5 x 0.9 = 2.25 in. above average, with a probable error of 1.05 in.

He used the example of predicting stature from the length of a thigh bone to explain the cause of regression to the mean in his popular account. He observed that "a *very* long thigh-bone should lead us to expect that the stature of the unknown man to whom it belongs was not *very* tall, but only tall" (Galton 1890b, 84). To explain this fact of regression, he observes that there are three groups of causes affecting thigh-bone length and stature: (1) those that affect both variables in the same way, though not necessarily to the same extent; (2) those that affect only the first variable; (3) those that affect only the second variable. The thigh bone is affected by (1) and (2), but "a large departure occurs very much more rarely than a small one, and therefore it is very much more likely that a given departure should be built up of two lesser departures acting in the same direction than by the excess of a large departure over a small one" (84). Hence the effect of (1) on stature will, on average, be moderate rather than large, and the effect of the independent factor (3) will *on the average* be nothing, "because the total effect of that set is just as often in the direction of diminution as in that of increase" (84). Hence, the unknown stature is probably more mediocre than the known thigh bone; the unknown stature of a man with a very long thigh bone is probably only tall rather than very tall. The degree of this regression toward mediocrity is measured by the correlation coefficient. Galton here came close to the true explanation of regression to the mean.

Thus Galton had succeeded in generalizing the idea of regression from the context of heredity in which he had discovered it, and he rightly foresaw that "there seems to be a wide field for the application of these methods to social problems" (86). He was also motivated by the idea that there are three groups of causes affecting a pair of variables to try to determine their relative importance. Though he never published directly on this question, it led to a correspondence with several mathematicians, of which the essence was published by H. W. Watson in 1891, which throws light on the reasons for his difficulty in communicating with them.

Two Concepts of Probability

Galton sought the help of several mathematicians to solve analytic problems that were beyond his ability. His collaboration with Hamilton Dickson, leading to the discovery of the bivariate normal distribution in 1886, was described above. Sir Donald McAlister derived the log-normal distribution in 1879 at Galton's suggestion to describe situations, such as those covered by the Weber-Fechner Law, which states that the response to a psychophysical stimulus is proportional to its intensity and for which the geometric mean is the best measure of central tendency. H. W. Watson helped him in 1874 to solve a problem in probability theory, that of the extinction of surnames, which was discussed in chapter 5, and in 1891 in connection with the interpretation of correlation. But Galton had a rather uneasy relationship with the mathematicians, particularly with regard to the last problem, and wrote in his autobiography: "The patience of some of my mathematical friends was tried in endeavouring to explain what I myself saw very clearly as a geometrical problem, but could not express in the analytical forms to which they were accustomed, and which they persisted in misapplying. It was a gain to me when I had at last won over Mr. Watson, who put my views into a more suitable shape" (1908, 305).

According to Watson (1891), Galton had put to him the following mathematical problem. Suppose a man to attempt to cover one foot by a hop, and then another man, starting from the end of the hop, to cover another foot by a stride, and suppose you know that the hop and the stride are normally distributed with probable errors a and b, respectively;

then you know that the resultant of the hop and the stride is normal with probable error h, where $h^2 = a^2 + b^2$. Suppose then that the conditions of the problem are modified, and you do not know the probable error of the stride (b), but you do know that of the resultant of the hop and the stride (h) as well as that of the hop, may you not infer that the probable error of the stride is b where $b^2 = h^2 - a^2$? In fact, may you not treat the formula $h^2 = a^2 + b^2$ as an ordinary equation, and deduce from it $b^2 = h^2 - a^2$?

Watson wrote: "My first impulse was to say, *decidedly No*, you must infer that the skill constant of the strider was $\sqrt{(h^2 + a^2)}$ and not $\sqrt{(h^2 - a^2)}$" (1891, 307). He argued that since the sum of stride and hop was normal with probable error h and the hop with probable error a, their difference should, by a well-known theorem, be normal with probable error $\sqrt{(h^2 + a^2)}$. (The fallacy in the argument is that the theorem only applies to the difference between two *independent* random variables.) Galton had previously been in communication with another mathematician who had arrived at the same conclusion; this mathematician was almost certainly Hamilton Dickson, who wrote a letter to Galton on this subject (Dickson 1890). It was in his dissatisfaction with this conclusion, and his desire to elicit one more in accordance with his own treatment, that Galton had appealed to Watson, who eventually though rather reluctantly came round to Galton's point of view.

Galton had encountered this problem in trying to interpret correlation. He had written to Watson, using the correlation between two brothers as an example:

> When two sets of occurrences are said to be correlated, the meaning that lies behind the phrase is that their variations are governed partly by causes common to both, and partly by causes special to each. I want to analyse and to separate these three sets of causes, and to learn their relative importance.... I want to go a step further and to use this r [the correlation or regression] as a means of determining the variability of that group of influences when taken by themselves, that are common to any pairs of correlated values, and consequently of determining the variability of the groups of influences that are special to each member of the pair; and it was when making this step that a certain problem occurred. (Watson 1891, 306)

Galton never published his conclusions, but we may reconstruct what he may have done. Write h, a, and b for the probable errors of z, x, and y in eq.[10], interpreting x as the deviation of the family mean from the population mean and y as the deviation of a particular individual from his family mean; thus x represents the hop, y the stride, and z their sum.

From eq.[11], the most probable value of x/z, for a known value of z, is a^2/h^2, and Galton remarks in *Natural Inheritance* that "this is also the value of fraternal regression" (1889, 127). The phenotypic probable error h can be measured, as can the fraternal regression r, so that we can estimate $a^2 = rh^2$; finally, if the formula $h^2 = a^2 + b^2$ can be treated as an ordinary equation, we can estimate $b^2 = (1 - r)h^2$. Thus the correlation coefficient r can be interpreted in this case as the proportion of the squared probable error (or variance) that is due to a factor common to the two brothers and $1 - r$ as the proportion due to factors special to each of them. This interpretation is only valid in the special case, such as the correlation between two relatives, in which the factors common to both variables affect them to the same extent, and the factors special to each of them have the same variability in each of them. Thus it cannot be extended to the correlations in table 6.6. It is likely that Galton was looking for an interpretation of correlation that was meaningful for any pair of variables, which may account for his failure to publish this work.

The interesting question is, why Watson at first doubted that the formula $h^2 = a^2 + b^2$ could be treated as an ordinary equation. He wrote: "Mr. Galton is in the habit ... of regarding groups of phenomena presenting themselves in certain physiological investigations as being as absolutely subject to these [statistical] laws as if they were objectively impressed upon them. He regards the [probable] error in these groups as properties inherent in the groups, which may for each group be determined by measurement, and, thus determined, form the basis of future measurement" (Watson 1891, 305). In other words, Galton had adopted a frequentist interpretation of statistical laws, and regarded them as empirical properties of the real world. On the other hand, Watson, like other contemporary mathematicians, had been trained in the older Laplacean tradition that tended to interpret probability as a logical relationship measuring the degree of belief that it was logical to place on the occurrence of a future event dependent on the evidence about it. He therefore thought at first, as did Hamilton Dickson, that Galton's use of the formula $h^2 = a^2 + b^2$ as an ordinary equation might be illegitimate because it was based on his empirical interpretation of statistical laws as properties of the real world rather than on a logical interpretation in which they might change according to their context.

The checkered history of the distinction between the two interpretations of probability, the logical interpretation as rational degree of belief and the empirical interpretation as relative frequency in the long run, is worth considering in more detail. The classical theory of probability, culminating in the work of Laplace at the beginning of the nineteenth century, did not distinguish clearly between these two interpretations, but gave primacy to the logical interpretation. In his *Essai philosophique sur les probabilités* in 1814, Laplace defined probability thus:

> Probability has reference partly to our ignorance, partly to our knowledge. We know that among three or more events, one, and only one, must happen; but there is nothing leading us to believe that any one of them will happen rather than the others. In this state of indecision, it is impossible for us to pronounce with certainty on their occurrence. It is, however, probable that any one of these events, selected at pleasure, will not take place; because we perceive several cases, all equally possible, which exclude its occurrence, and only one which favours it.
>
> The theory of chances consists in reducing all events of the same kind to a certain number of cases equally possible, that is, such that we are *equally undecided* as to their existence; and in determining the number of these cases which are favourable to the event of which the probability is sought. The ratio of that number to the number of all the possible cases, is the measure of the probability; which is thus a fraction, having for its numerator the number of cases favourable to the event, and for its denominator the number of all cases which are possible. (Laplace 1814, quoted in Mill 1846, 534)

In the first edition of his *System of Logic,* John Stuart Mill argued that the foundation of the doctrine of chances, as taught by Laplace in the above passage, was defective:

> To be able to pronounce two events equally probable, it is not enough that we should know that one or the other must happen, and should have no ground for conjecturing which. Experience must have shown that the two events are of equally frequent occurrence. Why, in tossing up a halfpenny, do we reckon it equally probable that we shall throw cross or pile? Because experience has shown that in any great number of throws, cross and pile are thrown about equally often; and that the more throws we make, the more nearly the equality is perfect. (Mill 1843, 1141)

Thus he rejected Laplace's definition of probability in favor of a definition based on long-term relative frequency, and he continued the attack: "It would indeed require strong evidence to persuade any rational person that by a system of operations upon numbers, our ignorance can be coined into science" (1142). But he was persuaded by a long letter from the philosopher of science John Herschel that the Laplacean view was valid (Porter 1986, 83), and he wrote in the second and subsequent editions:

> This view of the subject was taken in the first edition of the present work: but I have since become convinced, that the theory of chances, as conceived by Laplace and by mathematicians generally, has not the fundamental fallacy which I had ascribed to it.
>
> We must remember that the probability of an event is not a quality of the event itself, but a mere name for the degree of ground which we, or some one else, have for expecting it. The probability of an event to one person is a different thing from the probability of the same event to another, or to the same person after he has acquired additional evidence.... Every event is in itself certain, not probable: if we knew all, we should either know positively that it will happen, or positively that it will not. But its probability to us means the degree of expectation of its occurrence, which we are warranted in entertaining by our present evidence. (Mill 1846, 535)

These two passages are clear definitions of the two concepts of probability. We now realize that both concepts are valid, and that the question at issue is which concept is appropriate in a particular situation. This question is controversial today, and receives different answers from the frequentist and the Bayesian schools of statistical inference. The growing acceptance of statistical laws as scientific explanations in the nineteenth century was accompanied by a shift from the Laplacean to the frequentist concept of probability, since scientific laws were intended as empirical properties of the real world rather than as expressions of degrees of belief.

George Boole arrived at a frequentist interpretation of probability in *The Laws of Thought* (1854), but the most influential exposition of this interpretation was John Venn's *Logic of Chance* (1866). He stated his position in the Preface: "With what may be called the Material view of Logic as opposed to the Formal or Conceptualist—with that which regards it as taking cognisance of laws of things and not of the laws of our own minds in thinking about things—I am in entire concordance" (xiii). He rejected the idea that the science of probability was concerned with the degree of our certainty or belief about the things which we are supposed to contemplate. He argued instead that it has to do with limiting relative frequencies: "What, for instance, is the meaning of the statement that one cow in ten fails to suckle its young? It certainly does not declare that in any given herd of, say twenty, we shall find just two that fail; whatever might be the strict meaning of the words, this is not the import of the statement. It rather contemplates our examination of a large number, of a long succession of instances, and states that in such a succession we shall find a numerical proportion, not indeed accurate at first, but which tends in the long run to become accurate" (5). He concluded that probability was an objective property of a long series of similar events, and that it was not meaningful to try to calculate the probability of a unique event. In rejecting the subjective interpretation of probability, he made a withering attack on Laplace's Law of Succession, according to which, if an event has occurred n times in succession, the probability that it occurs on the next occasion is $(n + 1)/(n + 2)$; on the basis of this law, Laplace had calculated that, at the date of publication of his work, the probability of the sun rising tomorrow was 1,826,214 to 1.

As noted above, in discussing schemes of distribution, Galton adopted a frequentist interpretation of probability. In a letter to de Candolle in 1888, he wrote: "He [Venn] is the author of a most thoughtful book called the *Logic of Chance* which young statisticians ought to read, for it explains what statistics cannot as well as can do, in a very masterly way" (Pearson 1930b, 478). He described his difficulty in communicating with the mathematicians, particularly with respect to the problem discussed by Watson in 1891, in his autobiography:

> He [Watson] helped me greatly in my first struggles with certain applications of the Gaussian Law, which, for some reasons that I could never clearly perceive, seemed for a long time to be comprehended with difficulty by mathematicians, including himself. They were unnecessarily alarmed lest the well-known rules of Inverse Probability should be unconsciously violated, which they never were. I could give a striking case of this, but abstain because it would seem depreciatory of a man whose mathematical powers and ability were far in excess of my own. Still, he was quite wrong. The primary objects of the Gaussian Law of Error were exactly opposed, in one sense, to those to which I applied them. They were to get rid of, or to provide a just allowance for errors. But these errors or deviations were the very things I wanted to preserve and know about. This was the reason that one eminent living mathematician gave me. (Galton 1908, 305)

Thus Galton attributed his difficulty in communicating with the mathematicians to the different type of application of probability theory to which they were accustomed. They were used to applications in which variability was due to errors of measurement to be eliminated as far as possible, while Galton was interested in applications to human data whose variability was of biological significance. But the misunderstanding between them arose from a deeper distinction—that Galton used an empirical, frequentist interpretation of statistical laws while his mathematical colleagues adopted the older, Laplacean tradition. This is evident from Galton's comment that "they were unnecessarily alarmed lest the well-known rules of Inverse Probability should be unconsciously violated"; inverse probability refers to the use of Bayesian arguments such as were employed by Laplace in deriving the law of succession.

The Development of Statistics

Galton's geometrical intuition helped him to discover the basic ideas of regression and correlation, but his lack of analytic skills prevented him from developing these ideas. His reliance on graphical methods of estimating the location and dispersion of a distribution, rather than the sample mean and variance in common use at the time, was understandable but turned out to be inefficient. His method of fitting a regression line to the data by eye was clearly unsatisfactory. Most importantly, he never had a clear understanding of how to extend his ideas to three or more variables, and was frequently led into error when he tried to do so. It was left to more competent mathematicians inspired by his ideas, in particular Karl Pearson, to develop the theory of multiple regression on a sound mathematical basis.

Karl Pearson (1857–1936) was a strong mathematician, having been placed third wrangler in the Cambridge mathematical tripos in 1879, and was appointed in 1884 to the chair of applied mathematics and mechanics at University College London (UCL). In the early 1890s he became aware of the significance of Galton's work and devoted the rest of his life to the development of statistical methodology on a sound mathematical basis. He wrote in the foreword of his *Life of Galton*: "He was sixty-seven when his *Natural Inheritance* was published, the book which may be said to have created his school. For although his methods were developed in papers of the preceding decade, that book undoubtedly first made them known to us, and found him the lieutenants who built up the school of modern statistics" (1914, 2).

Pearson read *Natural Inheritance* within weeks of publication. His initial reaction was rather lukewarm, but he was soon stimulated by two people, Weldon and Edgeworth, to see the potential of Galton's methods (Stigler 1986). The zoologist Weldon, who had just begun his biometric studies motivated by Galton's work, moved from Oxford to UCL in 1891, and asked for Pearson's help in analyzing his data (see chapter 10).

Edgeworth, a mathematician and economist, gave a course of lectures on the uses and methods of statistics at UCL in 1892, in which he discussed Galton's work on correlation and its possible extension to three or more variables. In 1894 Pearson published the first of a long series of "Contributions to the mathematical theory of evolution," in which he developed statistical methodology for analyzing biological data. In this paper he showed how to dissect a skew distribution that Weldon had found for frontal breadth of crabs into a mixture of two normal distributions representing different races (see chapter 10). In the third paper, "Regression, heredity, and panmixia," published in 1896, he developed the general theory of correlation and multiple regression based on the assumption of multivariate normality, using an algebraic theorem due to Edgeworth. The algebraic details of his treatment are rather cumbersome; a simplified account of the aspects of multiple regression theory that will be used later is given in the appendix to this chapter.

Thus Karl Pearson was stimulated by Galton's work to develop the statistical theory of regression. His other major contribution was the introduction of the chi-square test of goodness of fit, but all his work depended on the assumption that large samples of data were available. In practice, scientists were often faced with the problem of interpreting experiments with limited data, and modern statistical techniques, associated particularly with R. A. Fisher, were developed to allow them to do this.

In the foreword to *Statistical Methods and Scientific Inference*, Fisher paid generous tribute to the pioneering statistical work of "that versatile and somewhat eccentric man of genius, Francis Galton." He wrote:

> Galton's great gift lay in his awareness, which grew during his life, of the vagueness of many of the phrases in which men tried to express themselves in describing natural phenomena.... That the methods he himself used were often extremely crude, and sometimes seriously faulty, is, indeed, the strongest evidence of the eventual value to the progress of science of his unswerving faith that objectivity and rationality were accessible, even in such elusive fields as psychology, if only a factual basis for these qualities were diligently sought. The systematic improvement of statistical methods and the development of their utility in the study of biological variation and inheritance were the aims to which he deliberately devoted his personal fortune, through the support and endowment of a research laboratory under Professor K. Pearson. (Fisher 1956, 1–2)

Fisher continued with a bitter attack on the work of Karl Pearson:

> The peculiar mixture of qualities exhibited by Pearson made this choice in some respects regrettable, though in others highly successful.... The terrible weakness of his mathematical and scientific work flowed from his incapacity in self-criticism, and the unwillingness to admit the possibility that he had anything to learn from others, even in biology, of which he knew little. His mathematics, consequently, though always vigorous, were usually clumsy, and often misleading.... His immense personal output of writings, his great enterprise in publication, and the excellence of production characteristic of the Royal Society and the Cambridge Press, left an impressive literature. The biological world, for the most part, ignored it, for it was indeed both pretentious and erratic. (Fisher 1956, 2–3)

Fisher then discussed the small sample methods which he himself did so much to develop:

> Though Pearson did not appreciate it, quantitative biology, especially in its agricultural applications, was beginning to need accurate tests of significance. So early as Darwin's experiments on growth rate the need was felt for some sort of test of whether an apparent effect "might reasonably be due to chance". I have discussed this particular case in *The Design of Experiments* (Chapter III). It was characteristic of the early period, and of Pearson, that such difficulties were habitually blamed on "paucity of data", and not ascribed specifically to the fact that mathematicians had so far offered no solution which the practitioner could use, and indeed had not been sufficiently aware of the difficulty to have discussed the problem. As is well known, it was a research chemist, W. S. Gosset, writing under the name of "Student", who supplied the test which [inaugurated] the first stage of the process by which statistical methods attained sufficient refinment to be of real assistance in the interpretation of data. (Fisher 1956, 3–4)

Darwin (1876) had described a series of experiments, mentioned above, showing that cross-fertilized plants were usually larger than self-fertilized plants. The data from a typical experiment are shown in table 6.7. Darwin had obtained seeds by self- and by cross-pollination and placed them on sand to germinate. As often as a pair of seedlings of different type germinated at the same time he planted them on opposite sides of a pot and measured their height (in eighths of an inch) as young plants, so that the observations on crossed and self-fertilized plants in the same row in the table form a matched pair; he also ensured as best he could that the crossed plants on one side of the pot enjoyed the same conditions of moisture and light as the selfed plants on the other. Fisher (1935) reanalyzed these data, which had originally been analyzed by Galton, and observed that because of the design of the experiment it was

Table 6.7. Heights of Crossed and Self-fertilized Zea mays Plants in Eighths of an Inch

	Crossed	Self-fertilized	Difference
Pot I	188	139	49
	96	163	–67
	168	160	8
Pot II	176	160	16
	153	147	6
	172	149	23
Pot III	177	149	28
	163	122	41
	146	132	14
	173	144	29
	186	130	56
Pot IV	168	144	24
	177	102	75
	184	124	60
	96	144	–48

Sources: Darwin 1876, Fisher 1935

natural to find the difference between each pair of observations, shown in the last column, and to test whether the mean difference was significantly different from zero by Student's t test. The mean difference is 20.93 with an estimated standard error of 9.75, giving $t = 20.93/9.75 = 2.15$ with 14 degrees of freedom, which is just on the borderline of significance at the 5 percent level. (The sampling distribution of the t statistic, allowing for sampling variability in both the mean and the estimated standard error, had been found empirically by Student and analytically by Fisher.) Similar results obtained by Darwin on six other species confirmed the superiority of crossed over self-fertilized plants.

Fisher praised Darwin's experimental design, but remarked that it was flawed by the absence of randomization. All the crossed plants were on one side of the pot and all the selfed plants on the other, so that some unkown factor might have affected all the crossed plants in a pot. His emphasis on randomization to underpin valid statistical inferences was one of Fisher's fundamental contributions to the design of experiments.

Darwin asked Galton to help him with the statistical analysis of these data, but his proposals can in retrospect be seen to be flawed (Darwin 1876, Fisher 1935). Galton suggested that the data for each type of plant should first be rearranged in rank order, either separately for each pot or over all four pots if differences between pots could be ignored, and that the differences between the ranked observations should then be analyzed. For example, ignoring differences between pots, the largest crossed plant measured 188 and the largest selfed plant 163, with a difference of 25; the second largest crossed and selfed plants measured 186 and 160, with a difference of 26; and so on. This proposal is in line with Galton's liking for ranked observations, but there are two problems with it. First, it ignores the key feature of Darwin's experimental design, that the pairs of plants in each row represent a matched pair; perhaps Darwin had not explained this adequately to Galton when he sent the data to him. Second, though it is possible to do a statistical analysis on ranked observations, it is not valid to do an analysis on differences between ranked observations; Fisher (1935) showed that this led to an underestimate of the standard error leading to a spuriously high significance level. Comparison of Galton's and Fisher's analysis of Darwin's data shows the great strides that had been made in the sixty years between them.

Appendix: Regression Theory

Consider first a pair of random variables, x and y, with mean values μ_x and μ_y and with standard deviations σ_x and σ_y. Another useful quantity is the covariance, defined as the average or Expected value of the product of the deviations from their respective means:

$$\mathrm{Cov}(x, y) = \mathrm{E}[(x - \mu_x)(y - \mu_y)]. \tag{14}$$

It is positive if they tend to vary in the same direction, and thus provides a starting point for estimating correlation. It can be estimated from the sample covariance

$$\mathrm{cov}(x, y) = \sum_{i=1}^{n} (x_i - m_x)(y_i - m_y)/(n - 1). \tag{15}$$

Suppose that the regression of y on x is linear with slope $\beta_{y|x}$, so that we may write

$$y = \mu_y + \beta_{y|x}(x - \mu_x) + e, \tag{16}$$

where e is the deviation about the regression line which by definition has zero Expected value. Multiplying both sides of this equation by $(x - \mu_x)$ and taking Expected values, we find that

$$\mathrm{Cov}(x, y) = \beta_{y|x}\mathrm{Var}(x) \qquad [17]$$

since the deviation e has zero Expected value for any fixed value of x. Hence

$$\beta_{y|x} = \mathrm{Cov}(x, y)/\mathrm{Var}(x). \qquad [18]$$

The correlation coefficient is the slope of the regression of y/σ_y on x/σ_x (or vice versa):

$$\rho = \frac{\mathrm{Cov}(x, y)}{\sigma_x \sigma_y}. \qquad [19]$$

These results only assume linearity of regression and do not depend on normality. This fact was first pointed out by G. U. Yule, as Pearson (1896) acknowledged in a footnote.

These results also suggest that linear regression slopes and correlation coefficients can be estimated from data by the corresponding sample quantities. For example, for the data in table 6.5, we find that

$$s_x^2 = 0.63, \quad s_y^2 = 5.53, \quad \mathrm{cov}(x, y) = 1.41. \qquad [20]$$

Hence the regression slopes and the correlation coefficient can be estimated as

$$\beta_{y|x} = \frac{1.41}{0.63} = 2.2, \quad \beta_{x|y} = \frac{1.41}{5.53} = 0.25,$$
$$r = \frac{1.41}{\sqrt{(0.63 \times 5.53)}} = 0.75. \qquad [21]$$

The product moment formula for calculating the correlation coefficient, $r = \mathrm{cov}(x, y)/s_x s_y$, was introduced by Pearson (1896), though there was a dispute over priority with Edgeworth (Stigler 1986).

To extend these results to the multivariate situation, consider a set of $n + 1$, possibly correlated, random variables, $y_0, y_1, y_2, \ldots, y_n$; they might, for example, be the sizes of different organs in the body, or the heights of related individuals. Write μ_i for the Expected value of y_i, V_{ij} for the covariance of y_i and y_j, and V_{ii} for the variance of y_i (the covariance of y_i with itself). Suppose that the regression of y_0 on the remaining variables is

$$y_0 - \mu_0 = \beta_1(y_1 - \mu_1) + \beta_2(y_2 - \mu_2) + \cdots + \beta_n(y_n - \mu_n) + e, \quad [22]$$

where e is normal with zero mean and constant variance, independent of the values of $y_1, y_2, \ldots, y_n$. Multiplying both sides of this equation by $(y_i - \mu_i)$, and taking Expected values, gives

$$V_{i0} = \beta_1 V_{i1} + \beta_2 V_{i2} + \cdots + \beta_n V_{in} \quad [23]$$

for $i = 1, 2, \ldots, n$. This gives n linear equations to solve for the regression coefficients, which can be expressed in matrix terminology

$$\mathbf{c} = \mathbf{V}\boldsymbol{\beta}, \quad [24]$$

where $\mathbf{c}$ is the vector of the covariances of y_0 with the y_i's, $c_i = V_{i0} = \mathrm{Cov}(y_i, y_0)$, $\mathbf{V}$ is the variance-covariance matrix of the y_i's ($i = 1, 2, \ldots, n$) whose (i, j)th element is V_{ij}, and $\boldsymbol{\beta}$ is the vector of the β_i's ($i = 1, 2, \ldots, n$). Hence

$$\boldsymbol{\beta} = \mathbf{V}^{-1}\mathbf{c}. \quad [25]$$

This is a standard result in multiple regression theory, which depends only on the linearity of the regression.

7

Statistical Theory of Heredity

> The outline of my problem of this evening is, that since the characteristics of all plants and animals tend to conform to the law of deviation [the normal distribution], let us suppose a typical case, in which the conformity shall be exact, and which shall admit of discussion as a mathematical problem, and find what the laws of heredity must then be to enable *successive* generations to maintain statistical identity.
>
> Galton, "Typical laws of heredity"

In *Hereditary Genius* (1869), Galton outlined the possibility of constructing a theory of quantitative genetics based on Darwin's theory of pangenesis, but he was prevented from developing this model-based approach by his mathematical weakness and by the inadequacy of the theory. In the 1870s he developed a physiological theory of heredity without considering its statistical consequences (chapter 4), while in the 1880s he turned his attention to constructing a purely statistical theory of the relationship between parent and child in which theoretical models of heredity were kept in the background. He outlined three main problems to be addressed by this statistical theory in the Introduction to *Natural Inheritance*: (1) Why was the statistical distribution of many quantitative characters constant from one generation to the next? (2) What was the average share contributed to the personal features of the offspring by each ancestor severally? Consideration of this question led to the law of ancestral heredity. (3) How could the nearness of kinship between different types of relatives be measured?

In this chapter I consider briefly the theory outlined in *Hereditary Genius* before turning to the more substantive statistical theory developed in the 1880s. The law of ancestral heredity is discussed in chapter 8.

A Theory Based on Pangenesis

> The doctrine of Pangenesis gives excellent materials for mathematical formulae, the constants of which might be supplied through averages of facts, like those contained in my tables, if they were prepared for the purpose. My own data are too lax to go upon; the averages ought to refer to some simple physical characteristic, unmistakeable in its quality, and not subject to the doubts which attend the appraisal of ability. (Galton 1869, 370)

In considering how a quantitative theory of heredity could be based on the hypothesis of pangenesis, Galton first considered the effect of what is today called mutation. He supposed that a child acquired a fraction r of its gemmules unchanged from the parents, the remainder, $1 - r$, being changed through individual variation; in other words, $1 - r$ is the mutation rate. The proportion of gemmules inherited unchanged from the grandparents is r^2, from the great-grandparents r^3, and so on. He admitted that he had no idea what the mutation rate was, but as an example he supposed that it was 0.1, so that the proportion of gemmules handed down unchanged from the parents was 0.9, from the grandparents 0.81, and from all the ancestors above the fiftieth degree only 0.005. He concluded that "the theory of Pangenesis ... appears to show that a man is wholly built up of his own and ancestral *peculiarities*, and only in an infinitesimal degree of characteristics handed down in an unchanged form, from extremely ancient times. It would follow that under a prolonged term of constant conditions, it would matter little or nothing what were the characteristics of the early progenitors of a race, the type being supposed constant, for the progeny would invariably be molded by those of its more recent ancestry" (1869, 371).

He then sketched how a theory of quantitative genetics might be constructed:

> The average proportion of gemmules, modified by individual variation under various conditions preceding birth, clearly admits of being determined by observation; and the deviations from that average may be determined by the same theory in the law of chances [the normal distribution], to which I have so often referred. Again, the proportion of the other gemmules which are transmitted in an unmodified form, would be similarly treated; for the children would, *on the average*, inherit the gemmules in the same proportions that they existed in the parents; but in each child there would be a deviation from that average....
>
> If the theory of Pangenesis be true, not only might the average qualities of the descendants of groups A and B, A and C, A and D, and every other

combination be predicted, but also the numbers of them who deviate in various proportions from those averages. Thus, the issue of F and A ought to result in so and so, for an average, and in such numbers, per million, of A, B, C, D, E, F, G, &c., classes. The latent gemmules equally admit of being determined from the patent characteristics of many previous generations, and the tendency to reversion into any ancient form ought also to admit of being calculated. In short, the theory of Pangenesis brings all the influences that bear on heredity into a form, that is appropriate for the grasp of mathematical analysis. (Galton 1869, 373)

This passage outlines an ambitious program for constructing a theory of quantitative genetics based on pangenesis; the groups A, B, C, and so on are the grades defined in table 6.3. But there is little indication about how it might be implemented in detail or about how empirical estimates might be made of the mutation rate or the tendency to reversion. Galton did not pursue the idea further.

We turn now to his substantive statistical theory of heredity, beginning with the question: Why is the statistical distribution of many quantitative characters constant from one generation to the next?

"Typical Laws of Heredity" (1877)

In February 1877 Galton delivered a lecture at the Royal Institution, in which he discussed a problem that had perplexed him for some years. In *Hereditary Genius,* he had suggested that the height of an isolated human population under constant conditions would remain constant from generation to generation; this would apply not only to the mean height but to the distribution of heights, which would be normal with the same mean and variance in each generation. The fossil record showed that this was true for many quantitative characters of plants and animals, and he concluded that "the processes of heredity are found to be so wonderfully balanced and their equilibrium to be so stable, that they concur in maintaining a perfect statistical resemblance so long as the external conditions remain unaltered" (1877, 492). He now asked, Why is this the case?

Some people might argue that "there is no wonder in the matter, because each individual tends to leave his like behind him, and therefore each generation must resemble the one preceeding" (492). But this is untrue. If we compare 100 giants with 100 men of medium height, the giants do not leave behind them their quota of giants in the next generation for several reasons: first, they are less fertile than average men and their taller offspring would be less likely to survive, or so Galton thought; second, their offspring would, on the average, be less tall than their fathers because of dilution by marriage and because the progeny of exceptional individuals tend to "revert" to mediocrity. On the other hand, the 100 medium men, being more fertile and breeding more truly to their like, leave more than their proportionate share of progeny. Thus one might expect fewer giants and more medium-sized men in the next generation. Thus the question remained, how do successive generations resemble each other in their statistical properties with great exactitude?

Galton now observed that nearly all quantitative characters followed, at least approximately, the normal distribution, acknowledging Quetelet as authority for this information. He then posed the problem in the epigraph of this chapter. He had called his lecture "Typical laws of heredity," by which he meant the statistical laws that heredity must follow in the typical case in which a normal distribution with constant mean and variance is maintained from one generation to the next.

He first pointed out that Quetelet and his followers had unaccountably overlooked an important fact, that "although characteristics of plants and animals conform to the [normal] law, the reason of their doing so is as yet totally unexplained. The essence of the law is that differences should be wholly due to the collective actions of a host of independent *petty* influences in various combinations.... Now the processes of heredity that limit the number of the children of one class such as giants, that diminish their resemblance to their fathers, and kill many of them, are not petty influences, but very important ones" (512). Thus his problem was to explain why biological characteristics should follow the normal law, as well as to explain how their mean and variability should remain constant from one generation to the next.

An Experiment with Sweet Peas

When Galton began to think about this question, he had decided to do some experiments to put him on the right track: "When the idea first occurred to me, it became evident that the problem might be solved by the aid of a very moderate amount of experiment. The properties of the law of deviation are not numerous and they are very peculiar. All, therefore, that was needed from experiment was suggestion. I did not want proof, because the theoretical exigencies of the problem would afford that. What I wanted was to be started in the right direction" (1877, 512).

He therefore looked for a measurable characteristic suitable for investigating the statistical relationship between successive generations and decided to experiment on seed size in plants. He did some work on cress (Pearson 1924, 392), but eventually selected sweet peas, which Joseph Hooker, director of Kew Gardens, and Charles Darwin suggested as having three advantages: they are self-fertilizing, so that only one parent needs to be considered; all the seeds in the pod are of similar size; and they are very hardy and prolific. A preliminary experiment at Kew in 1874 failed, and to minimize the chance of failure the next year, Galton persuaded nine of his friends and acquaintances in various parts of the United Kingdom, including Darwin, to grow plants for him from seed; seven of these nine experiments were successful. He procured a large number of seeds from the same bin from a seed merchant, sorted them into seven equally spaced size classes, and sent each of his friends seven little packets, each containing ten seeds of almost exactly the same size. The seeds from the different packets, representing different sizes, were planted in separate rows, and the seeds from their offspring were collected and returned to Galton for measurement and analysis. (See Galton 1877, 512–513; 1889, 79–82 and 225–226; 1908, 300–302 [though the date of 1885 given there is clearly wrong]; Pearson 1924, chap. 10.)

The results of this experiment are shown in table 7.1. Galton drew two conclusions from them. First, within each class (row), representing the offspring of parent seeds of the same size, there was an approximately normal distribution of the sizes of the filial seeds, with a probable error which was the same in each class. He called this probable error the "family variability," and he wrote: "I was certainly astonished to find the family variability of the produce of the little seeds to be equal to that of the big ones, but so it was, and I thankfully accept the fact, for if it had been otherwise I cannot imagine, from theoretical considerations, how the problem could be solved" (1877, 513). (Pearson [1930a, 7] later found the standard deviation to be slightly lower in the produce of the small seeds.)

The second conclusion is that there is reversion of the filial mean, which follows a simple linear law. In discussing the results in *Natural Inheritance*, by which time Galton had replaced the word "reversion" by "regression," he wrote: "It will be seen that for each increase of one unit on the part of the parent seed, there is a mean increase of only one-third of a unit in the filial seed; and again that the mean filial seed resembles the parental when the latter is about 15.5 hundredths of an inch in diameter. Taking 15.5 as the point towards which Filial Regression

Table 7.1. Diameters of Parent Seeds and Their Produce (hundredths of an inch)

Diameter of parent seed	Distribution of Diameters of Filial Seeds (percent)									Mean diameter of filial seeds
	< 15	15–16	16–17	17–18	18–19	19–20	20–21	> 21	Total	
21	22	8	10	18	21	13	6	2	100	17.5
20	23	10	12	17	20	13	3	2	100	17.3
19	35	16	12	13	11	10	2	1	100	16.0
18	34	12	13	17	16	6	2	0	100	16.3
17	37	16	13	16	13	4	1	0	100	15.6
16	34	15	18	16	13	3	1	0	100	16.0
15	46	14	9	11	14	4	2	0	100	15.3

Source: Galton 1889

points, whatever may be the parental deviation from that point, the mean Filial Deviation will be in the same direction, but only one-third as much" (1889, 225).

In other words, if $m_{y|x}$ is the mean size of seeds produced from seeds of size x, then

$$m_{y|x} - 15.5 = 0.33(x - 15.5), \text{ or}$$
$$m_{y|x} = 10.33 + 0.33x. \qquad [1]$$

This is the first regression line ever to have been calculated; the relationship is shown diagrammatically in fig. 7.1. It will also be seen that the mean filial size of 15.5 is considerably less than the mean parental size, which is about 18. Galton recognized this fact: "The point of convergence was considerably below the average size of the seeds contained in the large bagful I bought at a nursery garden, out of which I selected those that were sown, and I had some reason to believe that the size of the seed towards which the produce converged was similar to that of an average seed taken out of beds of self-planted specimens" (1885c, 246). This was probably an environmental effect; the plants raised by Galton's friends may have been cultivated under less ideal conditions than their parents, and Darwin wrote to him that they ought to have been planted much farther apart.

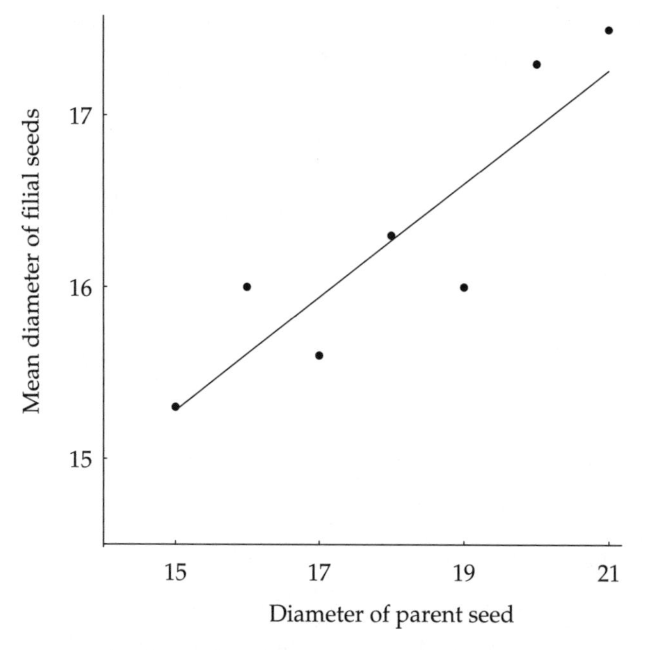

Fig. 7.1. Regression of filial on parental values for seed diameter (hundredths of an inch) in sweet peas (data from Galton 1885c). The fitted line is $y = 10.33 + 0.33x$.

Solution of the Problem

Galton used these results to answer the question, How do successive generations resemble each other in their statistical properties? Consider a character y which follows a normal distribution with mean μ (the racial mean) and modulus c. Consider first the simplest case of "simple descent" in which there is single parentage, as in sweet peas, with no selection on fecundity or viability. The two processes involved in producing the next generation are reversion, which decreases the variability, and family variability, which increases it. At equilibrium these two processes must balance. Suppose that reversion causes the mean value of offspring of parents with value y to be $\mu + r(y - \mu)$; it is assumed that reversion acts linearly, as found in sweet peas, and that it is directed to the racial mean, since environmental differences between generations are to be ignored. The offspring of parents with value y have deviation d from their mean value due to family variability, which on the basis of the sweet pea results is normally distributed with modulus v which does not depend on y. If the offspring value is denoted y^*, then

$$y^* = \mu + r(y - \mu) + d. \qquad [2]$$

Now $r(y - \mu)$ and d are normally and independently distributed with zero mean and moduli rc and v, respectively, so that y^* is normally distributed with mean μ and modulus c^*, whose square is the sum of the squares of the component moduli:

$$c^{*2} = r^2c^2 + v^2. \qquad [3]$$

This follows from the theorem known to Galton as "the law of the sum of two fallible measures" (see eq.[3] in chapter 6).

Hence the offspring values are normally distributed and have the same variance as the parental values if $c^* = c$, so that

$$c = v/\sqrt{(1 - r^2)}. \qquad [4]$$

For example, if $r = 1/3$, then at equilibrium $c = 1.061\ v$.

Galton was really interested in organisms with double parentage, in particular with human populations, and he considered human height as an example, though he had no data at this time. He suggested that female height should first be adjusted to have the same mean and modulus as male height, and that the average height of each couple should be calculated and treated as the equivalent of a single parent. (He later called this construct the mid-parent.) If male height is normally distributed with mean μ and modulus c, and if there is random mating for height, the average height of a couple is normal with mean μ and modulus $c/\sqrt{2}$, by the law of the sum of two fallible measures. If the regression of offspring value on mid-parental value is given by eq.[2], with y on the right-hand side reinterpreted as a mid-parental value, then by the same argument as before, the offspring value is normally distributed with mean μ and squared modulus

$$c^{*2} = \tfrac{1}{2}r^2c^2 + v^2. \qquad [5]$$

Setting $c^* = c$ at equilibrium gives

$$c = v/\sqrt{(1 - \tfrac{1}{2}r^2)}. \qquad [6]$$

Galton also considered the effects of selection on fecundity and viability. He first remarked that in the typical case in which he was interested, in which a normal distribution with the same mean and variance is preserved from one generation to the next, selection must be conformable with the normal distribution; it must also be symmetric about the mean in order not to change it. For viability selection he therefore proposed a fitness function equivalent to

$$w(y) = \exp \frac{-(y-\mu)^2}{s^2}, \qquad [7]$$

where $w(y)$ is the relative fitness of an individual with character value y. This represents selection for an optimal value equal to the mean. The parameter s measures the spread of fitness about the mean; the smaller s, the greater the intensity of selection. (Haldane [1954] later generalized this fitness function to selection for an optimal value θ, not necessarily equal to the mean, by replacing μ by θ; this is called the noroptimal model [Bulmer 1980].)

Suppose that in the parental generation, the character is normally distributed after selection with mean μ and modulus c. After random mating in a bisexual population, the character is normally distributed in the next generation before selection with mean μ and squared modulus given by eq.[5]. The character remains normal after selection with mean μ and squared modulus

$$c^{**2} = (s^2 + c^{*2})/s^2 c^{*2}. \qquad [8]$$

Given r, v, and s, the equilibrium variance after selection can be found by solving the equation $c^{**} = c$. Galton considered fecundity selection in a similar way.

From one point of view, this paper is a triumphant success, showing how the distribution of a character can remain normal with constant parameters from one generation to the next, and laying the foundations of many of the recurring themes of quantitative genetics. But it is less successful as an explanation of how the phenotypic variability takes its particular value. What Galton has shown, for example, in eq.[4] is the relationship that must exist between the modulus c, the family variability v, and the regression of offspring on parent r, under simple descent in order for the distribution to remain constant from one generation to the next. This is a purely descriptive result which does not explain why these parameters should take their observed values. Galton thought that he had done more than this, and in particular that he could predict the equilibrium value of the modulus from a recurrence relationship by treating r and v as constants:

> As regards the precise scale of deviation that characterises each population, let us trace, in imagination, the history of the descendants of a single medium-sized seed. In the first generation the differences are merely those due to family variability; in the second generation the tendency to wider dispersion is somewhat restrained by the effect of reversion; in the third, the dispersion again increases, but is more largely restrained, and the same process continues in successive generations, until the step-by-step progress of dispersion has been overtaken and exactly checked by the growing antagonism of reversion. Reversion acts precisely after the law of an elastic spring.... Its tendency to recoil increases the more it is stretched, hence equilibrium must at length ensue between reversion and family variability. (Galton 1877, 514)

He is arguing here that the phenotypic variance satisfies the recurrence relation

$$c_{t+1}^2 = r^2 c_t^2 + v^2 \qquad [9]$$

starting from $c_0 = 0$. This leads to the series

$$\begin{aligned} c_0 &= 0, c_1^2 = v^2, c_2^2 = (r^2 + 1)v^2, \\ c_3^2 &= (r^4 + r^2 + 1)v^2, \cdots, \end{aligned} \qquad [10]$$

a geometric series which converges to the limiting value given in eq.[4]. The hidden, and invalid, assumption is that v and r take the fixed values found in the original population. Current understanding of self-fertilizing populations (see below) shows that, if successive generations are propagated from a single seed, they all have a mean value determined by the mean value of the line from which that seed was obtained, with no reversion, and with modulus equal to that of the environmental variability, which is less than the modulus in the original, genetically variable population. Galton cannot be blamed for not anticipating this work; but this example illustrates the weakness of model-free, statistical theories.

Johannsen's Experiments with Beans

The Danish plant physiologist Wilhelm Johannsen (1857–1927) excelled in doing simple experiments that could be analyzed mathematically and statistically, resembling in these ways the experiments of Mendel. He greatly admired Galton's work, but he wanted to find the underlying biological reason for his purely statistical law of regression. In his book published in 1903, which was dedicated to Galton, he wrote of this law:

> A truly biological study of heredity cannot rest satisfied with such essentially statistical researches. A race, a population ... is by no means always, from the point of view of the biologist, to be treated as homogeneous, even when the individual variations are grouped round an average value, presumably 'typical', in the manner prescribed by the law of error. Such a population may ... contain a number of independent types, differing markedly from one another, which may not be discoverable at all by direct observation of the empirical frequency curve or table.... Before such a population is treated as homogeneous, it should therefore be biologically analysed in order to be clear as to its elements, that is to have some knowledge of the independent types already existing in the population. (Johannsen 1903, 4–5)

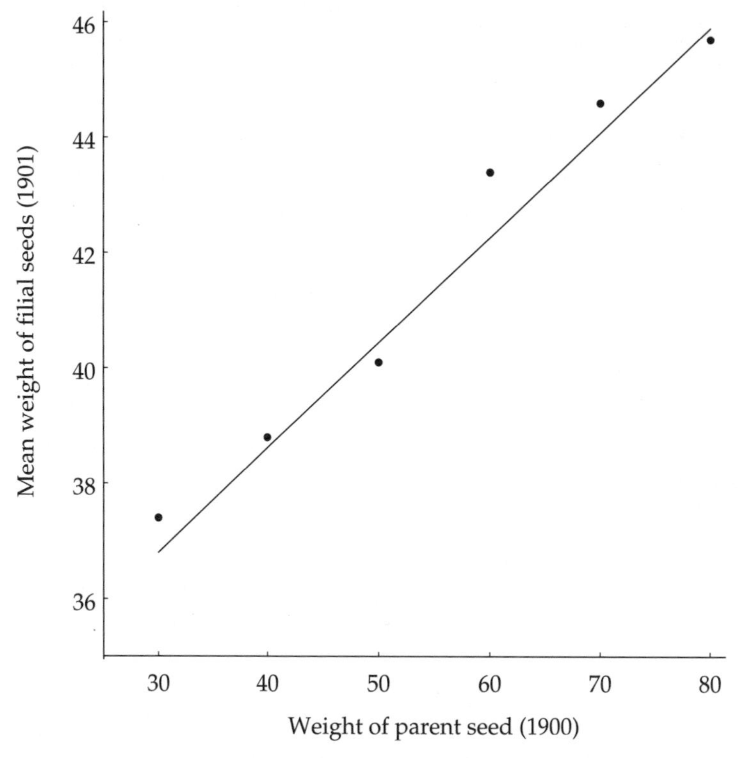

Fig. 7.2. Regression of filial on parental values for seed-weight (centigrams) in a population of haricot beans (data from Johannsen 1903). The fitted line is $y = 31.4 + 0.18x$.

Johannsen realized that a self-fertilizing population of plants differed essentially from a cross-fertilizing population of plants or animals. A self-fertilizing population was composed of a number of "pure lines," the posterity of a single individual, while there was continual crossing between the lines in a cross-fertilizing population. His aim of studying the different types within a population could be accomplished most simply in a self-fertilizing population, and he presented the results of three such investigations, on the weight and the length-breadth ratio of the seeds of the bean *Phaseolus vulgaris* and on the phenomenon of relative sterility in barley.

The best-known investigation was on the seed-weight of beans. Single beans were weighed and sown, the resulting plants were allowed to self-fertilize (in a net-covered enclosure), and their seeds were harvested. A sample of these seeds was sown in their turn, and allowed to self-fertilize in similar conditions. He bought a bag of beans from a seed-merchant in 1900 and compared the seed-weights of their offspring in 1901 with the parental seed-weights, as Galton had done for seed diameter in sweet peas. The results in fig. 7.2 show the same law of regression to the mean as Galton found, though the regression coefficient was only 0.18 compared with Galton's value of 0.33.

To investigate whether regression was due to the presence of different types within the population, he generated nineteen pure lines by sowing nineteen beans in 1900, weighing the seeds of their offspring in 1901, and sowing a sample of these offspring to produce plants in 1902. The result of plotting the seed-weight of these plants in 1902 against that of their parents in 1901 is shown in fig. 7.3 for two of the lines with the most extensive data. There has been complete regression to the mean, that is to say the regression coefficient is zero. There was also a substantial difference between the mean weights of the lines in the same year, and between the two years. The mean weight of line B (solid circles), in hundredths of a gram, was 52 in 1901 and 56 in 1902, compared with 40 in 1901 and 45 in 1902 for line J (open circles). Thus seed-weight was heavier in line B than in line J, and in 1902 than in 1901. There was also substantial variability within a line in each year, but this variability was not heritable.

Johannsen concluded that the population of beans was composed of a large number of "pure lines," each with its own type, which he later called a genotype. The genotype, together with environmental factors such as the weather in a particular year, determined its average seed-weight. There was also variation about this average value which he

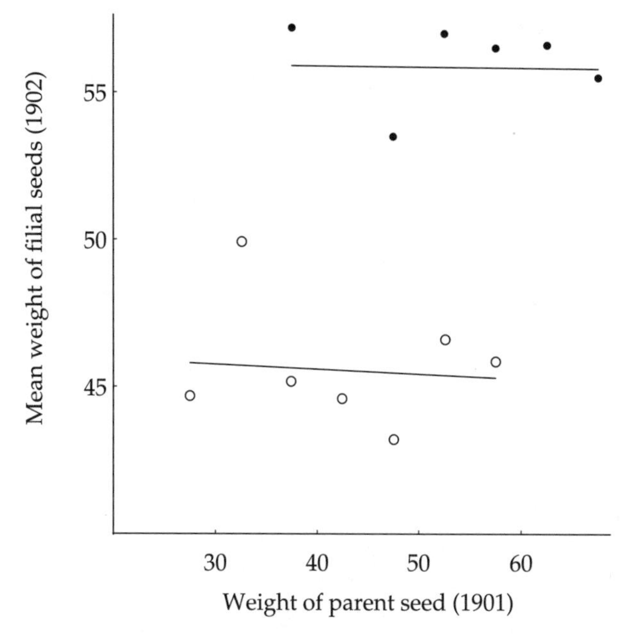

Fig. 7.3. Regression of filial on parental values for seed-weight (centigrams) in two pure lines of haricot beans (data from Johannsen 1903). The fitted lines are $y = 56 - 0.003\,x$ for line B (solid circles) and $y = 46.2 - 0.02\,x$ for line J (open circles).

called "fluctuating variability," which had a standard deviation of about 8 centigrams and was not heritable. (De Vries's and Johannsen's use of this term must be carefully distinguished.) The incomplete regression to the mean shown in fig. 7.2, in which data for the whole population were lumped together, was due to the mixture of different genotypes; a large bean was more likely to come from a line with a high genotypic value than a small bean and was therefore more likely to have offspring with large seeds.

There is unfortunately a complication. Weldon and Pearson (1903) published an immediate critical review of Johannsen's book from a biometrical viewpoint in their newly founded journal *Biometrika*. They argued that, if his pure line theory were true, the regression of the mean weight of each line in 1902 on that in 1901 ought to be unity, since it should only be displaced by a constant year effect, whereas it was in fact only 0.59. The explanation of this fact is that the year effect is not constant but varies from line to line; this phenomenon is known today as

Table 7.2. Mean Seed-Weights (centigrams) of 19 Lines of Beans in 1901 and 1902

Line	1901	1902	Line	1901	1902
A	60	64	L	36	45
B	52	56	M	34	43
C	57	55	N	31	41
D	60	55	O	31	35
E	51	51	P	39	45
F	40	48	Q	44	49
G	40	47	R	41	45
H	38	46	S	40	49
J	40	45	T	40	51
K	51	45			
			Average	43	48

Source: Johannsen 1903

genotype x year interaction. Table 7.2 shows the mean weights of the nineteen lines in the two years; each of these means is quite accurate, having a standard error of less than 1 cg. On average the seeds were about 5 cg heavier in 1902 than in 1901, but three lines (C, D, and K) did better in 1901 than in 1902, one line (E) did the same in both years, and two lines (N and T) did better by more than 10 cg in 1902 than in 1901.

Johannsen (1911, 145) was familiar with this type of interaction:

> It is well known to breeders that some strains of wheat yield relatively much better than others on rich soil, while the reverse is realized on poorer soils. In four subsequent years two pure lines of barley, both characterized by a considerable degree of disposition to produce vacant spikelets (aborted grains) in the heads, presented the phenotypes here indicated in percentages of such vacancies.

Pure line L:	30	33	27	29
Pure line G:	5	45	3	28

In modern terminology, we may write $y = g + e$, where y is the phenotypic value (the observed weight of a bean), g is the genotypic value (the average weight in a particular line in a particular year), and e is the deviation from the mean (Johannsen's fluctuating variability). If g_i is the genotypic value in year i, the regression of the line means in year i on those in the previous year is

$$\mathrm{Cov}(g_{i-1}, g_i)/\mathrm{Var}(g_{i-1}) \qquad [11]$$

from eq.[18] in chapter 6. This is unity when there is no genotype x year interaction, since both the numerator and the denominator reduce to Var(g), the genetic variance, which is the same from year to year; but otherwise it may be substantially less than unity.

Similarly, the regression of the phenotypic values of offspring on parent (see figs. 7.1 and 7.2), in a mixed population with several genotypes, is

$$\mathrm{Cov}(g_{i-1}, g_i)/[\mathrm{Var}(g_{i-1}) + \mathrm{Var}(e)]. \qquad [12]$$

When there is no genotype x year interaction, this becomes

$$\mathrm{Var}(g)/[\mathrm{Var}(g) + \mathrm{Var}(e)], \qquad [13]$$

so that the regression is a measure of the proportion of the total variance within the population that is due to variability between genotypes. But otherwise it may be substantially less than this, and may vary from year to year. Indeed, the regression of offspring in 1901 on parents in 1900 was 0.18 (see fig. 7.2), but the regression of offspring in 1902 on parents in 1901, calculated from Johannsen's table 3, was 0.27. This may explain why Weldon and Pearson (1903) found that the regression of offspring bean in 1902 on grandmaternal bean in 1900 was substantially less than that of offspring bean in 1902 on maternal bean in 1901, which was another of their reasons for rejecting Johannsen's pure line theory.

But this complication does not affect the main conclusion from Johannsen's pure line theory, that there was no correlation between offspring and parent within a pure line and that selection within a line would therefore be ineffective (see fig. 7.3). There was no justification for Weldon and Pearson's criticism of the pure line theory. (Provine [1971, 96] is even more scathing in his criticism of Johannsen: "Pearson and Weldon calculated this ... correlation [within a pure line] and found it was 0.3481, scarcely the zero Johannsen claimed." The figure of 0.3481 cited by Weldon and Pearson in fact refers to the correlation between offspring in 1902 and parents in 1901 *within the whole population*, not to the correlation within a pure line; neither they nor Johannsen suggested that it should be zero. Provine has misunderstood Weldon and Pearson's argument.)

Johannsen's work was very influential in elucidating the distinction between genotype and phenotype, and in demonstrating the role of nonheritable environmental variability ("fluctuating variability") about the genotypic value. He also provided an explanation of Galton's law of partial regression to the mean in self-fertilizing populations. If the population comprised a mixture of several different genotypes, parents with a large phenotypic value of some character were more likely to have a genotype conferring a larger than average phenotypic value on their offspring. It was not possible to determine by inspection whether the population comprised a single genotype or a mixture of several genotypes, since a mixture of several normal distributions could easily mimic a single normal distribution, and he showed how the genotypes could be separated experimentally in a self-fertilizing population.

But all of this was unknown to Galton. We must now return to his efforts to show that the law of regression that he had demonstrated in sweet peas, though he had not understood its underlying cause, also held in man.

The Inheritance of Human Height

Galton was mainly interested in the inheritance of human characters, but he had no human data on which to develop his statistical theory until he devised his *Record of Family Faculties* (RFF) in 1884. This was an extensive questionnaire in which respondents were asked to provide information for themselves, their parents, grandparents, and great-grandparents, and any brothers and sisters thereof, about their height, color of hair and eyes, other physical and mental characteristics, and medical history. To encourage response, Galton provided £500 in prize money for the most complete records. He received 150 replies, mostly from professional people, and awarded prizes to 84 of them (£7 to 40 of them and £5 to the remainder). He used these records most extensively to study the inheritance of height, but he also discussed the inheritance of eye color, of good and bad temper, of the artistic faculty, and of disease (Galton 1886c, 1887b, and 1889). He reported his results on height in four papers: the presidential address to the Anthropology Section of the British Association (1885b), "Regression towards mediocrity in hereditary stature" (1885c), "Family likeness in stature" (1886a), and a presidential address to the Anthropological Institute (1886b); he summarized and consolidated his conclusions in *Natural Inheritance* (1889).

The Advantages of Height

He argued that height had many advantages for the statistical investigation of heredity. First, it was easy to measure, was practically constant during thirty-five or forty years of middle life, depended little on differences of bringing up, and had little influence on mortality.

Another advantage of height was that it followed a normal distribution very closely, so that Galton was able to use the well-known properties of this distribution. He attributed this fact to the central limit theorem:

> Human stature is not a simple element, but a sum of the accumulated lengths or thicknesses of more than a hundred bodily parts....
>
> This multiplicity of elements, whose variations are to some degree independent of one another, ... , corresponds to an equal number of sets of rows of pins in the apparatus [fig. 6.4a] by which the cause of variability was illustrated. The larger the number of these variable elements, the more nearly does the variability of their sum assume a "Normal" character.... The beautiful regularity in the Statures of a population, whenever they are statistically marshalled in the order of their heights, is due to the number of variable and quasi-independent elements of which Stature is the sum. (Galton 1889, 83–85)

In fact, the central limit theorem does not, strictly speaking, apply to this situation, since the lengths of the different bones are correlated; Galton acknowledged this problem by writing about "quasi-independent elements." The normal distribution is nevertheless a very good empirical approximation for height. In *Natural Inheritance,* he suggested that the data might have been better approximated by the log-normal distribution, in which case every measurement would be replaced by its logarithm, and these logarithms treated as if they had been the observed values (1889, 118–119). But he found the gain to be so small that it was not worth pursuing, since he was indisposed to do anything that was not really necessary, which might further confuse the reader.

In discussing how he would analyze data on human height in "Typical laws of heredity," Galton had suggested that female height should first be adjusted to have the same mean and modulus as male height. From the RFF data, he found that this could be achieved by multiplying female height by the factor 1.08, and he called these the transmuted heights. (An insight into his method of work is provided by his remark that the exact value of this factor made little difference, "for it happened that, owing to a mistaken direction, the computer to whom I first

Table 7.3. Effect upon Adult Children of Differences in Height of Their Parents

Difference between heights of parents	Proportion per 50 of cases in which the heights of the children deviated to various amounts from the mid-filial stature of their respective families					No. of children
	< 1 in.	< 2 in.	< 3 in.	< 4 in.	< 5 in.	
< 1 in.	21	35	43	46	48	105
1–2 in.	23	37	46	49	50	122
2–3 in.	16	34	41	45	49	112
3–5 in.	24	35	41	47	49	108
> 5 in.	18	30	40	47	49	78

Source: Galton 1889

Original note: Every female height has been transmuted to its male equivalent by multiplying it by 1.08, and only those families have been included in which the number of adult children amounted to six, at least.

entrusted the figures used a somewhat different factor, yet the result came out closely the same" [1885b, 1207].)

He had then suggested that, to simplify the analysis of the relationship between parents and offspring, the average height of each couple should be calculated and treated as the equivalent of a single parent; he called the average of the father's height and the transmuted height of the mother the mid-parental value. To justify the method of dealing with mid-parentages instead of with single parents, he argued: "If the Stature of children depends only upon the *average* Stature of their two Parents, that of the mother having been first transmuted, it will make no difference in a Fraternity whether one of the Parents was tall and the other short, or whether they were alike in Stature. But if some children resemble one Parent in Stature and others resemble the other, the Fraternity will be more diverse when their Parents had differed in Stature than when they were alike" (1889, 89). He therefore used the RFF data to determine whether the variability within a sibship depended on the difference in height between the parents, with the results shown in table 7.3. He concluded that differences in height between the parents had little or no effect on their offspring. In other words, height was an example of a character that showed blending inheritance, so that the height of the offspring depended only on the mid-parental height and not separately on the heights of the two parents.

The analysis of the relationship between offspring and mid-parent is simpler if parents marry at random, with no tendency for like to marry like. Galton found this to be the case for the characters on which he had information from RFF: height, good and bad temper, eye-color, and artistic tastes; and he remarked that this was not extraordinary, "for though people may fall in love for trifles, marriage is a serious act, usually determined by the concurrence of numerous motives. Therefore we could hardly expect either shortness or tallness, darkness or lightness in complexion, or any other single quality, to have in the long run a large separate influence" (85). In the case of height, he took the 205 pairs of parents on whom he had information from RFF, and divided the husbands and wives into three groups, tall, medium, and short, with medium individuals having heights (or transmuted heights) between 67 and 70 inches. The cross-tabulation is shown in table 7.4, from which he concluded: "We may therefore regard the married folk as couples picked out of the general population at haphazard when applying the law of probabilities to heredity of stature" (206). Pearson (1896, 270) reexamined Galton's data by modern statistical methods, confirming that the correlation between husband and wife was barely different from zero. But later studies of the inheritance of height have shown that there is substantial assortative mating, with a tendency of like to mate with like, which needs to be taken into account; for example, Pearson and Lee (1903) found a correlation of 0.28.

Table 7.4. Marriage Selection in Respect to Stature

Wives	Husbands		
	Short	Medium	Tall
Tall	12	20	18
Medium	25	51	28
Short	9	28	14

Short and tall, 12 + 14 = 26 cases;
short and short or tall and tall, 9 + 18 = 27 cases.

Source: After Galton 1889

Galton summarized this discussion:

> The advantages of stature as a subject in which the simple laws of heredity may be studied will now be understood. It is a nearly constant value that is frequently measured and recorded, and its discussion is little entangled with considerations of nurture, of the survival of the fittest, or of marriage selection. We have only to consider the mid-parentage and not to trouble ourselves about the parents separately. The statistical variations of stature are extremely regular, so much so that their general conformity with the results of calculations based on the abstract law of frequency of error is an accepted fact by anthropologists. I have made much use of the properties of that law in cross-testing my various conclusions, and always with success. (Galton 1885b, 1209)

He added that the only drawback to the use of height was its small variability. As we saw in the last chapter, its probable error was 1.7 inches. On the other hand, he thought that the precision of the data was small, partly due to uncertainty in some cases whether height was measured with the shoes on or off, and he estimated that the probable error of measurement of a single observation was two-thirds of an inch. He does not state how he obtained the latter figure, but it would, in modern terminology, reduce the heritability by 15 percent since $(0.67/1.7)^2 = 0.15$.

He had carefully validated all his claims about the suitability of height for the study of heredity with one exception: the claim that it was almost unaffected by nurture. The Anthropometric Committee of the British Association, of which he was an active member, drew attention to environmental differences in height in the British Isles, and he cited them as one of the reasons for greater variability among cofraternities than among fraternities: "There are three reasons why Co-Fraternals should be more diverse among themselves than brothers.... Thirdly, because the nurture or rearing of Co-fraternals is more various than that of Fraternals.... The large differences between town and country-folk, and those between persons of different social classes, are conspicuous in the data contained in the Report of the Anthropological Committee of the British Association in 1880, and published in its Journal" (1889, 94). (A co-fraternity consists of all the offspring of a group of mid-parents with the same height.) Thus by his own admission there may have been some influence of nurture on the variability of height, though any influence common to all members of the same family would not have reduced, and may even have increased, the correlations between relatives.

These provisos about measurement error and the effect of nurture are important, because Galton ignored them in interpreting the significance of the regression of offspring on mid-parent for height and assumed that it could be explained entirely in terms of hereditary factors. He also regarded his results as being general laws of heredity: "It is needless to say that I look upon this inquiry into stature as a representative one. The peculiarities of stature are that the paternal and maternal contributions blend freely, and that selection, whether under the aspect of marriage selection or of the survival of the fittest, takes little account of it. My results are presumably true, with a few further reservations, of all qualities or faculties that possess these characteristics" (1886b, 491).

The Regression of Offspring on Mid-Parent

Galton's estimate of the regression of offspring on mid-parental height was 2/3, as we saw in the last chapter. He admitted in *Natural Inheritance* that his first estimate was 3/5, but that he afterwards substituted the value of 2/3 "because the data seemed to admit of that interpretation also, in which case the fraction of two-thirds was preferable as being the more simple expression" (1889, 98). He argued that the regression of offspring on a single parent ought to be 1/3 since the two parents contribute equally and the contribution of either of them could only be one half of that of the mid-parent. When he tried to test this prediction, he encountered a complication "that the height of the children of both sexes, but especially that of the daughters, takes after the height of the father more than that of the mother. My present data are insufficient to deter-

Table 7.5. Parent-Offspring Correlations for Height

	Father and		Mother and	
	Son	Daughter	Son	Daughter
Galton's data	0.40	0.36	0.30	0.28
Pearson and Lee's data	0.51	0.51	0.49	0.51
Pearson and Lee's data, corrected for assortative mating	0.40	0.40	0.39	0.40

Sources: Pearson 1896, 1930a, Pearson and Lee 1903

mine the ratio satisfactorily" (1885b, 1208). Karl Pearson determined the uniparental correlations from Galton's original data and showed that this conclusion was correct, though the effect was small (first row in table 7.5). He also showed that this asymmetry disappeared in the large data set collected by himself and Alice Lee, in which great care had been taken with the accurate measurement of height (second row in table 7.5); the correlations shown at the bottom of table 7.5 have been corrected for the effect of assortative mating by dividing the original correlations by 1.28. He concluded that the asymmetry in Galton's data was due to measurement error: "I think it may well have been due to amateur measuring of stature in women, when high heels and superincumbent chignons were in vogue; it will be noted that the intensity of heredity decreases as more female measurements are introduced. Daughters would be more ready to take off their boots and lower their hair knots, than grave Victorian matrons" (Pearson 1930a, 18). It can be inferred that if this source of measurement error were removed, all the uniparental regressions would be about 0.4, and the regression of offspring on mid-parent about 0.8.

Despite this complication, Galton took the best estimate of the regression of offspring on mid-parent as 2/3, and he applied the theory developed in "Typical laws of heredity" by substituting $r = 2/3$ in eqs.[5] and [6]. He was encouraged to have confirmed for human height the same phenomenon of regression toward mediocrity that he had first demonstrated for seed-weight in the sweet pea.

He also thought that he had developed a better understanding of the reason for regression:

> I was then [in 1877] blind to what I now perceive to be the simple explanation of the phenomenon ... [which] is as follows. The child inherits partly from his parents, partly from his ancestry. Speaking generally, the further his genealogy goes back, the more numerous and varied will his ancestry become, until they cease to differ from any equally numerous sample taken at haphazard from the race at large. Their mean stature will then be the same as that of the race ... or, to put the same fact in another form, the most probable value of the mid-ancestral deviates in any remote generation is zero.
>
> For the moment let us confine our attention to the remote ancestry and to the mid-parentages, and ignore the intermediate generations. The combination of the zero of the ancestry with the deviate of the mid-parentage is that of nothing with something, and the result resembles that of pouring a uniform proportion of pure water into a vessel of wine. It dilutes the wine

to a constant fraction of its original alcoholic strength, whatever that strength may have been.

The intermediate generations will each in their degree do the same. The mid-deviate of any one of them will have a value intermediate between that of the mid-parentage and the zero value of the ancestry. Its combination with the mid-parental deviate will be as if, not pure water, but a mixture of wine and water in some definite proportion had been poured into the wine. The process throughout is one of proportionate dilutions, and therefore the joint effect of all of them is to weaken the original wine in a constant ratio....

The average regression of the offspring to a constant fraction of their respective mid-parental deviations, which was first observed in the diameters of seeds, and then confirmed by observations on human stature, is now shown to be a perfectly reasonable law which might have been deductively foreseen. (Galton 1885b, 1207–1210)

The argument depends on the idea that the offspring of very distinguished, or very tall, or very short, parents would tend to revert to more remote ancestors, who would on average be less extreme. This idea was familiar to Galton from the time he wrote his first paper on heredity, "Hereditary talent and character," in 1865. The wine and water metaphor added the essential idea that regression should be linear because the wine is diluted "to a constant fraction of its original alcoholic strength, whatever that strength may have been."

The idea that statistical regression was due to reversion, that is, to the expression of hereditary elements which had been latent since they had been last expressed in a more or less distant ancestor, was very different from Johannsen's idea that it depended on the masking of the genotype of the parent by environmental variability. It also led Galton to consider the question, What was the average share contributed to the personal features of the offspring by each ancestor severally? This in turn led him to the law of ancestral heredity which is discussed in chapter 8.

Kinship

We come now to the third problem that Galton set out to solve in *Natural Inheritance*: "The last of the problems that I need mention now, concerns the nearness of kinship in different degrees. We are all agreed that a brother is nearer akin than a nephew, and a nephew than a cousin, and so on, but how much nearer are they in the precise language of numerical statement?" (1889, 2). He proposed to measure the nearness of kinship between two related individuals by the corresponding regression coefficient, β. He was also interested in calculating the prediction error about the regression line, Q_e, given by

$$Q_e = \sqrt{(1 - \beta^2)}Q, \qquad [14]$$

where Q is the population probable error (see chapter 6). If an individual had deviation x, then the deviation of a relative would be normally distributed with mean βx and probable error Q_e, so that the entire distribution can be calculated.

His conclusions about the nearness of kinship, calculated from the appropriate regression, together with the prediction error for height, with $Q = 1.7$, are shown in table 7.6. As we have seen, he had estimated from his data on height that the regression of offspring on mid-parent was 2/3, from which he inferred that the regression of child on a single parent (or vice versa) was 1/3. He also estimated from his data on height for brothers that the fraternal regression was 2/3; we discuss this conclusion shortly. He then assumed that the other regressions could be calculated from these two values by appropriate multiplication. A nephew is the son of a brother, so that his regression on an uncle is 1/3 x 2/3 = 2/9; a grandchild is the child of a child, so that the regression of grandchild on grandfather is 1/3 x 1/3 = 1/9; cousins are the offspring of two siblings, so that the regression is 1/3 x 2/3 x 1/3 = 2/27. These calculations are plausible but mistaken, since regressions cannot be calculated in this way. But Galton must be given credit for the fruitful idea of trying to calculate a numerical coefficient of kinship between different types of relatives.

Table 7.6. Nearness of Kinship between Relatives

Relationship	Regression, β	Prediction Error $= 1.7\sqrt{(1 - \beta^2)}$
Parent or child	1/3	1.60
Brother or sister	2/3	1.27
Uncle or nephew	2/9	1.66
Grandparent or grandchild	1/9	1.69
Cousin	2/27	1.70

Source: After Galton 1889

Fraternal Regression

Galton explained fraternal regression in a slightly different way from parent-offspring regression: "The *rationale* of the regression from father to son is largely to be ascribed . . . to the double source of the child's heritage. That heritage is derived partly from a remote and numerous ancestry, who are on the whole like any other sample of the past population, and therefore mediocre, and partly only from the persons of the parents. . . . The *rationale* of the regression from a known man to his unknown brother is due to a compromise between two conflicting probabilities: the one that the unknown brother should differ little from the known man, the other that he should differ little from the mean of his race" (1886b, 492–493). It will be remembered that he thought that brothers had identical stirps, and that the differences between them were due to different hereditary elements becoming patent, that is to say, to developmental variability. An empirical estimate of fraternal regression was needed since it could not be inferred from parent-offspring regression.

Galton obtained data on brothers from two sources: from the records of family faculties already described, and from a special survey of records of height among brothers. For the RFF data he used only the heights of the adult sons, ignoring the transmuted daughters, and excluding families of six or more brothers; for the special data he excluded families of five or more brothers. In each case he took the height of each man in turn and recorded the heights of all his brothers, creating a cross-tabulation like table 6.5 of the heights of the men and of their brothers. For men of fixed height (x) he then calculated the median height of their brothers, plotted it against x, and found the best-fitting line by eye (fig. 7.4). It was

$$\begin{aligned} y &= 68.25 + 0.48(x - 68.25) \text{ for the RFF data} \\ y &= 68 + 0.67(x - 68) \text{ for the special data.} \end{aligned} \qquad [15]$$

(In *Natural Inheritance* Galton did not estimate the slope from the RFF data explicitly, saying only that it could not exceed one-half; the value of 0.48 has been inferred from his figure 13.)

Galton considered the special data to be more accurate than the RFF data, and he therefore accepted the figure of 2/3 as the true fraternal regression, and he attributed the lower figure for the RFF data to errors of measurement. (See discussion above.) However, he observed that the probable error of a single value was the same (1.7 inches) for both sets of data, and he speculated that the inaccuracy in the RFF data was due to two factors: a measurement error, such as doubt about whether the shoes were on or off, and a tendency on the part of his correspondents to record medium statures when they were in doubt. The first would increase the probable error while the second would reduce it, but both factors would tend to reduce the fraternal regression. He concluded that the RFF regression should be increased by a factor of one-third to give the correct value of 2/3, and he wondered whether the regression of child on mid-parent estimated from the RFF data should be corrected in the same way to give 8/9, but he could not believe such a high value to be correct. He therefore provisionally adopted the value of 2/3 from the RFF data for the mid-parental regression as being near enough for the time being, until more accurate data could be obtained, while accepting the value of 2/3 from the special data for fraternal regression. He drew the conclusion that an individual was related twice as closely to a brother (kinship = 2/3) as to one of his parents (kinship = 1/3).

Karl Pearson (1896) was stimulated by *Natural Inheritance* to develop the theory of regression in relation to heredity, and in the course of this work he reexamined Galton's original, raw data by his new methodology. Using all the brothers, not excluding large families, he found the fraternal correlation to be 0.39 for the RFF data and 0.60 for the special data, rather lower than Galton's estimates of 0.48 and 0.67, but with the same difference between the two data sets. Pearson also noted a peculiarity in part of the special data, which he called the Essex contribution,

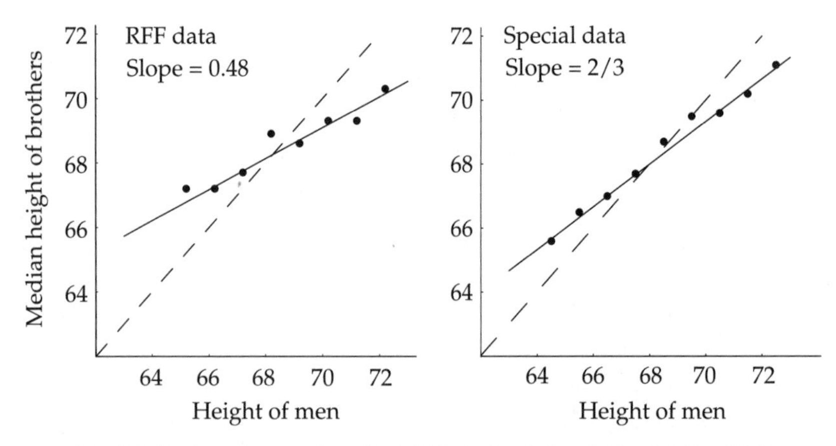

Fig. 7.4. Fraternal regression of height in inches (after Galton 1886a, 1889)

whose members were uniformly short. They were drawn from brothers in a volunteer regiment, and Pearson suggested that there might be unconscious selection as to height by those who join the volunteers. The correlation was reduced from 0.60 to 0.56 by excluding the Essex contribution. He also thought that the difference between the RFF and the special data might reflect a difference between classes, since the former were largely drawn from the upper middle and the latter from the working class. He concluded that it seemed unlikely that fraternal correlation was twice as large as parent-offspring correlation, as Galton supposed. He later found, in a large and carefully measured data set, that the fraternal correlation for height was 0.54, compared with 0.51 for the parent-offspring correlation (Pearson and Lee 1903; see table 7.5).

Variability in Fraternities and Co-Fraternities

Galton distinguished carefully between what he called fraternities and co-fraternities: "As all the Adult Sons and Transmuted daughters of the *same* Mid-Parent, form what is called a Fraternity, so all the Adult Sons and Transmuted Daughters of a *group* of Mid-Parents who have the same Stature (reckoned to the nearest inch) will be termed a Co-Fraternity" (1889, 94). Each row in table 6.4 represents a co-fraternity, and he found empirically that the probable error of a co-fraternity was $Q_{y|x} = 1.5$ inches, where x and y are the heights of mid-parents and of their offspring, compared with $Q_y = 1.7$ inches in the population. This value is in almost exact agreement with the theoretical value for $Q_{y|x}$ calculated from eq.[9] in chapter 6, with $Q_x = 1.7/\sqrt{2}$ and $\beta = 2/3$.

He realized that the variability in fraternities should be less than that in co-fraternities: "There are three reasons why Co-Fraternals should be more diverse among themselves than brothers. First, because their Mid-Parents are not of identical height, but may differ even as much as one inch. Secondly, because their grandparents, great-grandparents, and so on indefinitely backwards, may have differed widely. Thirdly, because the nurture or rearing of Co-Fraternals is more various than that of Fraternals" (94). He therefore tried to estimate the variability within fraternities, both empirically and theoretically. Estimates of co-fraternal and fraternal variability obtained from the special data and the RFF data are shown in table 7.7. As expected, co-fraternal is greater than fraternal variability.

Galton developed four methods for estimating fraternal variability, Q_F, which demonstrate the ingenuity and depth of his statistical think-

Table 7.7. Co-fraternal and Fraternal Variability, Q, Estimated by Different Methods for Two Data Sets

	Special data	RFF data
Co-fraternal, empirical		1.50
Co-fraternal, theoretical		1.50
Fraternal, empirical	1.07	1.38
Fraternal, theoretical	0.98	1.23

Source: After Galton 1889

ing. Two of the methods are empirical estimates from data, and the other two are based on theoretical calculations. I describe one method of each type since the two empirical estimates are numerically similar, and the two theoretical estimates are algebraically identical, though the methods look different.

To obtain an empirical estimate of fraternal variability from his data on brothers, Galton took all the sibships of the same size n, found the median value and the deviations from this median for each sibship, collected these deviations together, and determined their probable error Q from the half-interquartile range. To allow for the bias introduced by using the sample median rather than its true value, he estimated the fraternal variability from the equation

$$Q_F^2 = \frac{n}{n-1} Q^2. \qquad [16]$$

For the special data he found four independent values of 1.01, 1.01, 1.20, and 1.08 for sibships of size 4, 5, 6, and 7, with a mean value of 1.07.

In his theoretical calculation, he concentrated on the second reason why co-fraternals should be more variable than fraternals; this reason is that co-fraternal variability is derived from two sources, the variability of brothers and sisters about the mean value of the fraternity, and variability of the fraternal means in different fraternities with the same mid-parental values. He used the solution to the second problem that he had put to Hamilton Dickson, which was described in chapter 6 and which was posed in terms of bullets aimed at a target. In the present context, the problem is formulated in terms of the model

$$z = x + y, \qquad [17]$$

where x is the deviation of the true sibship mean from the population mean, y is the (independent) deviation of an individual from his true sibship mean, and z is his deviation from the population mean. If z is known, the predicted value of x is βz, where

$$\beta = \frac{Q_x^2}{Q_x^2 + Q_y^2} \qquad [18]$$

is the fraternal regression (see eq.[11] in chapter 6). Hence the fraternal variability Q_y can be calculated from the equation

$$Q_y^2 = (1 - \beta)\, Q_z^2. \qquad [19]$$

This equation is used today for calculating within family variability (Falconer and Mackay 1996, Bulmer 1980).

8

The Law of Ancestral Heredity

A second problem regards the average share contributed to the personal features of the offspring by each ancestor severally. Though one half of every child may be said to be derived from either parent, yet he may receive a heritage from a distant progenitor that neither of his parents possessed as *personal* characteristics. Therefore the child does not on the average receive so much as one half of his *personal* qualities from each parent, but something less than a half. The question I have to solve, in a reasonable and not merely in a statistical way, is, how much less?

Galton, *Natural Inheritance*

Galton's ancestral law followed naturally from his statistical theory of heredity combined with his ideas about the mechanism of inheritance; but he made a number of mistakes in deriving the law because of his lack of mathematical sophistication. Karl Pearson corrected these mistakes, and in so doing developed the theory of multiple regression. After 1900, the ancestral law played an important part in the controversy between the Mendelians and the biometricians.

The ancestral law has often been misunderstood because Galton did not state clearly what it meant. A major source of confusion arose from his failure to distinguish between two interpretations of the law, as a representation of the contributions of different ancestors to an individual and as a regression formula for predicting the value of a trait from ancestral values. He also failed to make clear what the "contribution of an ancestor" meant under the first interpretation. Pearson ignored the idea of ancestral contributions and interpreted the law as a statistical prediction formula, but some of the biometricians, in particular Weldon, took the idea of ancestral contributions seriously. Misunderstanding also stemmed from Pearson's mistaken view that Galton had this law in mind in 1865 (see chapter 4); in fact, he first formulated the law in 1885 as part of his statistical theory of heredity.

In this chapter, which is partly based on Bulmer (1998), I try to clarify some of the confusion and misunderstanding surrounding the ancestral law. I build on previous explications by Swinburne (1965) and Froggatt and Nevin (1971a,b), but I ignore the interpretation of Provine (1971), which I do not understand.

Galton's Formulation of the Ancestral Law

In the Introduction to *Natural Inheritance*, Galton outlined three main questions to be addressed by his statistical theory of heredity. His answers to the first and third questions were discussed in chapter 7. The second question was posed very clearly in the epigraph of this chapter, which was briefly discussed in chapter 4. This passage introduces the distinction between the heritage or genotype and the personal features or phenotype. But his concept of the reason underlying this distinction was very different from the modern concept, since his ideas about the mechanism of inheritance were so different. For Galton, the heritage or stirp consisted of all the hereditary particles, patent and latent, inherited from the parents, while the personal features were determined by the patent or expressed particles. Thus it was natural for him to ask: How many of the patent particles in a particular individual were patent in a parent? How many were last patent in a grandparent? How many were last patent in a great-grandparent? and so on. Consideration of this question led to the law of ancestral heredity.

The ancestral law has a dual interpretation, as a representation of the separate contributions of each ancestor (on average and interpreted as above) to the expressed phenotype of the offspring, and as a prediction formula for predicting the value of a trait from ancestral values. Galton stated the law in its first sense, as a representation of the separate contributions of each ancestor: "The influence, pure and simple, of the mid-parent may be taken as 1/2, of the mid-grandparent 1/4, of the mid-great-grandparent 1/8, and so on. That of the individual parent would therefore be 1/4, of the individual grandparent 1/16, of an individual in the next generation 1/64, and so on" (1885c, 261).

This is a statement of the ancestral law as a representation of the separate contributions of each ancestor, on average, to the expressed phenotype of the offspring. The law can also be interpreted as a prediction formula for predicting the offspring value y_0 given the values of the mid-parent y_1, of the mid-grandparent y_2, and of all the more remote mid-ancestors from the regression formula

$$E(y_0 \mid y_1, y_2, y_3, \cdots) = \tfrac{1}{2}y_1 + \tfrac{1}{4}y_2 + \tfrac{1}{8}y_3 + \cdots . \quad [1]$$

Galton assumed, incorrectly, that these two interpretations of the ancestral law were equivalent.

Galton made several attempts to study inheritance over several generations to verify the law as a prediction formula, from data on human eye-color, and by breeding moths and mice. The most extensive data are on coat color in basset hounds, obtained from pedigree records. There are two coat colors in these hounds, tricolor or T (white, yellow, and black) and non-tricolor or N (white and yellow). He applied the prediction formula to this all-or-nothing character by substituting p_i, the proportion of T hounds in the *i*th generation, for y_i; this is equivalent to coding T as 1 and N as 0. For example, the predicted proportion of tricolor offspring in a pedigree with two T parents ($p_1 = 1$) and three T grandparents ($p_2 = 0.75$) is 0.5 + 0.1875 + 0.1467 = 0.8342; 0.5 ($p_1/2$) is the effect of the parents, 0.1875 ($p_2/4$) is the effect of the grandparents, and 0.1467 is the estimated effect of the probable more remote ancestry, given three T and one N grandparents (made up of 0.0408 for each T and 0.0243 for

Table 8.1. Observed and Predicted Number of Tricolor (T) Basset Hounds According to the Law of Ancestral Heredity

Number of T grandparents	Number of T parents	Number of T offspring observed	Number of T offspring predicted	Total offspring
4	2	106	108	119
3	2	101	99	119
2	2	24	21	28
1	2	8	8	11
4	1	20	24	37
3	1	79	92	158
2	1	36	30	60
1	1	4	3	6
2	0	7	5	18
1	0	2	1	6

Source: After Galton 1897a

each N grandparent from a rather ad hoc argument). The predicted number of T offspring is 119 x 0.8342 = 99, which agrees well with the observed number of 101. The good agreement between observed and predicted numbers shown in the third and fourth columns of table 8.1 encouraged Galton and his followers, particularly Karl Pearson, to believe in the validity of the law.

(Current knowledge suggests that coat color in basset hounds is determined by a pair of alleles at the agouti locus, with N dominant to T [Burns and Fraser 1966, Willis 1989, Robinson 1990]. In this case, all offspring from a mating between two T parents should be T, whereas only 86 percent of them are. The discrepancy may be due to inadequacy of the model or of the data. When he reanalyzed the data, Pearson (1900a) found that the dam's coat color had a greater effect than the sire's in determining the offspring's color, and he suggested that this might be due to the inaccuracy of kennel hands in recording the sire's coat color. The situation is complicated by the fact that the headings "dam" and "sire" are transposed in some of Galton's tables as he subsequently acknowledged [1897c], but this makes no difference to table 8.1 in which parents of both sexes are combined. Inspection of the data shows little evidence of an effect of the number of T grandparents when the number of T parents is fixed, contrary to the ancestral law.)

How did Galton come to formulate a law of such remarkable simplicity and generality? He first derived it by an ingenious, semi-empirical argument in 1885, which he repeated in *Natural Inheritance* in 1889. This argument had several weaknesses due to his failure to understand multiple regression theory. When he returned to the subject in 1897, he replaced this semi-empirical argument by two a priori arguments, which are much less convincing. It is tempting to dismiss the law as a meaningless aberration, but it was in fact a brave attempt to answer a question which was perfectly sensible given Galton's ideas about the mechanism of heredity, but which he lacked the technical expertise to solve. I conclude this section by trying to reconstruct how Galton might have formulated the ancestral law if he had articulated his theory of heredity explicitly and if he had understood the theory of multiple regression that Pearson, stimulated by the ancestral law, developed in 1896.

Galton's Derivation of the Law in 1885

Galton's first attempt to derive the law was given in an appendix to his paper on "Regression towards mediocrity in hereditary stature" (1885c);

this argument was repeated in *Natural Inheritance* (1889). The argument can be rephrased in modern terminology. Suppose that the regression of an individual on all his ancestors is

$$d_0 = \beta_1 d_1 + \beta_2 d_2 + \beta_3 d_3 + \cdots + e, \qquad [2]$$

where d_0 is the deviation from the mean of the individual, d_1 the mid-parental deviation, d_2 the mid-grandparental deviation, and so on, e is the prediction error, and the β's are regression coefficients to be determined. Taking Expected values conditional on the mid-parental deviation yields

$$E(d_0|d_1) = \beta_1 d_1 + \beta_2 E(d_2|d_1) + \beta_3 E(d_3|d_1) + \cdots . \qquad [3]$$

Galton had found that the regression of offspring on mid-parent was 2/3, while the regression of mid-parent on offspring was 1/3. The regression of mid-grandparent on mid-parent must also be 1/3, and he argued by analogy that the regression of mid-great-grandparent on mid-parent was 1/9, and so on. Hence

$$\tfrac{2}{3} d_1 = \beta_1 d_1 + \tfrac{1}{3}\beta_2 d_1 + \tfrac{1}{9}\beta_3 d_1 + \cdots \qquad [4]$$

so that

$$\tfrac{2}{3} = \beta_1 + \tfrac{1}{3}\beta_2 + \tfrac{1}{9}\beta_3 + \cdots . \qquad [5]$$

It was to be expected that $\beta_1 < 2/3$ because the total regression coefficient of 2/3 was influenced not only by the direct effect of the mid-parent but also by the indirect effects of the more remote ancestors (above-average parents were themselves likely to have above-average parents, grandparents of the offspring, and so on); the partial regression coefficient β_1 reflected only the direct influence of the mid-parent.

To evaluate the partial regression coefficients in eq.[5], Galton considered two limiting hypotheses. Under the constant hypothesis, $\beta_i = \beta$ for all i, so that

$$\begin{aligned} \tfrac{2}{3} &= (1 + \tfrac{1}{3} + \tfrac{1}{9} + \cdots)\beta = \tfrac{3}{2}\beta, \text{ or} \\ \beta &= \tfrac{4}{9}. \end{aligned} \qquad [6]$$

Under the geometric decrease hypothesis, $\beta_i = \beta^i$, so that

$$\tfrac{2}{3} = (1 + \tfrac{1}{3}\beta + \tfrac{1}{9}\beta^2 + \cdots)\beta = 3\beta/(3 - \beta), \text{ or}$$
$$\beta = \tfrac{6}{11}. \qquad [7]$$

Galton now remarked that the two estimates of β were nearly the same, and that their average was nearly 1/2, and he concluded that $\beta_1 = 1/2$, $\beta_2 = 1/4$, $\beta_3 = 1/8, \cdots$, which led him to the law of ancestral inheritance:

$$E(d_0 | d_1, d_2, d_3, \cdots) = \tfrac{1}{2}d_1 + \tfrac{1}{4}d_2 + \tfrac{1}{8}d_3 + \cdots . \qquad [8]$$

(Since the mean value is constant and the regression coefficients sum to unity, eqs.[1] and [8], expressed in terms of phenotypic values and of deviations from the mean respectively, are equivalent.)

Unfortunately, there are several problems in this derivation of the law. First, there is no reason why the coefficients of the β's in eq.[4] should follow a geometric series, which assumes that if the regression of parent on offspring is 1/3, then that of grandparent on offspring is 1/9, and so on; we shall see later that these coefficients, starting from a parent-offspring regression of 1/3 under either Galton's model of inheritance or a simple Mendelian model, are not 1/3, 1/9, 1/27, $\cdots$ but 1/3, 1/6, 1/12, $\cdots$. Second, the two results of $\beta = 4/9$ and $\beta = 6/11$ are obtained under different models, so that there is little logic in averaging them to obtain a value of 1/2; furthermore, Galton abandoned the constant hypothesis in favor of the geometric decrease hypothesis as soon as he had obtained the average value of 1/2. Third, there is no particular reason to suppose that the geometric decrease hypothesis, $\beta_i = \beta^i$, should hold for the partial regression coefficients in eq.[2]. Finally, Galton assumed that the regression coefficients in this prediction formula were also measures of the direct effects of the different ancestral generations; we shall show later that this is incorrect.

Nevertheless, Galton thought that he had discovered a general law of heredity, that the direct influence of a single parent was one quarter, that of a single grandparent one-sixteenth, and so on; and he also thought that these contributions could be used to find the regression of offspring values on those of their ancestors. In deriving this law in 1885, he had used data from his investigation on the inheritance of stature, but next year he extended the law to the inheritance of eye color:

> Stature and eye-colour are not only different as qualities, but they are more contrasted in hereditary behaviour than perhaps any other simple qualities. Speaking broadly, parents of different statures transmit a blended heritage to their children, but parents of different eye-colours transmit an

> alternative heritage. If one parent is as much taller than the average of his or her sex as the other parent is shorter, the statures of their children will be distributed in much the same way as those of parents who were both of medium height. But if one parent has a light eye-colour and the other a dark eye-colour, the children will be partly light and partly dark, and not medium eye-coloured like the children of medium eye-coloured parents. The blending in stature is due to its being the aggregate of the quasi-independent inheritances of many separate parts, while eye-colour appears to be much less various in its origin. If then it can be shown, as I shall be able to do, that notwithstanding this two-fold difference between the qualities of stature and eye-colour, the shares of hereditary contribution from the various ancestors are in each case alike, we may with some confidence expect that the law by which those hereditary contributions are governed will be widely, and perhaps even universally, applicable. (Galton 1886c, 402–403)

For a character such as stature with blended inheritance, Galton assumed that an individual received, on average, one quarter of the peculiarity of each parent, one-sixteenth of that of each grandparent, and so on, giving rise to the ancestral law in eqs.[1] or [8]. For a character like eye color with alternative inheritance, he implicitly assumed that an individual inherited uniquely the character of a particular ancestor, having a probability of one quarter of inheriting directly from each of the parents, a probability of one-sixteenth of inheriting directly from each of the grandparents, and so on. (This is what Weldon [1899] thought Galton to mean.) For a dichotomous character, such as light versus dark eyes, this gives rise to the ancestral law in eq.[1], with p_i, the proportion of light-eyed ancestors in the ith generation, substituted for y_i. Galton collected data on about 200 human families in which the eye colors of the children and of their parents and grandparents were known. After making an allowance for the effect of the probable more remote ancestry and for intermediate (hazel) eye color, which he thought was due to blending inheritance, he found that the ancestral law successfully predicted the average frequency of light-eyed children from the numbers of light-eyed parents and grandparents.

Derivation of the Law in 1897

Galton returned to the subject in 1897 with two new, a priori, arguments, when he presented data verifying the validity of the law as a formula for predicting coat color in basset hounds (see table 8.1). In the opening paragraph he wrote:

> In the following memoir the truth will be verified in a particular instance, of a statistical law of heredity that appears to be universally applicable to bisexual descent. I stated it briefly and with hesitation in my book 'Natural Inheritance' because it was then unsupported by sufficient evidence. Its existence was originally suggested by general considerations, and it might, as will be shown, have been inferred from them with considerable assurance. Consequently, as it is now found to hold good in a special case, there are strong grounds for believing it to be a general law of heredity. (Galton 1897a, 401)

After stating the law—in the form that the two parents contribute between them, on the average, one half of the total heritage of the offspring, the four grandparents one quarter, and so on—he produced a new argument to support it:

> It should be noted that nothing in this statistical law contradicts the generally accepted view that the chief, if not the sole, line of descent runs from germ to germ and not from person to person. The person may be accepted on the whole as a fair representative of the germ, and, being so, the statistical laws which apply to the persons would apply to the germs also, though with less precision in individual cases. Now this law is strictly consonant with the observed binary subdivisions of the germ cells, and the concomitant extrusion and loss of one-half of the several contributions from each of the two parents to the germ cell of the offspring. The apparent artificiality of the law ceases on these grounds to afford cause for doubt; its close agreement with physiological phenomena ought to give a prejudice in *favour* of its truth rather than the contrary. (Galton 1897a, 403)

He was appealing to recent discoveries about the reduction division of the germ cells, but the argument is wrong. He appears to have been misled by a completely false analogy between the halving of the number of chromosomes in the reduction division and the coefficients in the ancestral law. The reduction division is the mechanism whereby each parent transmits only half of his or her hereditary particles to the offspring, thus keeping constant the total number of particles in successive generations; Galton had realized the need for the number of particles to be halved since he developed his theory of the stirp in the 1870s (see chapter 4). Thus the reduction division explains how "one half of every child may be said to be derived from either parent" while maintaining the total size of the stirp unchanged. But it does not address the main feature of the ancestral law, that "the child does not on the average receive so much as one half of his *personal* qualities from each parent, but something less than a half." We saw in chapter 5 that Galton had, at the suggestion of

Weldon, studied recent work on the reduction division and the significance of polar bodies in writing his unpublished paper on "The service of sex" in 1896, but had misunderstood it.

Galton added a second a priori argument:

> Again, a wide though limited range of observations assures us that the occupier of each ancestral place *may* contribute something of his own peculiarity, apart from all others, to the heritage of the offspring. Therefore there is such a thing as an average contribution appropriate to each ancestral place, which admits of statistical valuation, however minute it might be. It is also well known that the more remote stages of ancestry contribute considerably less than the nearer ones. Further, it is reasonable to believe that the contributions of parents to children are in the same proportion as those of the grandparents to the parents, of the great-grandparents to the grandparents, and so on; in short, that their total amount is to be expressed by the sum of the terms in an infinite geometric series diminishing to zero. Lastly, it is an essential condition that their total amount should be equal to 1, in order to account for the whole of the heritage. All these conditions are fulfilled by the series of $1/2 + (1/2)^2 + (1/2)^3$ + &c., and by no other. These and the foregoing considerations were referred to when saying that the law might be inferred with considerable assurance *a priori*; consequently, being found true in the particular case about to be stated, there is good reason to accept the law in a general sense. (Galton 1897a, 403)

In other words, he argued that it was plausible to assume the geometric relationship $\beta_i = \beta^i$, and that the terms must sum to unity. Hence $\beta = 1/2$, giving the ancestral law. The argument rests on the unsupported assumption of an exact geometric relationship. The assumption that the terms must sum to unity is justified if they are interpreted as the separate contributions of the different ancestral generations, but seems less obvious if they are interpreted as regression coefficients in a prediction formula. There is a strong hint in the final sentence that Galton had this a priori argument in mind in his original derivation of the law in 1885. It would explain why he pooled the results from the constant model and the geometric decrease model to obtain an average value of one-half, and then immediately abandoned the constant model.

In conclusion, Galton first tried to derive the law in 1885 by a statistical argument based on his recent discovery of regression toward mediocrity. His attempt was flawed, largely due to his ignorance of multiple regression theory. He tried to justify the law in 1897 by two a priori arguments based on little or no foundation. A logical reconstruction of how he should have formulated the ancestral law may help to clarify what he may have had at the back of his mind.

Galton's Law As It Should Have Been

It has been argued in chapter 4 that the following model was implied in *Natural Inheritance*, though it was not explicitly stated. Heredity is mediated through particulate elements, and in bisexual inheritance each parent transmits half of his or her elements to the offspring, thus maintaining the total number of elements in successive generations. These elements may be latent or patent, only the patent elements being expressed, but a latent element may become patent in a subsequent generation. Both patent and latent elements can be transmitted from one generation to the next. The simplest model incorporating these components is that a proportion p of the elements is patent in each individual, that patent and latent elements have the same chance of one-half of being transmitted to the offspring, and that, if transmitted, all elements have the same chance p of becoming patent regardless of their status in the parent and more remote ancestors.

Under this model, the chance that a patent element in a parent is present and patent in a child is $p/2$ (because it has a one-half chance of being present and independently a chance p of being patent); by a similar argument, the chance that a patent element in a grandparent is present and patent in a grandchild is $p/4$; and so on. For a continuous character such as height, these probabilities are also the respective correlations, so that the correlation coefficient r_i between a child and a single ancestor i generations back is

$$r_i = (\tfrac{1}{2})^i p. \qquad [9]$$

(Galton mistakenly assumed that $r_i = (1/3)^i$.) It is shown in the appendix to this chapter that the partial regression coefficients β_i in the linear regression of offspring on all the mid-ancestors given these correlations is

$$\beta_i = (1 - \beta)\beta^{i-1}, \qquad [10]$$

where β is given in eq.[42] of the appendix with $\alpha = p$. Some numerical values are shown in table 8.2.

Table 8.2. Values of β as a Function of p, the Proportion of Patent Elements

p	0.010	0.100	0.500	0.600	0.667	0.900	0.990
β	0.990	0.908	0.586	0.500	0.438	0.169	0.020

The proportion of patent elements, p, can be estimated as $2r_1$, which is the regression of offspring on mid-parent. For his data on stature, Galton estimated that the mid-parental regression was 2/3, so that he should have taken $\beta = 0.438$, giving the multiple regression equation

$$E(d_0 | d_1, d_2, d_3, \cdots) = .56d_1 + .25d_2 + .11d_3 + \cdots . \qquad [11]$$

The ancestral law in eq.[8] is obtained as a special case when $p = 0.6$, so that $\beta = 0.5$. We may call the multiple regression arising from the correlations in eq.[9] the generalized ancestral law; it reduces to the special law considered by Galton when $p = 0.6$.

It is of interest to observe that Galton would have obtained eq.[11] by the method he used in 1885 if he had assumed that $r_i = (2/3)(1/2)^i$ instead of $r_i = (1/3)^i$ for the ancestral correlations and if he had also assumed that the partial regression coefficients were of the modified geometric form $\beta_i = (1 - \beta)\beta^{i-1}$, which sum to unity for all values of β. In this case eq.[5] becomes

$$\begin{aligned} \tfrac{2}{3} &= \beta_1 + \tfrac{1}{3}\beta_2 + \tfrac{1}{6}\beta_3 + \cdots \\ &= (1 - \beta) + \frac{\beta(1 - \beta)}{3(1 - \frac{1}{2}\beta)} . \end{aligned} \qquad [12]$$

This gives a quadratic equation whose relevant solution is $\beta = 0.438$. If he had estimated the regression of offspring on mid-parent as 0.6 rather than 2/3, which was his first thought, he would have found $\beta = 0.5$ by this method, leading to the ancestral law in eq.[8].

The ancestral law can also be interpreted as a representation of the separate contributions of each ancestor to the expressed phenotype of the offspring. Under Galton's model of heredity, the probability that an element patent in the offspring was last patent in an ancestor i generations ago can be calculated as $p(1 - p)^{i-1}$. This seems a reasonable representation of the contribution of ancestors i generations ago:

$$\text{direct contribution from the } i\text{th ancestral generation} = p(1-p)^{i-1}. \quad [13]$$

But it is not the same, as Galton plausibly, but wrongly, assumed, as the corresponding coefficient in the multiple regression coefficient. The series 1/2, 1/4, 1/8, etc., is recovered for the contributions of the parents, grandparents, great-grandparents, and so on, when $p = 1/2$, corresponding to a mid-parental regression of 1/2; but in this case the multiple regression equation is

$$E(d_0 | d_1, d_2, d_3, \cdots) = .41d_1 + .24d_2 + .14d_3 + \cdots . \quad [14]$$

Alternative inheritance can be represented in a Galtonian setting by supposing that each individual has two elements for the character, of which one is randomly chosen to be transmitted to each child. The elements are of two kinds, representing light and dark eye color or tricolor or non-tricolor coat color; one of the two elements in each individual is chosen randomly to become patent and determine the expressed phenotype. This is equivalent to a Mendelian model in which one of the two alleles is chosen at random to be phenotypically expressed. It is also equivalent to the preceding model with $p = 1/2$. The ancestral contributions follow the Galtonian recipe of one quarter from each parent, one-sixteenth from each grandparent, and so on, and it is transparent in this case why this is so. If we trace back from an individual expressed allele to the most recent generation in which a direct ancestor of that allele was expressed, the probabilities of going back to the parental, grandparental, great-grandparental generations and so on are 1/2, 1/4, 1/8, etc., which may be taken as the direct contributions of those generations. But, contrary to what Galton supposed, the regression of offspring on mid-parent under this model is not 2/3 but 1/2; and the multiple regression on mid-parents, mid-grandparents and so on is not given by eq.[8] but by eq.[14] (with p_i substituted for d_i). This model of alternative inheritance can be generalized by supposing that each individual has $2n$ elements for the character, of which one is randomly chosen to be expressed; n of the elements are randomly chosen to be transmitted to the offspring. In this case the proportion of patent elements is $1/2n$.

Thus the law of ancestral heredity has a precise interpretation under the model of heredity that Galton implicitly assumed; in fact, it has two different interpretations: as a prediction equation, and as a representation of the separate contributions of each ancestor to the phenotype of the offspring. The main conclusions are, first, that the law depends on a free parameter p, the proportion of patent elements, which can be estimated from the regression of offspring on mid-parent; and second, that the ancestral contributions are not the same as the coefficients in the prediction equation.

Galton fell into error through his lack of mathematical sophistication, through his failure to formulate a precise model of heredity, and through his failure to distinguish between the two interpretations of the law. He tried to justify the law in several ways, none of them very convincing today, and he may have been misled into seeking too simple (i.e., too reasonable) an answer to the second question that he posed in the Introduction to *Natural Inheritance*: "The child does not on the average receive so much as one half of his *personal* qualities from each parent, but something less than a half. The question I have to solve, in a reasonable and not merely in a statistical way, is, how much less?" (1889, 2).

It is hoped that this logical reconstruction has clarified how he might have formulated his model of heredity more precisely and where it would have led him if he had been able to appreciate the mathematical results that his work led Pearson to develop. In particular, the idea of ancestral contributions has a clear meaning under Galtonian models with latent and patent elements, though it loses this meaning under Mendelian models.

Karl Pearson's Interpretation of the Ancestral Law

We saw in chapter 6 that in 1896 Pearson was stimulated by Galton's statistical work to derive the general theory of multiple regression. In particular, he calculated the joint regression of offspring on the parents and more remote ancestors on Galton's geometric assumption that if the offspring-parent correlation is r, then the offspring–grandparent correlation is r^2, the offspring–great-grandparent correlation r^3, and so on. He showed that, under this assumption, if the values of both parents are known, then the regression coefficients on the grandparents and any more remote ancestors are all zero, and he concluded (his italics): "On the theory with which we are concerned, a knowledge of the ancestry beyond the parents in no way alters our judgement as to the size of organ or degree of characteristic probable in the offspring, nor its variability. *An exceptional father is as likely to have exceptional children if he comes of a mediocre stock as if he comes of an exceptional stock.*... This result seems to me somewhat surprising, but I cannot see how it is to be escaped" (1896, 306).

This conclusion flatly contradicts the law of ancestral heredity in eqs.[1] and [8]. In the multiple regression of the offspring on the mid-parental, mid-grandparental values, and so on, the coefficient for the mid-parental value is $2r$ (that is to say 2/3 if $r = 1/3$, as Galton supposed) and all the other coefficients are zero. In the next year Galton (1897a) published his paper verifying the ancestral law as a formula for predicting coat color in basset hounds. Pearson was convinced by this paper that the ancestral law was basically correct (though there was little evidence of the effect of ancestry beyond the parents), and he concluded that Galton's assumption that the correlation coefficients form the geometric series $r_i = r^i$, with $r = 1/3$, must be wrong. In 1898 he published a paper "On the law of ancestral heredity," with the dedication "A New Year's Greeting to Francis Galton, January 1, 1898." In the Introduction to this paper he wrote:

> In Mr. Galton's 'Natural Inheritance' we find a theory of regression based upon the "mid-parent." This formed the starting point of my own theory of biparental inheritance [1896].... In that memoir I pretty fully developed the theory of multiple correlation as applied to heredity ... but I reached the paradoxical conclusion that "a knowledge of the ancestry beyond the parents in no way alters our judgement as to the size of organ or degree of characteristic probable in the offspring." ... The recent publication of Mr. Galton's paper on Basset hounds has led me back to the subject, because that paper contains facts in obvious contradiction with the principle above cited.... I have come to the conclusion that what I shall in future term *Galton's Law of Ancestral Heredity*, if properly interpreted, reconciles the discrepancies in 'Natural Inheritance' and between it and my memoir of [1896]. It enables us to predict *a priori* the values of all the correlations of heredity, and forms, I venture to think, the fundamental principle of heredity from which all the numerical data of inheritance can in future be deduced, at any rate, to a first approximation. (Pearson 1898, 386)

Pearson therefore set out to find what correlations r_i would lead to the ancestral law. He found that the modified geometric law

$$r_i = 0.6\left(\tfrac{1}{2}\right)^i \qquad [15]$$

would lead to this law, that is to say to a multiple regression on mid-parent, mid-grandparent, and so on in which the partial regression coefficients are 1/2, 1/4, 1/8, and so on. (See the appendix to this chapter. I have ignored Pearson's generalization of the law, allowing for different phenotypic means and variances in different generations.) That is to say, the correlations are not 1/3, 1/9, 1/27, and so on, as Galton supposed, but 0.3, 0.15, 0.075, and so on. Pearson remarked that Galton's first estimate of the regression of offspring on mid-parent for height was 0.6, in exact agreement with the value of $2r_1$ predicted from eq.[9], and that he afterward changed it to 2/3, which was in less good agreement.

Thus Pearson showed that the correlations r_i could be calculated from Galton's ancestral law, and that in particular the parent-offspring correlation should be about 0.3. He wrote: "The confidence I put in the truth of the law is not measured by Mr. Galton's researches on stature or on colour in Basset hounds, however strong evidence these may provide, but rather on the fact that the theory gives *a priori* the correlation between parents and offspring, and that this correlation is practically identical with the value I have myself determined from these and other observations" (1898, 387). (In modern terminology, the law predicts a heritability of 0.6, which is not far from the estimated heritability for many characters.)

Galton's law gives the partial regression coefficients when *all* the mid-ancestors are known. Pearson showed that it could also be used to find the partial regression coefficients when only some of the mid-ancestors are known, from the correlations in eq.[15] and the standard formula in eq.[25] in chapter 6. This would, for example, provide a more accurate way of calculating the predicted values in table 8.1 given the mid-parental and mid-grandparental values. (He also thought in 1898 that the law could be used to calculate correlations between brothers and other collateral relatives. I have not been able to follow his argument, which he later repudiated: "The reader may enquire whether there exists no general relation between … the parental and fraternal correlations. I do not think this is so, unless we make some additional hypothesis" [Pearson 1930c, 25]).

Pearson concluded: "In short if Mr. Galton's law can be firmly established, *it is a complete solution, at any rate to a first approximation, of the whole problem of heredity*" (1898, 393). However, he was worried that the inflexibility of the law made the strength of inheritance an absolute constant. It seemed more likely that the strength of inheritance would vary from one character to another. In particular, he had recently been working with Alice Lee on the inheritance of fertility in man, which seemed to be heritable to some extent, though the correlation between parent and offspring was scarcely one-tenth of that given by Galton's law. He observed that Galton had obtained his law in 1897 by supposing that the partial regression coefficients could be represented by the geometric law $\beta_i = \beta^i$, which together with the requirement that they sum to unity gave the unique solution $\beta_i = (1/2)^i$. Pearson proposed that the law could be generalized by representing these coefficients by the relationship $\beta_i = \gamma\beta^i$, where γ measured the strength of inheritance. If they are constrained to sum to unity, so that $\gamma = (1 - \beta)/\beta$, this leaves one free parameter to be estimated from the data, allowing a greater scope for variety of inheritance in different species and in different characters. (He made an algebraic slip at this point, which he corrected in 1910, on which the appendix to this chapter is based.) This leads to the generalized ancestral law considered above. With the reservation that γ might for some characters differ from unity, he concluded that "the law of ancestral heredity is likely to prove one of the most brilliant of Mr. Galton's discoveries; it is highly probable that it is the simple descriptive statement which brings into a single focus all the complex lines of hereditary influence. If Darwinian evolution be natural selection combined with *heredity*, then the single statement which embraces the whole field of heredity must prove almost as epoch-making to the biologist as the law of gravitation to the astronomer" (1898, 412).

But the two men differed in their interpretation of what the ancestral law meant. To Galton it had a dual interpretation, as a representation of the separate contributions of each ancestor to the offspring, and as a formula for predicting the value of a trait from ancestral values. He assumed wrongly that the two representations should be identical, but he also thought that the first one was primary. To Pearson, only the second interpretation as a prediction formula was important. In discussing the ancestral law he wrote:

> The correlation of a somatic character in a great grandparent, say, and great grandchild is not in any sense a real measure of what the former contributes to the latter, nor is the corresponding multiple regression coefficient such a measure. We are testing what *on the average* we can predict of the somatic characters of the offspring from a knowledge of what the germ-plasms of the "stirp" have produced in the past. In other words the term "contribution of an ancestor" should be interpreted as, or be replaced by, "contribution of the ancestor to the prediction formula." *It is in no sense a physical contribution to the germ-plasms on which the somatic characters of the offspring depend....* The fact, I think, is that Galton's own ideas at this time were obscured by his belief that the ancestors actually did contribute to the heritage; he regarded the incipient structure of the new being to be the

> result of a clash of elements contributed from many ancestral sources, and the resulting building up out of more or less opposing elements of a particulate individual inheritance as the result of chance. (Pearson 1930a, 60–61)

Pearson's interpretation of the ancestral law purely as a statistical prediction formula arose from his phenomenalist philosophy of science, which was influenced by the views of the Austrian physicist Ernst Mach, inventor of the unit of speed relative to the speed of sound and a precursor of the logical positivists who formed the Vienna circle in the 1920s. Pearson wrote an influential book, *The Grammar of Science*, expounding his position, and he summarized his philosophy in the preface to the second edition: "Mechanism is not at the bottom of phenomena, but is only the conceptual shorthand by aid of which [men of science] can briefly describe and [summarize] phenomena.... All science is description and not explanation" (1900b, 5). He was therefore attracted to a model-free statistical theory that would provide an economical description of the facts of heredity.

Pearson included two extra chapters on evolution in the second edition of *The Grammar of Science*, in the second of which he summarized his views on the ancestral law. Assuming constant mean and variance, the law can be written

$$\mathrm{E}(d_0 | d_1, d_2, d_3, \cdots) = \beta_1 d_1 + \beta_2 d_2 + \beta_3 d_3 + \cdots, \qquad [16]$$

where the partial regression coefficients follow the modified geometric law $\beta_i = \gamma\beta^i$. It was natural to require that they sum to unity, so that $\gamma = (1 - \beta)/\beta$. To give this result, the ancestral correlations must be of the form $r_i = \alpha(1/2)^i$, where α can be calculated from β or vice versa (see the appendix). This is the generalized ancestral law. In the important special case when $\alpha = 0.6$, he found that $\beta = 1/2$, so that the generalized law reduced to the special law in eq.[8]. Pearson thought that this held approximately for many continuous characters, such as stature, that showed blending inheritance, though a different parameter would sometimes be needed.

For characters which did not blend but which showed alternative inheritance, such as eye color in man or coat color in dogs, he proposed a different law, the law of reversion, which he expounded in a paper dedicated as "A New Year's Greeting to Francis Galton—January 1, 1900" (Pearson 1900a). He proposed that an individual inherited the character directly from one of his ancestors, with probability β of following either parent, and probabilities $\gamma\alpha, \gamma\alpha^2, \gamma\alpha^3$, and so on of following a grandparent, great-grandparent, great-great-grandparent, and so on. Since the individual follows one of these ancestors, the probabilities sum to unity, so that

$$2\beta + 4\gamma\alpha + 8\gamma\alpha^2 + 16\gamma\alpha^3 + \cdots = 2\beta + 4\gamma\alpha / (1 - 2\alpha) = 1. \qquad [17]$$

Thus the model has two free parameters to be estimated from the data. The parameter β is the probability of inheriting directly from a parent, and the remaining terms represent the probabilities of reversion to more remote ancestors. Galton's model, in which each parent contributed on average one-quarter of the heritage, each grandparent one-sixteenth, and so on, is recovered by setting $\alpha = \beta = \gamma = 1/4$.

Pearson then found equations for the ancestral correlations r_1, r_2, r_3, and so on under this model, taking into account that they depended not only on direct inheritance from a particular ancestor but also on indirect inheritance; for example, the correlation between offspring and father depends not only on direct inheritance from the father (β) but also on inheritance from one of the paternal grandfathers multiplied by their correlation with the father, and so on, so that

$$r_1 = \beta + 2\gamma\alpha r_1 + 4\gamma\alpha^2 r_2 + \cdots . \qquad [18]$$

Hence he showed how the parameters of the model and the ancestral correlations r_3, r_4, r_5, and so on could be calculated under this model given r_1 and r_2. He advocated the measurement of the correlations by his newly-developed method of tetrachoric correlation, assuming that eye colors, for example, can be put in rank order (light blue, dark blue, gray, ... , dark brown, black), and that there is a normally distributed variable underlying this ranking. Pearson fitted this rather strange model to Galton's data on coat color in basset hounds, but he was unable to fit it with meaningful parameter values to other data showing alternative inheritance.

Pearson gave some biological meaning to the parameters in his law of reversion, but when he returned to the subject three years later, he took a completely empirical approach, which was more natural to him:

> In all cases as those of man, horse and dog, where parents of identical character do not produce identical offspring, the theory of statistics shows us that closer prediction may be obtained when we predict from many instead of few relatives.... Attention is therefore properly paid to ancestry in such cases.... The law of ancestral heredity in its most general form is

> not a biological hypothesis at all, it is simply the statement of a fundamental theorem in the statistical theory of multiple correlation applied to a particular type of statistics. If statistics of heredity are themselves sound the results deduced from this theorem will remain true whatever biological theory of heredity be propounded. (Pearson 1903a, 226)

Accordingly, he thought that the proper method to proceed in heredity was by the statistical theory of multiple regression, and that empirical knowledge of the nearer coefficients of correlation would suggest the more distant ones, which would probably be expressible as a geometric series. Hence the multiple regression coefficients could be found, which would also form a geometric series, though he dropped the assumption that these coefficients should sum to unity. He illustrated the methodology from data on eye color in man and coat color in horses. Since these were discrete rather than continuous characters, he measured the correlations by his method of tetrachoric correlation. He found the correlations shown in table 8.3, and he observed that in both cases the geometric series $r_i = 0.75(2/3)_i = 0.5, 0.33, 0.22, 0.15, \ldots$ was a close approximation to the observed series, and he calculated the multiple regression on all the mid-ancestors based on these correlations. The result is given in the appendix; the coefficients do not sum to unity because the correlations do not halve in each generation. (Pearson calculated this regression wrongly both in 1903a and in 1910.)

Pearson's interpretation of the ancestral law as an empirical formula for predicting the phenotypic value of an individual from ancestral values was based on the philosophy that "all science is description and not explanation." His evaluation of the coefficients in the regression formula

Table 8.3. Tetrachoric Correlations

Relationship	Eye color in man	Coat color in horses	$\frac{3}{4}\left(\frac{2}{3}\right)^i$
Parental	0.49	0.52	0.50
Grandparental	0.32	0.30	0.33
Great-grandparental	0.19	0.19	0.22
Great-great-grandparental	—	0.15	0.15

Source: Pearson 1903

evolved through time. In 1896 he accepted Galton's assertion that the correlations between relatives were of the form $r_i = r^i$, with $r = 1/3$, and concluded that the coefficients in the regression formula were $\beta_1 = 2r = 2/3$, $\beta_2 = \beta_3 = \cdots = 0$. In 1898 he was persuaded that Galton's derivation of the coefficients as $\beta_i = \beta^i$, with $\beta = 1/2$, was usually correct, from which he concluded that the correlations must be $r_i = 0.6(1/2)^i$. But he suggested that the coefficients might sometimes be of the more general form $\beta_i = \gamma\beta^i$, though he required that they sum to unity so that $\gamma = (1 - \beta)/\beta$; in this case the correlations must be of the form $r_i = \alpha(1/2)^i$, giving rise to the generalized law. In 1900 he developed an alternative law of reversion for discrete characters, though this model had little success. By 1903 he had dropped the assumption that the regression coefficients in the ancestral law must sum to unity, so that the correlations might be of any modified geometric type $r_i = \alpha\rho^i$. This may be called the overgeneralized law since it has found little application; the factor of one-half arises from the equal contributions of each parent in biparental inheritance.

Throughout the evolution of his thought about the constants in the ancestral law, Pearson remained firm in his conviction that it should be interpreted as an empirical formula that summarized our knowledge of heredity; the only exception was his brief flirtation with the biological idea of reversion in 1900. He concluded that the ancestral law made any theory of the mechanism of heredity, such as Mendel's laws, redundant. As late as 1930 he wrote: "It has often been suggested that the Ancestral Law is contradicted by the discoveries of Mendel and his fellows; it is needless to say that this cannot be the case, for the law does not depend on any mechanism of the germ plasma" (1930c, 4). This statement is disingenuous because it ignores the requirement in the ancestral law that the multiple regression is linear. Pearson himself showed in 1904 that, under a Mendelian model with dominance, the regression of offspring on both parents is nonlinear (see below).

The Ancestral Law and Mendelism

Mendelism was simultaneously "rediscovered" in 1900 by de Vries, Correns, and Tschermak, and was introduced to England in a lecture to the Royal Horticultural Society by the Cambridge zoologist William Bateson (1900b). He began by praising Galton's law as a statistical law of wide application, but he pointed out that there were many cases involving dominance to which it would not apply; for example, the offspring of a polled Angus cow and a shorthorn bull is almost invariably polled (having no horns), despite the presence of horns in all the shorthorn ancestors. All these cases could be explained by Mendel's work, which he described in enthusiastic terms.

The conflict between the Mendelians (headed by Bateson) and the biometricians (headed by Pearson and Weldon) has been ably discussed, by Provine (1971) and MacKenzie (1981) among others. It was largely a clash of personalities and of modes of thought.

Bateson was a strong naturalist but a weak mathematician. He was convinced of the importance of Mendel's ideas, but he distrusted statistical arguments that he did not understand. (E. B. Ford [1980, 340] wrote: "Most of the earlier Darwinian zoologists ... were extremely unmathematical, and they felt Mendelism could be an intrusion of mathematics into biology. I once spent part of an afternoon trying to explain $p^2{:}2pq{:}q^2$ to William Bateson. Not only could he not understand it but he could see no possible point in it.") Pearson, on the other hand, was a weak naturalist but a strong mathematician. His phenomenalist philosophy of science predisposed him to favor model-free statistical theories, and he distrusted Mendelian models.

Weldon and Bateson were close friends at Cambridge in the 1880s, but they went in different directions in the 1890s (see chapters 9 and 10). Bateson became convinced of the evolutionary importance of discontinuous variation, which predisposed him favorably to Mendelism as the source of such variation; Weldon was stimulated by Galton's statistical work to study characters that varied continuously, to which Mendelism was not so clearly relevant, and was thus drawn into the biometrical camp.

The disagreement about evolutionary discontinuity is discussed in chapter 9. I confine myself here to their disagreement over the ancestral law. I have already discussed the dual interpretation of the ancestral law, as a representation of the contributions of each ancestor to the expressed phenotype of the offspring, and as a prediction formula for predicting the value of a trait from ancestral values. Weldon and Pearson, the leading supporters of biometry, took different views of the law. The biologist Weldon took Galton's idea of ancestral contributions seriously, and he was reluctant to accept Mendelism because it did not seem to incorporate this idea; Pearson, as we have seen, treated the law as a prediction formula, and his main concern was to determine whether it could be reconciled with Mendelism as a statistical rule. Two papers by Yule showed how the ancestral law, interpreted as a prediction formula, could be reconciled with Mendelism, but they were ignored by Pearson.

Weldon and Mendelism

Weldon (1902) reviewed Mendel's work in the first volume of the biometrical journal *Biometrika*. He summarized Mendel's theory in two laws: the law of dominance and the law of segregation. He understood the law of dominance to state that, for example, when a variety of garden peas with yellow cotyledons was crossed with a race with green cotyledons, all the offspring in the F_1 generation would be yellow, regardless of the varieties of yellow and green peas used. The law of segregation stated that, when these hybrids were allowed to self-fertilize, three quarters of the progeny would be yellow and one quarter green. He showed that the law of dominance was not universally true for all crosses, and he concluded:

> The fundamental mistake which vitiates all work based upon Mendel's method is the neglect of ancestry, and the attempt to regard the whole effect upon offspring, produced by a particular parent, as due to the existence in the parent of particular structural characters; while the contradictory results obtained by those who have observed the offspring of parents apparently identical in certain characters show clearly enough that not only the parents themselves, but their race, that is their ancestry, must be taken into account before the result of pairing them can be predicted. (Weldon 1902, 252)

Bateson (1902) was incensed by Weldon's review and immediately published a polemical defense of Mendel's principles of heredity. He argued that Mendel had not stated a law of dominance, and that Mendelism allowed the heterozygote to be either identical with one of the homozygotes, or intermediate between them, or different from either of them. He agreed with Weldon that yellow was usually but not universally dominant to green (for example, the green variety Telephone was fully or partially dominant in crosses with yellow varieties of peas), but he argued that this was consistent with Mendelism. (The green gene in Telephone presumably differs from the green gene in other varieties; whether this reflects ancestry is a question of terminology.) Bateson considered the fundamental principle of Mendelism to be what he called the purity of the gamete, which was reflected in the law of segregation. But Weldon was unconvinced. At the time of his early death from pneumonia in 1906, he was engaged in trying to disprove the Mendelian inheritance of coat color in horses from the *General Stud Book of Racehorses* (Provine 1971). His thinking is illustrated by an alternative theory that he had been trying to develop.

In 1905, Weldon had been working with Pearson on a "mathematical theory of determinantal inheritance," that would incorporate Mendelian ideas but at the same time retain the effect of distant ancestry, Weldon providing the biological ideas and Pearson the mathematical analysis. After Weldon's death in 1906, Pearson (1908) wrote up the work from notes Weldon had left. Pearson indicated the lines on which Weldon had been working:

> He was not inclined to accept the theory of unit characters, of allelomorphs and of pure gametes as capable of fully describing even the inheritance of the simplest characteristics. At the same time he recognised the importance of the segregation first pointed out by Mendel in the offspring of hybrids, though even here he was not prepared to make the segregating classes so distinct and so wanting in continuous variation, as some Mendelians have held them to be. He was convinced that in a sufficiently general theory of inheritance some place must be left for a normally arising percentage, however small, of variants, specially related to the distant ancestry. He was thus seeking for a mechanical explanation of latent characters, or in other aspects of reversions and even mutations. (Pearson 1908, 81)

Weldon proposed that an individual might possess $2p$ determinants for a particular character, and that the gametes were formed by choosing p determinants at random, without replacement, from the total $2p$ determinants. When $p = 1$ the model reduces to Mendelism. When $p = 2$, consider a cross between a pure race with genotype *AAAA* and another with genotype *aaaa*. The offspring in the F_1 generation are all *AAaa*, who produce gametes *AA*, *Aa*, and *aa* with frequencies, 1/6, 2/3, and 1/6 by random sampling without replacement. If the F_1 generation is allowed to self-fertilize, as in Mendel's experiments, or to mate at random, the frequencies in the next F_2 generation are shown in table 8.4. If the two genotypes at each end are indistinguishable, so that the phenotypic classes can be collapsed into "preponderance of *A*," "balanced," and "preponderance of *a*," the Mendelian 1:2:1 ratios are recovered.

Pearson wrote: "The peculiar suggestiveness of this result lies in the exact Mendelian properties arising on a simple view of dominance apart from any hypothesis of the pure gamete. There exists a latent [*a*] determinant in the heterogenic chromosomes of a large percentage of the 25 per cent with dominant [*A*] character. This, if judicious cross-breeding

Table 8.4. Predicted Frequencies in the F_2 Generation under Weldon's Model with $p = 2$

Genotype	*AAAA*	*AAAa*	*AAaa*	*Aaaa*	*aaaa*
Frequency	$\frac{1}{36}$	$\frac{2}{9}$	$\frac{1}{2}$	$\frac{2}{9}$	$\frac{1}{36}$
	$\frac{1}{4}$			$\frac{1}{4}$	

were adopted, might be rendered manifest in some, if only a small number, of the grandchildren of the offspring of the hybrids" (1908, 85). Being the mathematician he was, Pearson worked out the consequences of the model for arbitrary values of p, and illustrated them for $p = 3$ and 4. It might be thought that this is irrelevant today because of the inadequacy of the model, but it is in fact equivalent to autopolyploidy with random chromosomal segregation, which is valid for genes very close to the centromere (Li 1955); the results for $p = 1, 2, 3, \ldots$ are applicable to diploidy, tetraploidy, hexaploidy, and so on.

The point of the model was that it avoided the conclusion, so repugnant to Weldon and other ancestrians, that extracted recessives are identical to recessives from the original pure race. If a yellow pea from a pure race is crossed to a green pea from a pure race, all the offspring are yellow, but one quarter of the offspring in the F_2 generation are green. Under Mendelism, these extracted green peas breed true, despite their yellow grandparents. Under Weldon's model with $p = 2$ their yellow ancestry may become apparent from appropriate crosses. It was later shown by Darbishire (1909) that extracted green peas behaved in crosses in exactly the same way as green peas from a pure race, irrespective of the number of yellow peas in their ancestry, so that there was nothing like ancestral contributions in the literal sense envisaged by Weldon.

Pearson and Mendelism

Pearson (1904a) published an important paper, "On a generalised theory of alternative inheritance, with special reference to Mendel's laws." The generalization lay in considering a character determined by an arbitrary number of loci. He acknowledged Weldon as providing the incentive for the memoir, and he wrote that he had not called it a generalized theory of Mendelian heredity because Mendel's original theory had been replaced by what were termed Mendelian Principles. "The fundamental principles propounded by Mendel are given up, and for each case a pure gamete formula of one kind or another is suggested as describing the facts. This formula is then emphasised, modified or discarded, according as it fits well, badly, or not at all with the growing mass of experimental data. It is quite clear that it is impossible while this process is going on to term anything whatever Mendelian as far as theory is concerned" (1904a, 53). He and Weldon were, not entirely without reason, uncomfortable with the proliferation of post hoc Mendelian formulas propounded by Bateson and his followers.

Pearson considered a quantitative character determined by n loci, with alleles A_i and a_i at the ith locus, with A_i dominant to a_i. He supposed that a race homozygous for A, with genotype $A_1A_1.A_2A_2 \ldots A_nA_n$, was crossed with a race homozygous for a, with genotype $a_1a_1.a_2a_2 \ldots a_na_n$; all the offspring in this F_1 generation were perfect heterozygotes, with genotype $A_1a_1.A_2a_2 \ldots A_na_n$. He supposed that there was now random mating in this and all subsequent generations. He showed that the frequencies of the genotypes A_iA_i, A_ia_i and a_ia_i were 1/4, 1/2, and 1/4, respectively and that the frequencies at different loci were statistically independent of each other in all subsequent generations. The first result is a special case of the Hardy-Weinberg law with a gene frequency of 1/2 (see chapter 10). The second result expresses the fact that the population is in linkage (gametic phase) equilibrium, which follows from the fact that all individuals were perfect heterozygotes in the F_1 generation.

Pearson then calculated that the mean number of recessive loci (a_ia_i) among offspring of fathers with s recessive loci was

$$\tfrac{1}{6}n + \tfrac{1}{3}s. \qquad [19]$$

His method of finding this result was rather complicated. A simpler method is this. If a father is recessive at the ith locus, then half his children are recessive at that locus, since all of them receive a_i from him and half of them from their mother, the gene frequency being one half. On the other hand, if the father is not recessive, then he has a chance of one third of being A_iA_i, (in which case none of his children can be recessive) and a chance of two thirds of being A_ia_i (in which case one quarter of his children are recessive, since they have a chance of one half of receiving a_i from him and the same chance of receiving a_i from their mother); thus one-sixth (= 2/3 x 1/4) of his children are recessive. Hence if he has s

recessive and $(n - s)$ nonrecessive loci, the number of recessive loci in his children is on average $s/2 + (n - s)/6$, leading to eq.[19].

Thus a quantitative character determined by the effects of any number of loci showing complete dominance should show a linear parent-offspring regression with a coefficient of 1/3. Pearson remarked that this was numerically identical with the value obtained by Galton in his original investigation on the inheritance of stature. This would seem "a great step forward, as linking up perfectly definite inheritance results with a physiological theory of heredity" (1904a, 64). Unfortunately, more recent investigations had shown that the parental correlation appeared to be markedly greater than 1/3, nearer to .45 to .5 (he was perhaps failing to take assortative mating into account; see table 7.4), and that it also varied from character to character and also between species. He concluded: "We can only say, at present, that a generalised theory of the pure gamete leads to precisely the same general features of regression as have been observed by the biometricians, but it appears numerically too narrow to describe the observed facts" (65).

Pearson also found the regression of offspring on more remote ancestors under this model. The grandparent-offspring regression was 1/6, the great-grandparent–offspring regression 1/12, and so on, decreasing by one half in each generation. "The results show . . . that a general theory of the pure gamete, embracing the simpler forms of the Mendelian principle, leads us directly to a series of ancestral correlations decreasing in a geometrical progression. Thus . . . ancestry is of the utmost importance, and the population follows laws identical in form with those propounded in the biometrical theory on the basis of a linear regression multiple correlation. Only the values of the constants deduced for the law of ancestral heredity from the present theory of the pure gamete . . . are sensibly too small to satisfy the best recent observations on inheritance" (73).

Finally, he considered the simultaneous regression on both parents, supposing that the father has s and the mother t recessive loci. He pointed out that the regression could not be linear, since the appropriate linear prediction with a slope of one-third is $n/12 + (s + t)/3$. (The constant $n/12$ ensures the correct prediction of $n/4$ when $s = t = n/4$, the mean value.) When both parents are pure recessives so that $s = t = n$, this formula predicts $3n/4$ for the average offspring number, but the correct prediction is n, since all the offspring must be pure recessives. Pearson then showed that the expected number of recessive loci in the offspring is

$$\frac{(n + 2s)(n + 2t)}{9n}. \qquad [20]$$

This can be put in a more meaningful form if s and t are expressed as deviations from the mean, $s' = s - n/4$, $t' = t - n/4$; the expected deviation in the offspring is

$$\frac{s' + t'}{3} + \frac{4s't'}{9n}. \qquad [21]$$

When the number of loci is large, the second term becomes negligible, and the formula reduces to the Galtonian linear regression on the mid-parental value with a slope of 2/3, but when n is small there is substantial non-linearity.

Pearson concluded that "in the theory of the pure gamete there is nothing in essential opposition to the broad features of linear regression, skew distribution, the geometric law of ancestral correlation, etc., of the biometric description of inheritance in populations. But it does show that the generalised theory here dealt with is not elastic enough to account for the numerical values of the constants of heredity hitherto observed" (86).

In 1909 Pearson extended his model in two ways. He considered a single locus with alleles A and a, but he allowed arbitrary gene frequencies, p and q, following Hardy (1908), and he calculated the gametic as well as the somatic correlations. To find the somatic correlation of an individual with an ancestor i generations ago, assuming complete dominance of A over a, he first found the average number of dominant (AA or Aa) offspring dependent on the ancestor being dominant or recessive. The difference between these two averages gives the regression of the offspring on the ancestral character, coding dominant as 1 and recessive as 0. This regression is also the correlation coefficient, which he determined as

$$r_i = \left(\tfrac{1}{2}\right)^{i-1} \frac{q}{1 + q}. \qquad [22]$$

If $q = 1/2$, this gave the correlation that he had found previously

$$r_i = \tfrac{1}{3}\left(\tfrac{1}{2}\right)^{i-1}. \qquad [23]$$

As q approached unity, so that there were very few homozygous dominants in the population, the correlations became

$$r_i = \left(\tfrac{1}{2}\right)^i. \quad [24]$$

In this case the correlations were of the form $r_i = r^i$, which as he had shown in 1896 led to an ancestral regression in which the regression coefficients on the grandparents and any more remote ancestors, given the parental values, are all zero; in other words, the ancestry does not matter. But when $q < 1$ the ancestry does matter since the correlations are of the form

$$r_i = \alpha\left(\tfrac{1}{2}\right)^i, \quad [25]$$

with $\alpha < 1$.

Pearson also found the corresponding gametic correlations, by which he meant the correlation between the number of A alleles in offspring and ancestor. Write x_o and x_a for the number of A alleles in the offspring and the ancestor (x = 2, 1, or 0 for AA, Aa or aa). He found the average value of x_o for given values of x_a. Regardless of the gene frequency,

$$\begin{aligned} &\mathrm{E}(x_o|x_a = 2) - \mathrm{E}(x_o|x_a = 1) \\ &\quad = \mathrm{E}(x_o|x_a = 1) - \mathrm{E}(x_o|x_a = 0) \\ &\quad = \left(\tfrac{1}{2}\right)^i. \end{aligned} \quad [26]$$

Thus the regression of offspring on ancestral gametic value is linear with coefficient $(1/2)^i$. This is also the correlation between offspring and ancestral values (compare eq.[24]), so that given the parental genotypes the higher ancestry is of no predictive value.

Pearson drew a wider conclusion:

> There is, however, I venture to think, another aspect of these results which is worthy of fuller consideration. Namely, the fairly close accordance now shown for the first time to exist between the ancestral gametic correlations in a Mendelian population and the observed ancestral somatic correlations suggests that the accordance between gametic and somatic constitutions is for at least certain characters possibly more intimate than is expressed by an absolute law of dominance. If (Aa) were a class, or possibly on a wider determinantal theory a group of several classes [see above], marked by an individual somatic character—not invariably identical with the somatic character of (AA)—there would be little left of contradiction between biometric and Mendelian results as judged by populations sensibly mating at random. It is the unqualified assertion of the principle of dominance which appears at present as the stumbling block. (Pearson 1909, 229)

The gametic correlations are the correlations expected under Mendelism in the absence of dominance, and Pearson was suggesting that a relaxation of the requirement of complete dominance would give a Mendelian theory in agreement with his empirical findings of ancestral correlations.

At this point Pearson could have effected a reconciliation of the ancestral law with Mendelism, if he had thought of incorporating environmental variability into the model. Furthermore, he completely ignored two papers in 1902 and 1906 by George Udny Yule (1871–1951) which had laid the foundations for this reconciliation (though he had discussed Yule's 1902 paper in an appendix to Pearson 1903a). Yule was a demonstrator at University College London under Pearson in the 1890s and maintained his association with the college in the 1900s, but relations between Pearson and Yule had cooled during the latter decade (MacKenzie 1981, Stigler 1999). Yule was a practical statistician with a strong understanding of the underlying cause of regression, and his interpretation of the law is cogent today.

Yule's Reconciliation of the Law with Mendelism

In 1902, Yule first defined what he meant by ancestral heredity: "Supposing the character of the parent known, ... will a knowledge of the grandparent's character enable one to increase the accuracy of the estimate? If the answer to the question be in the affirmative, ... then there is what may be termed a *partial* heredity from grandparent as well as parent, and it is to the existence of such *partial* heredity that statistical writers generally refer when they speak of ancestral heredity" (201). He went on to say that further knowledge of the great-grandparental, great-great-grandparental, and so on, characters would be likely to increase the accuracy of prediction even further, though it would never become exact. He concluded: "This law then, that *the mean character of the offspring can be calculated with the more exactness, the more extensive our knowledge of the corresponding characters of the ancestry,* may be termed the Law of Ancestral Heredity" (202). Yule's statement of the law is very wide, allowing for any values of the partial regression coefficients on the grandparents and more remote ancestors, provided they are not zero, but it depends on the statistical interpretation of the law as a prediction formula.

Yule contrasted this statement of the law with Galton's formulation that "the two parents contribute between them on average one half ... of the total heritage of the offspring, the four grandparents one quarter, ... and so on." Apart from the fixed values of the coefficients, he had difficulty in understanding the conception of the "heritage" and of "ancestry" contributing thereto. (Johannsen [1911, 138] wrote: "Ancestral influence! As to heredity, it is a mystical expression for a fiction. The ancestral influences are the 'ghosts' in genetics, but generally the belief in ghosts is still powerful.") Yule pointed out that "the law of regression . . . and the law of ancestral heredity are both susceptible of a very simple physical explanation on totally different lines," arguing that "the somatic character of an individual is not . . . an absolute guide to the character of the ovum from which he sprang nor, *a fortiori*, to the mean character of the germ cells which he produces" (1902, 205–206). Today we should say that the genotype cannot usually be inferred from the phenotype.

Yule considered two reasons for this fact. The first was environmental variability, from which it followed that "the odds are, therefore, that a given abnormal somatic type is an abnormal development of a mediocre germ cell rather than a mediocre (or subnormal) development of an abnormal cell" (206). It followed that regression would occur. (This extends to cross-fertilizing populations Johannsen's explanation of regression in a mixture of genotypes in a self-fertilizing population.) It also followed that the ancestry could contribute further information about the genotype of the offspring. "If ancestry as well as parents be abnormal it is more probable that the parents are an average development of a really abnormal type of germ cell, and hence more probable that the offspring will follow, and not regress from, the parental type, i.e. we have 'ancestral heredity' " (206).

Another reason the phenotype cannot always be inferred from the genotype under Mendelian inheritance arises from dominance. If *A* is dominant to *a*, so that *AA* and *Aa* possess the dominant and *aa* the recessive character, then an individual with the dominant character may be either *AA* or *Aa*, and information about the phenotypes of his ancestors increases the chance of predicting his genotype correctly and hence of predicting the phenotype of his offspring; that is to say, there is ancestral heredity. On the other hand, the genotype of an individual with the recessive character is known for certain, in the absence of environmental variability, and there is no ancestral heredity.

Yule concluded that "although the theory of ancestral contributions to a heritage implies the law of ancestral heredity, the converse is not true: the law of ancestral heredity need not in any way imply actual physical contributions of the ancestry to the offspring. The ancestry of an individual may serve as guides to the most probable character of his offspring simply because they serve as indices to the character of his germplasm as distinct from his somatic characters" (206).

Yule's second paper in 1906 was a response to Pearson's conclusion (1904a) that Mendelian theory was not sufficiently elastic to cover the observed facts. Yule suggested that the theory could be made more elastic by dropping the requirement of complete dominance. He imagined a length to be made up of a number of distinct segments, the length of each of which is determined by an independent allelomorphic pair. Let each segment take the length l_1, l_2, l_3, according as the corresponding *AA*, *Aa*, or *aa* genotype is present; if the numbers of these genotypes are m_1, m_2, and m_3, then the total length L is:

$$L = m_1 l_1 + m_2 l_2 + m_3 l_3. \qquad [27]$$

With equal gene frequencies, he found the coefficient of correlation between parent and offspring for such a character to be:

$$r = \frac{(l_1 - l_3)^2}{2(l_1 - l_3)^2 + (l_1 - 2l_2 + l_3)^2}. \qquad [28]$$

He observed that if $l_1 = l_2$ or $l_2 = l_3$, there was complete dominance and the correlation was 1/3, as Pearson found. But if l_2 was exactly intermediate between l_1 and l_3, so that $l_2 = (l_1 + l_3)/2$, then $r = 1/2$, which was the maximum value it could take in the absence of assortative mating. He concluded: "There is therefore no difficulty in accounting for a coefficient of 0.5 on the theory of segregation, but such a value probably indicates an absence of the somatic phenomenon of dominance. In the case of characters like stature, span, &c. in man this does not seem very improbable" (1906, 141). He also found the coefficients of correlation with the higher ancestry, and found that they halved in each generation. He observed that in the absence of dominance, the correlations were 1/2, 1/4, 1/8, and so on, which implied a complete absence of ancestral inheritance "in the proper sense of the term," since the partial coefficients of correlation between the offspring and the higher ancestry were zero.

Yule did not say how he had obtained these results, except that he had used methods which were relatively much simpler than those employed by Pearson. He probably calculated r as the ratio of the covariance to the variance. A little algebra shows that the variance at each locus under this model is one sixteenth of the denominator in eq.[28], and that the covariance between parent and offspring is one-sixteenth of the numerator.

Yule also remarked that a complete theory of heredity should take into account the effect of the environment in modifying the soma, "an effect which is hardly likely to be negligible in the case of such a character as stature" (141). He supposed that the standard deviations of the contributions of the three genotypes were s_1, s_2, and s_3, respectively; it can be calculated that the value of r in eq.[28] is reduced by the addition to the denominator of the term

$$4s_1^2 + 8s_2^2 + 4s_3^2, \qquad [29]$$

which is sixteen times the average environmental variance. Unfortunately, Yule made a mistake in calculating the coefficients in eq.[29], giving them as 3, 4, and 3 rather than 4, 8, and 4, but this does not affect his conclusions qualitatively. He remarked that the common ratio of the ancestral correlations remained unaltered at 1/2, so that as far as these correlations went, it was impossible to distinguish between the dominance and nonheritable environmental effects. He concluded: "The case taken is a limited one, but the results are sufficient to show that the theory of the pure gamete, as applied to compound characters, is much more flexible than would appear from Professor Pearson's work, and can hardly be summarily dismissed as inapplicable to cases in which the coefficients of correlation approximate to 0.5" (142).

This remarkable paper foreshadowed the major results of theoretical quantitative genetics in a random-mating population under a Mendelian model without epistasis (Fisher 1918). The phenotypic variance $\mathrm{Var}(y)$ can be decomposed into genetic and environmental components:

$$\mathrm{Var}(y) = \mathrm{Var}(g) + \mathrm{Var}(e). \qquad [30]$$

The genetic variance can in turn be decomposed into additive and dominance components:

$$\mathrm{Var}(g) = \mathrm{Var}(a) + \mathrm{Var}(d). \qquad [31]$$

The additive genetic variance represents the variance due to the additive effects of the two genes at each locus; the dominance variance represents the variance due to departures from additivity, that is, to statistical interaction between the effects of the two genes. For a locus with two equally frequent alleles and with genotypic effects l_1, l_2, and l_3 as postulated by Yule, the genetic components of variance are

$$\mathrm{Var}(a) = \tfrac{1}{8}(l_1 - l_3)^2$$
$$\mathrm{Var}(d) = \tfrac{1}{16}(l_1 - 2l_2 + l_3)^2. \qquad [32]$$

The formulas for unequal gene frequencies and for loci with several alleles are similar but slightly more complicated. The total genetic variance for a phenotypic character, and its additive and dominance components are obtained by summing the contributions from individual loci affecting that character. This idea underlies the modern theory of quantitative genetics based on Mendelian theory, which is discussed further in chapter 10. The resemblance between lineal relatives is determined by the additive genetic variance. The heritability of a character, h^2, is the proportion of the phenotypic variance that is due to the additive genetic variance:

$$h^2 = \frac{\mathrm{Var}(a)}{\mathrm{Var}(y)}. \qquad [33]$$

The correlation between parent and child is $h^2/2$. The correlation coefficient r_i between a child and a single ancestor i generations back is

$$r_i = (\tfrac{1}{2})^i h^2. \qquad [34]$$

This is equivalent to the correlation in eq.[9] if p is replaced by h^2; thus Galton's model is formally equivalent to Mendelism, with latency replaced by dominance and environmental variability. Thus the generalized ancestral law is valid under Mendelism provided that the regression is linear. As Pearson (1904a) showed, the regression of offspring on both parents is not linear when there is dominance, but the departure from linearity becomes negligible as the number of loci increases. This is the basis of the infinitesimal model of quantitative genetics, which shows that a quantitative character determined by a very large number of loci with very small effects has a multivariate normal distribution for any number of relatives (Bulmer 1980). The hope is that this mode provides a reasonable approximation for many quantitative characters. Linear regression formulas based on the correlations in eq.[34], and their extension to the multivariate case, underpin the modern theory of selection on quantitative characters (see chapter 10). Thus the generalized ancestral law, interpreted as a prediction formula, lies at the heart of modern evolutionary theory, though its interpretation as a representation of the ancestral contributions to the offspring has been abandoned along with Galton's theory of heredity.

Despite Yule's successful attempt to reconcile the ancestral law with Mendelism, the dispute between the two sides rumbled on for some time. In 1908 William Bateson's close collaborator, Reginald Punnett, read a paper to the newly formed Royal Society of Medicine on "Mendelism in relation to disease." Dr. H. M. Vernon sent a written contribution to the discussion, hoping that medical men "would not be carried away by the idea that Mendelism was the one all-important question of heredity" (Punnett 1908, 161). He pointed out that the three diseased conditions quoted by Punnett as examples of Mendelism were very rare, and that among normal characters only eye color had been shown to conform to the law. He continued:

> All the other measurable characters in man and cases of hereditary transmission of disease ... had nothing to do with the law, as far as could be seen. The gametes corresponding to such characters were able to blend and form blended zygotes, which gave rise to blended gametes and not segregated alternative ones, as was required by Mendel's law. The vast amount of work done by Galton, Pearson and others on the transmission of such blended characters and their relation to the characters of the parents, grandparents, &c., was practically ignored by the Mendelians. For the average medical man a knowledge of the laws of ancestral heredity, as defined by the workers mentioned, appeared more important than a knowledge of the segregated transmission of a few very rare diseases, interesting as such cases were. (Dr. Vernon in Punnett 1908, 161–162)

He seems to have thought that blended inheritance involved physical fusion of the hereditary material (see chapter 4).

Later in the discussion, Mr. Major Greenwood said that "as a pupil of Karl Pearson, he ought to say something with regard to the Mendelian school ... there being a tendency, on the part of the Mendelians, to sing a Te Deum on the slightest provocation.... It was desirable to know what meaning the Mendelians attached to the word 'proof'. A statistician recently ... said that approximations could be classified into three groups: close approximations, rough approximations, and Mendelian approximations" (Punnett 1908, 162–163).

Punnett remained firm in his reply: "Dr Vernon's letter raised the old controversy between the Mendelians and the biometricians, and dwelt upon the practical value of the law of ancestral heredity as defined by Pearson and others. But it did not seem to him that a law which utterly collapsed before such simple facts as the production of colour from two pure strains of poultry or sweet peas was likely to be of much value to the average medical man or to anybody else" (167–168). This exchange of views illustrates the bitterness of the dispute between the ancestrians and the Mendelians in the first decade of the twentieth century. Punnett was of course correct in asserting that the ancestral law could not be applied to characters determined by a single locus with dominance because it ignored the non-linearity in the biparental regression. The lack of understanding between the Mendelians and the biometricians is partly explained by the fact that the two sides studied different sorts of characters. The Mendelians were interested in single locus characters in which Mendelian segregation could be demonstrated but to which the ancestral law could often not be applied because they showed dominance; the biometricians were interested in continuous, polygenic characters to which the generalized ancestral law provided a satisfactory approximation.

Appendix: The Regression on Mid-Ancestral Values

This appendix gives a method, based on the modified geometric series assumption, for finding the linear multiple regression of an individual on all the mid-ancestral values. The method is based on the treatment by Pearson (1910).

If d_0 is the offspring deviation, d_1 the mid-parental deviation, and so on, the covariance between d_i and d_j is $r_k V/2^m$, where $k = |i-j|$, $m = \min(i, j)$, and V is the variance of a single observation. Premultiplying eq.[25] in chapter 6 by $\mathbf{V}$, we obtain the equations for determining the partial regression coefficients β_i:

$$\begin{aligned} r_1 &= \tfrac{1}{2}\beta_1 + \tfrac{1}{2}r_1\beta_2 + \tfrac{1}{2}r_2\beta_3 + \cdots \\ r_2 &= \tfrac{1}{2}r_1\beta_1 + \tfrac{1}{4}\beta_2 + \tfrac{1}{4}r_1\beta_3 + \cdots \\ r_3 &= \tfrac{1}{2}r_2\beta_1 + \tfrac{1}{4}r_1\beta_2 + \tfrac{1}{8}\beta_3 + \cdots \qquad \text{and so on.} \end{aligned} \tag{35}$$

Suppose that

$$r_i = \alpha\rho^i \tag{36}$$

and that

$$\beta_i = \gamma\beta^i. \tag{37}$$

Substituting these expressions in eq.[35], we find that

$$\alpha\rho = \tfrac{1}{2}\gamma\beta + \tfrac{1}{2}\alpha\rho\gamma\beta^2/(1-\rho\beta)$$

$$\alpha\rho^2 = \tfrac{1}{2}\alpha\rho\gamma\beta + \tfrac{1}{4}\gamma\beta^2 + \tfrac{1}{4}\alpha\rho\gamma\beta^3/(1-\rho\beta) \qquad [38]$$

$$\alpha\rho^3 = \tfrac{1}{2}\alpha\rho^2\gamma\beta + \tfrac{1}{4}\alpha\rho\gamma\beta^2 + \tfrac{1}{8}\gamma\beta^3)(1-\beta) + \tfrac{1}{8}\alpha\rho\gamma\beta^4/(1-\rho\beta)$$

and so on.

Multiply the first equation by $\beta/2$ and take it from the second equation, giving

$$\alpha\rho(\rho - \tfrac{1}{2}\beta) = \tfrac{1}{2}\alpha\rho\gamma\beta. \qquad [39a]$$

Multiply the second equation by $\beta/2$ and take it from the third equation, giving

$$\alpha\rho^2(\rho - \tfrac{1}{2}\beta) = \tfrac{1}{2}\alpha\rho^2\gamma\beta. \qquad [39b]$$

In all cases,

$$(\rho - \tfrac{1}{2}\beta) = \tfrac{1}{2}\gamma\beta. \qquad [40]$$

Making this substitution in the first equation in eq.[38], we obtain

$$\alpha\rho(1-\rho\beta) = (\rho - \tfrac{1}{2}\beta)[1 - \rho\beta(1-\alpha)]. \qquad [41]$$

Eqs.[40] and [41] provide a pair of equations for calculating γ and β from α and ρ, or *vice versa*. In calculating β from the quadratic equation, the root between 0 and 1 should be taken. The fact that this method leads to a unique solution shows that the assumptions in eqs.[36] and [37] are consistent with each other.

Example 1. $\alpha = 1, \rho = 1/3$. These are the correlations assumed by Galton. From eq.[41] we find that $\beta = 0$ and from eq.[40] that $\gamma\beta = 2/3$. Hence $\beta_1 = 2/3$, $\beta_2 = \beta_3 = \ldots = 0$.

Example 2. $\gamma = 1$, $\beta = 1/2$. These are the coefficients in Galton's ancestral law. The solution is $\alpha = 0.6$, $\rho = 0.5$. The correlations required to give Galton's ancestral law are $r_1 = 0.3$, $r_2 = 0.15$, $r_3 = 0.075$, and so on.

Example 3. $\rho = 1/2$. Segregation requires this value either under Galton's implied model of heredity or under Mendelism. It follows from eq.[41] that

$$\beta = 1 + \frac{1 - \sqrt{[1 + 4\alpha(1-\alpha)]}}{2(1-\alpha)}. \qquad [42]$$

From eq.[40],

$$\gamma = (1 - \beta)/\beta, \qquad [43]$$

which ensures that the β_i's sum to unity.

Example 4. $\alpha = 3/4$, $\rho = 2/3$. These are the values obtained by Pearson from empirical evidence. The partial regression coefficients are found from $\gamma = .431$, $\beta = .785$, so that $\beta_i = .431 \times .785^i = \{.338, .265, .208, \cdots\}$. These coefficients sum to 1.57. They differ from the coefficients calculated by Pearson (1910) for these parameters.

9

Discontinuity in Evolution

> *Evolution not by Minute Steps Only.*—The theory of Natural Selection might dispense with a restriction, for which it is difficult to see either the need or the justification, namely, that the course of evolution always proceeds by steps that are severally minute, and that become effective only through accumulation. That the steps *may* be small and that they *must* be small are very different views; it is only to the latter that I object, and only when the indefinite word 'small' is used in the sense of 'barely discernible,' or as small compared with such large sports as are known to have been the origins of new races.
>
> Galton, *Natural Inheritance*

Darwin had taken with him on the *Beagle* the first volume of Charles Lyell's *Principles of Geology* (1830), and had been converted to Lyell's theory that geology could be better explained by gradual changes acting continuously over long periods of time under forces that could be observed today, rather than by occasional catastrophic events. He applied the same gradualist principle to the operation of natural selection. In chap. 4 of *The Origin of Species*, he wrote: "Natural selection can act only by the preservation and accumulation of infinitesimally small inherited modifications, each profitable to the preserved being; and as modern geology has almost banished such views as the excavation of a great valley by a single diluvial wave, so will natural selection, if it be a true principle, banish the belief of the continued creation of new organic beings, or of any great and sudden modification in their structure" (1859, 95–96). He epitomized the gradualness of evolution in the canon *Natura non facit saltum*, "Nature does not make a leap."

Darwin recognized three types of heritable variability: individual differences, the very slight differences such as appear among offspring of the same parents; sports of nature, exhibiting large deviations from the usual type; and modifications due to the physical conditions of life, such as thicker fur of mammals reared in cold conditions, which might be transmitted to their offspring through the inheritance of acquired characters. He accepted that all three types of variability might play a role in bringing about evolutionary change, but he argued that natural selection of small individual differences was by far the most important evolutionary mechanism, and he largely discounted the role of sports.

However, Darwin's gradualist stance was rejected by some of his contemporaries. His great supporter T. H. Huxley, in his otherwise enthusiastic review of *The Origin of Species* in *The Times*, upheld the evolutionary importance of sports: "As a general rule, the extent to which an offspring differs from its parent is slight enough; but, occasionally, the amount of difference is much more strongly marked, and then the divergent offspring receives the name of a Variety.... In each [case] the variety appears to have arisen in full force, and, as it were, *per saltum*.... There seems to be, in many instances, a prepotent influence about a newly-arisen variety which gives it what one may call an unfair advantage over the normal descendants from the same stock" (1860, 35–37). As examples of varieties or sports he gave the Ancon sheep, a variety with short legs valued by farmers because they could not jump over low walls, and hexadactyl fingers and toes in man. He concluded that "Mr Darwin's position might, we think, have been even stronger than it is if he had not embarrassed himself with the aphorism, *'Natura non facit saltum,'* which turns up so often in his pages. We believe, as we have said above, that Nature does make jumps now and then, and a recognition of the fact is of no small importance in disposing of many minor objections to the doctrine of transformation" (77).

Galton's Theory of Discontinuous Evolution

Throughout his life, Huxley believed that the discontinuities in the fossil record revealed real discontinuities in evolution, whereas Darwin attributed them to the imperfection of the geological record. Galton also rejected Darwin's gradualist theory of evolution in favor of a saltationist model. He argued that there was only a small number of stable equilibria in which the system could rest, so that evolution required a leap from one stable equilibrium to another. After his discovery of regression, he reinforced this argument by maintaining that there must be perpetual regression back to the original position when selection was relaxed. He concluded that small deviations from the type (individual differences) would not lead to evolutionary change since they would be unstable and would also be subject to regression back to the central type. Evolution could only occur through the occurrence of large deviations (sports), which would represent a new stable position and a new focus of regression.

Stability of Type

Galton introduced his idea of stability of type in a short section of the final chapter of *Hereditary Genius*:

> I will now explain what I presume ought to be understood, when we speak of the stability of types, and what is the nature of the changes through which one type yields to another. Stability is a word taken from the language of mechanics; it is felt to be an apt word; let us see what the conception of types would be, when applied to mechanical conditions. It is shown by Mr. Darwin, in his great theory of "The Origin of Species," that all forms of organic life are in some sense convertible into one another, for all have, according to his views, sprung from common ancestry, and therefore A and B having both descended from C, the lines of descent might be remounted from A to C, and redescended from C to B. Yet the changes are not by insensible gradations; there are many, but not an infinite number of intermediate links; how is the law of continuity to be satisfied by a series of changes in jerks? The mechanical conception would be that of a rough stone, having, in consequence of its roughness, a vast number of natural facets, on any one of which it might rest in "stable" equilibrium. That is to say, when pushed it would somewhat yield, when pushed much harder it would again yield, but in a less degree; in either case, on the pressure being withdrawn, it would fall back into its first position. But, if by a powerful effort the stone is compelled to overpass the limits of the facet on which it has hitherto found rest, it will tumble over into a new position of stability, whence just the same proceedings must be gone through as before, before it can be dislodged and rolled another step onwards. The various positions of stable equilibrium may be looked upon as so many typical attitudes of the stone, the type being more durable as the limits of its stability are wider. We also see clearly that there is no violation of the law of continuity in the movements of the stone, though it can only repose in certain widely separated positions. (Galton 1869, 368–369)

After giving another metaphor, of people jammed in a great crowd which moves forward by fits and starts through a narrow passage, he concluded: "It is easy to form a general idea of the conditions of stable equilibrium in the organic world, where one element is so correlated with another that there must be an enormous number of unstable combinations for each that is capable of maintaining itself unchanged, generation after generation" (370).

A little later, in discussing his analysis of the effect of mutation (see chapter 7), he entered a caveat: "It must be understood that I am speaking of variations well within the limit of stability of the race, and also that I am not speaking of cases where individuals are selected for some peculiarity, generation after generation. In this event a new element must be allowed for, inasmuch as the average value of [the mutation rate] cannot be constant. In proportion as the deviation from the mean position of stability is increased, the tendency of individual variation may reasonably be expected to lie more strongly towards the mean position than away from it" (372).

This passage is suggestive of Galton's later idea of regression toward the mean. It is also reminiscent of the first objection of Fleeming Jenkin to Darwin's theory, that variability in any species was confined within strict limits, so that transformation of one species into another was impossible:

> A given animal or plant appears to be contained, as it were, within a sphere of variation; one individual lies near one portion of the surface, another individual, of the same species, near another part of the surface; the average animal at the centre. Any individual may produce descendants varying in any direction, but is more likely to produce descendants varying towards the centre of the sphere, and the variations in that direction will be greater than the variations towards the surface. Thus, a set of racers of equal merit indiscriminately breeding will produce more colts and foals of inferior than of superior speed, and the falling off of the degenerate will be greater than the improvement of the select. A set of Clydesdale prize horses would produce more colts and foals of inferior than superior strength. More seedlings of [the rose] 'Senateur Vaisse' will be inferior to him in size and colour than superior. The tendency to revert, admitted by Darwin, is generalized in the simile of the sphere here suggested. (Jenkin 1867, 282)

Jenkin contrasted this idea with Darwin's view "that there is no typical or average animal, no sphere of variation, with centre and limits" (285). His analogy of the sphere of variation elaborates a standard anti-transformationist argument, expressed for example by Lyell: "There are fixed limits beyond which the descendants from common parents can never deviate from a common type" (1832, 23).

It seems that Galton accepted Huxley's interpretation of gaps in the fossil record as evidence of discontinuities in evolution: "The changes are not by insensible gradations; there are many, but not an infinite number of intermediate links; how is the law of continuity to be satisfied by a series of changes in jerks?" (1869, 369). His idea of stability of type was an attempt to explain these discontinuities, and to reconcile the anti-transformationist idea of reversion toward a stable fixed type with the possibility that a large deviation could occur into a different stable type leading to transformation of the species.

But he did not give a satisfactory reason in 1869 that there should be a limited number of stable organic equilibria. The correlation of growth was accepted by Darwin in *The Origin of Species* as placing constraints on evolution, in that morphological characters do not change independently of one another; but it does not entail discontinuous change. Biased mutation was seen by Galton as the consequence, not the cause, of the existence of a stable equilibrium position.

He provided a fuller account in the third chapter of *Natural Inheritance*, on "Organic Stability," from which one can obtain a clearer idea of what he had in mind. In discussing the changes that occur during development, he wrote: "Every particle must have many immediate neighbours.... We may therefore feel assured that the particles which are still unfixed must be affected by very numerous influences acting from all sides and varying with slight changes of place, and that they may occupy many positions of temporary and unsteady equilibrium, and be subject to repeated unsettlement, before they finally assume the positions in which they severally remain at rest" (1889, 21). The "particles" presumably correspond with what he elsewhere calls patent elements, which are destined to take part in development (chapter 4). In this passage he appears to be referring to developmental constraints on the adult form. He lists several examples from everyday life to illustrate that there is only a limited number of stable groupings. For example, there are only a few stable forms of government, such as autocracies, constitutional monarchies, oligarchies, and republics. Even in cookery, there is only a limited number of satisfactory recipes.

In the same chapter, he repeated his metaphor of a rough stone, which he illustrated with a model of a polygonal slab, and as another example of stable mechanical types he discussed "the three kinds of public carriages that characterise the streets of London; namely, omnibuses, hansoms, and four-wheelers.... Attempted improvements in each of them are yearly seen, but none have as yet superseded the old familiar patterns, which cannot, as it thus far appears, be changed with advantage, taking the circumstances of London as they are" (26). A little later, he wrote: "The hansom cab was originally a marvellous novelty. In the language of breeders it was a sudden and remarkable 'sport,' yet the suddenness of its appearance has been no bar to its unchanging hold on popular favour. It is not a monstrous anomaly of incongruous parts, and therefore unstable, but quite the contrary" (30). (The hansom cab, invented by the English architect Joseph Hansom in 1847, was a light two-wheeled covered carriage with the driver's seat elevated behind.) In discussing the infertility of mixed types, he wrote: "It is not difficult to see in a general way why very different types should refuse to coalesce, and it is scarcely possible to explain the reason why, more clearly than by an illustration. Thus a useful blend between a four-wheeler and a hansom would be impossible; it would have to run on three wheels and the half-way position for the driver would be upon its roof" (31). In this example, he seems to equate stability with utility; the analogy in nature would be that selection favors only a small number of discrete types. (Pearson wrote: "The reader may be pardoned a little vexation when he finds such important topics ... only discussed by reference to the analogy of hansom cabs and the impossibility of their useful blend with four-wheelers!" He added in a footnote: "I find my copy of the *Natural Inheritance*, read and annotated forty years ago, defaced by many marginal notes expressing anger at Galton's analogies in this chapter. But these notes were written before I had read and grasped the value of much of the later work in the book" [1930a, 61].) Galton concluded his argument with the epigraph to this chapter.

Galton's idea of stability of type incorporated two ideas that are accepted today. First, there may be developmental constraints on evolution, encapsulated in the definition: "A developmental constraint is a bias on the production of variant phenotypes or a limitation on phenotypic variability by the structure, character, composition, or dynamics of the developmental system" (Maynard Smith et al. 1985, 266). Galton expressed this idea in the passage quoted above, beginning: "Every particle must have many immediate neighbours." Second, many combinations of characters, such as winged pigs, would clearly be maladptive because pigs would be bad at flying; this corresponds to a blend between a four-wheeler and a hansom cab. But he was misled, by pursuing too far his mechanical metaphor of a stone or slab with a number of facets on which it could rest in stable equilibrium, into believing that small deviations from the type (individual differences) would not lead to evolutionary change because they would be unstable and that only a large deviation to a new stable equilibrium (a sport) would lead to permanent change.

Perpetual Regression

When Galton discovered regression, he interpreted the complement of the regression coefficient as a measure of the strength of the stability of the type to small perturbations. In his paper on "Regression towards mediocrity" (1885c), he wrote:

> The stability of a Type, which I should define as "an ideal form towards which the children of those who deviate from it tend to regress," would, I presume, be measured by the strength of its tendency to regress; thus a mean regression from 1 in the mid-parents to 2/3 in the offspring would indicate only half as much stability as if it had been to 1/3.
>
> The limits of deviation beyond which there is no regression, but a new condition of equilibrium is entered into, and a new type comes into existence, have still to be explored. (Galton 1885c, 258)

It was natural for him to conclude (wrongly, as we show below) that there would be perpetual regression back to the type in response to weak selection, so that it could have no permanent effect. He expressed this idea in the Preface to the second edition of *Hereditary Genius*:

> Another topic would have been treated at more length if this book were rewritten—namely, the distinction between variations and sports....
>
> It is impossible briefly to give a full idea, in this place, either of the necessity or of the proof of regression.... Suffice it to say, that the result gives precision to the idea of a typical centre from which individual variations occur in accordance with the law of frequency.... The filial centre falls back further towards mediocrity in a constant proportion to the distance to which the parental centre has deviated from it.... All true variations are (as I maintain) of this kind, and it is in consequence impossible that the natural qualities of a race may be permanently changed through the action of selection upon mere variations. The selection of the most serviceable *variations* cannot even produce any great degree of artificial and temporary improvement, because an equilibrium between deviation and regression will soon be reached, whereby the best of the offspring will cease to be better than their own sires and dams....
>
> The case is quite different in respect to what are technically known as "sports". In these, a new character suddenly makes its appearance in a particular individual, causing him to differ distinctly from his parents and

> from others of his race. Such new characters are also found to be transmitted to descendants. Here there has been a change of typical centre, a new point of departure has somehow come into existence, towards which regression has henceforth to be measured, and consequently a real step forward has been made in the course of evolution. When natural selection favours a particular sport, it works effectively towards the formation of a new species, but the favour that it simultaneously shows to mere variations seems to be thrown away, so far as that end is concerned. (Galton 1892a, xvii–xix)

This echoes what he had written in *Inquiries into Human Faculty* when discussing eugenic improvement of the human race:

> So long as the race remains radically the same, the stringent selection of the best specimens to rear and breed from, can never lead to any permanent result. The attempt to raise the standard of such a race is like the labour of Sisyphus in rolling his stone uphill; let the effort be relaxed for a moment, and the stone will roll back. Whenever a new typical centre appears, it is as though there were a facet upon the lower surface of the stone, on which it is capable of resting without rolling back. It affords a temporary sticking-point in the forward progress of evolution. (Galton 1883, 198)

Galton (1897b) formulated mathematically his view that regression would severely limit the change in a character that could be produced under continued selection, in the absence of a change of type. As an example, he considered a human population with a mean of 68.5 inches and a probable error of 1.7 inches, female height having been corrected to its equivalent male value. He supposed that persons were selected as parents in each generation who lay at the 90th percentile of the distribution in that generation, that is, at 1.9 probable errors above the mean, assuming normality. (The probable error of the standard normal distribution is 0.6745 and the 90th percentile is 1.2816, so that the 90th percentile is 1.2816/0.6745 = 1.90 probable errors above the mean.) In the first generation, the selected parents have a height 1.9 x 1.7 = 3.23 inches above the mean, and their children are on average 2/3 x 3.23 = 2.15 inches above the mean of the race, with a probable error of 1.5 inches since they form a co-fraternity. The selected parents in this generation are 1.9 x 1.5 = 2.85 inches above the mean of the co-fraternity and 2.15 + 2.85 = 5 inches above the mean of the race, and their offspring are 2/3 x 5 = 3.33 inches above the mean of the race, with a probable error of 1.5 inches. In general, if the offspring in the *n*th generation of selection are on average d_n inches above the mean of the race with a probable error of

1.5 inches since they form a co-fraternity, the selected parents in this generation are $d_n + 2.85$ inches above the racial mean, and their offspring are on average $2/3 \times (d_n + 2.85)$ inches above this mean, with a probable error of 1.5 inches. Thus the deviations satisfy the recurrence relation

$$d_{n+1} = \tfrac{2}{3}(d_n + 2.85). \qquad [1]$$

Starting from $d_1 = 2.15$, we can calculate the deviations in successive generations as 2.15, 3.33, 4.12, 4.65, 5.00, etc. The deviations tend to the limiting value $2 \times 2.85 = 5.7$ inches, calculated by equating d_{n+1} to d_n.

Thus regression to the mean limits the effect of selection for height in this example to increasing the average height of the population from 68.5 to 74.2 inches. The argument can be generalized to any intensity of selection from the formula

$$d_{n+1} = w(d_n + tq'), \qquad [2]$$

where w is the coefficient of regression (2/3 in the example), t is the deviation corresponding to the intensity of selection expressed in probable errors (1.9 in the example), and q' is the probable error of a co-fraternity. The deviations tend to the limiting value $[w/(1 - w)]tq'$. Furthermore, any improvement under selection is lost when selection ceases. "There is no stability in a breed under the supposed conditions; but … as soon as selection ceases it will regress to the level of the rest of the population through stages in which the deviation at starting, sinks successively to $w, w^2 \ldots w^n$ of its value" (606). In other words, the deviations regress to zero according to the recurrence

$$d_{n+1} = wd_n \qquad [3]$$

when selection ceases.

The exception to this rule occurs if there is a change of type: "It may, however, happen that a stable form will arise during the process of high breeding, that shall afford a secondary focus of regression, and become the dominant one, if the ancestral qualities that interfere with it be eliminated by sustained isolation and selection. Thus a new variety would, as I conceive, arise.… We can thus understand the facility with which races of butterflies acquire mimetic forms, the severity of selection in their case being very great, while one of their generations occupies only a year" (606).

Selection Experiments

Galton planned two selection experiments to test his theory of perpetual regression. In 1887 he began a selection experiment on the Thorn moth (*Selenia*), in collaboration with an entomologist. This moth produces two generations a year, so that results could be expected fairly quickly. Some preliminary results were published in 1888, but the moth proved unsuitable as an experimental animal, and the study was abandoned in 1889. Nevertheless, it is interesting to note what he had hoped to achieve.

Galton (1887a) had planned a selection experiment on wing length in which three lines would be established: a high line bred from long-winged moths for six generations, a low line bred from short-winged moths, and a control line bred from medium-winged moths (medium-winged meaning with reference to the brood in question, and not with reference to the original brood). After six generations, selection would cease and all three lines would be bred from medium-winged moths. "After the sixth generation or thereabouts has been reached in each of the three lines of descent, it is further desired to proceed conversely, by breeding from medium specimens in each of the three lines, and again from medium specimens of their several broods, and so on until all trace of the [high and low] peculiarities shall have disappeared from their respective descendants" (21).

The design is similar to a modern selection experiment, except that the control line is maintained by mating intermediate rather than random animals. Galton had two aims, to test his law of ancestral heredity from pedigrees going back several generations, and to test his ideas about regression and stability of type. When selection was relaxed he predicted that the selected lines would regress to the original value, as the above quotation shows, but he qualified this prediction at the end of the paper: "The point towards which Regression tends cannot, as the history of Evolution shows, be really fixed. Then the vexed question arises whether it varies slowly or by abrupt changes, coincident with changes of organic equilibrium, which may be transmitted hereditarily; in other words, with small or large changes of type" (28). His planned experiment would test this point.

Galton (1897d) returned to the question of testing his ideas on selection experimentally, in a letter to *The Gardeners' Chronicle* entitled "Retrograde selection." He asked advice from horticulturalists as to suitable plants to use in an experiment to select backwards from a stable variety V, say a dwarf variety, to the original stock R from which it had been obtained. Three lines would be maintained: R and V, each maintained by cross-pollinating average plants of the two types; and the experimental line X in which plants as close as possible to R would be crossed, starting from the largest plants of the dwarf variety V. A record would be kept of the speed of response, and of whether it occurred continuously or in jumps. The letter did not lead to any actual experiments, though several suggestions were made in the *Chronicle*.

The Fallacy of Perpetual Regression

Galton thought it self-evident that the law of perpetual regression followed inevitably from the law of regression which he had discovered in a population not subject to selection; he did not realize that it was in conflict with the law of ancestral heredity. Suppose for simplicity that selection is practiced for one generation by choosing individuals with deviation d from the mean; the parents of these individuals on average have deviation d', their grandparents d'', and so on. By the ancestral law, their offspring on average have deviation

$$d^* = \tfrac{1}{2}d + \tfrac{1}{4}d' + \tfrac{1}{8}d'' + \cdots . \qquad [4]$$

If these offspring mate at random without further selection, the average deviation among their offspring (the grandchildren of the selected individuals) is, by the ancestral law,

$$d^{**} = \tfrac{1}{2}d^* + \tfrac{1}{4}d + \tfrac{1}{8}d' + \cdots = \tfrac{1}{2}d^* + \tfrac{1}{2}d^* = d^* \qquad [5]$$

as before. In other words, no further regression toward the mean has occurred. It is straightforward to show that this result holds good in subsequent generations, and after any number of generations of selection. If the ancestral law is to be taken at face value, Galton was wrong in supposing that there would be perpetual regression toward the mean after selection was relaxed, and that regression would limit the change in a character that could be produced under continued selection.

Karl Pearson made a similar point (1930a, 82–84). He also argued that Galton had made an invaluable contribution in discovering regression but that he did not really understand how it worked:

> Galton I think reached his views owing to a misinterpretation of the statistical phenomena of regression. It was a misfortune that he really did not get beyond the idea of regression in two variates, because to be clear as to the true relation between his "midparental heredity" and his "Law of Ancestral Heredity" a knowledge of multiple regression is essential. But it

> was the greatest good fortune that he got as far as he did; he blazed the track, which many have followed since, and if he left unsolved or half-solved problems, his disciples ought to be grateful that the master has provided the problem as well as the tool, rather than be stern critics of his pioneer work. (Pearson 1930a, 78)

It is not surprising that Galton fell into the error of thinking that there must be perpetual regression toward the mean since this seems to be a natural consequence of the idea of regression (Nesselroade, Stigler, and Baltes 1980; Stigler 1999). To understand the true situation, write r_i for the correlation between an individual and a direct ancestor i generations ago, and β_i for the regression coefficient of an individual on the average values of all his ancestors in that generation. If y_0 is the value of an individual and y_i is the average value of the 2^i ancestors i generations ago, then

$$\beta_i = \mathrm{Cov}(y_0, y_i)/\mathrm{Var}(y_i) = 2^i r_i \,. \qquad [6]$$

Note that β_i is a total, not a partial, regression coefficient.

Suppose that selection is practiced for a single generation, so that the parents from that generation all have a deviation d from the racial mean. The average deviation of their children is $\beta_1 d$, that of their grandchildren $\beta_2 d$, and so on. Two extreme models can be contrasted. If $\beta_1 = \beta_2 = \cdots = \beta$, all the regression occurs in the first generation and none afterward; we may call this instantaneous regression. But if $\beta_1 = \beta$, $\beta_2 = \beta^2, \ldots, \beta_n = \beta^n$, regression to the mean continues indefinitely; we may call this perpetual regression. An intermediate situation may occur in which regression continues for several generations, but does not proceed all the way back to the mean. Which of these situations is likely to hold?

We first note that if, as Galton supposed, the correlations r_i form a geometric series, $r_i = r^i$, then the regression coefficients also form a geometric series, $\beta_i = (2r)^i$. There would then be perpetual regression, but Galton gave no reason why the correlations should form a geometric series. We now consider what would happen under some different genetic models.

Consider first the model of patent and latent elements considered in the last chapter, in which a proportion, p, of the elements is patent in each individual, patent and latent elements are equally likely to be transmitted to the offspring, and all elements have the same chance, p, of being patent regardless of their status in the parent. Under this model, the correlation between an individual and an ancestor i generations back is

$$r_i = p\left(\tfrac{1}{2}\right)^i \qquad [7]$$

so that the regression coefficient on the mid-ancestral value is

$$\beta_i = 2^i r_i = p. \qquad [8]$$

There is instantaneous regression, which ceases after the first generation.

We can extend this model by supposing that patent and latent elements differ in their chance of being patent if they are present in the next generation. Suppose that a patent element has probability π of being patent if it is present in the next generation, while a latent element has probability λ of being patent in this case, with $\lambda \leq \pi$. Then $r_1 = \pi/2$, since the probability that a patent element in a parent is present in the offspring is 1/2, and the probability that it is patent is π. Similarly, $r_2 = [\pi^2 + \lambda(1 - \pi)]/4$; the probability that a patent element in a grandparent is present in the grandchild is 1/4, and it can be patent in the grandchild in two ways, patent in both the intervening parent and the grandchild (with probability π^2), or latent in the intervening parent and patent in the grandchild (with probability $\lambda(1 - \pi)$). Correlations with more remote ancestors can be calculated in a similar way, though the formulas becoming increasingly complicated. But in the extreme case when λ is so small that it can be ignored in the formulas, the correlations take the simple form

$$r_i = \left(\tfrac{1}{2}\pi\right)^i \qquad [9]$$

so that

$$\beta_i = 2^i r_i = \pi^i. \qquad [10]$$

The regression coefficients form a geometric series, so that there is perpetual regression to the mean. In the intermediate case when λ is not negligible but is less than π, regression is intermediate between perpetual and instantaneous; it continues indefinitely, but not all the way back to the starting value.

Instantaneous regression is expected under this model when the chance that an element is patent is independent of its status in previous generations ($\lambda = \pi$), so that the genetic system has no memory, but this ceases to hold when there is some memory of the past ($\lambda < \pi$). A similar

Table 9.1. Genetic Model with Two Loci Showing Epistatic Interaction

	BB	Bb	bb
AA	0	1	2
Aa	1	1	1
aa	2	1	0

conclusion holds under Mendelian models: regression is instantaneous when there are no epistatic interactions between loci, and when environmental effects are independent between generations, but this is no longer true when either of these conditions is violated. As an extreme example of epistatic interaction, in which the effect of an allele at one locus depends on an allele at a second locus, consider the genetic model in table 9.1. Suppose that the population starts in equilibrium under random mating with a gene frequency of 1/2 at both loci, so that the mean value is 1. Selection of individuals with value 2 produces equal numbers of the genotypes Ab/Ab and aB/aB, so that the gene frequencies remain 1/2 at both loci. If the genes are unlinked, the linkage (gametic phase) disequilibrium induced by selection is halved in each generation when selection is relaxed, and the population mean in successive generations is 2, 1.5, 1.25, 1.125, ··· ; in other words, there is perpetual regression to the mean.

"Discontinuity in Evolution" (1894)

In 1894 William Bateson published a massive book, *Materials for the Study of Variation Treated with Especial Regard to Discontinuity in the Origin of Species.* He was more concerned to collect facts than to formulate theories, but he briefly gave his reasons for collecting them in the introduction. He accepted the Darwinian theory that all living things are related to each other through common descent, and that they are more or less adapted to the environments in which they are placed through natural selection. This theory implies that there has been a transition from one form to another in a temporal series. Only the last term in the series is known, but if the whole series were before us, should we find that the transition had been brought about by minute and insensible differences between successive terms in the series (continuous evolution), or should we find distinct and palpable gaps in the series (discontinuity)?

> The Series may be wholly continuous; on the other hand it may be sometimes continuous and sometimes discontinuous; we know however by common knowledge that it is never wholly discontinuous. It may be that through long periods of the Series the differences between each member and its immediate predecessor and successor are impalpable, while at certain moments the series is interrupted by breaches of continuity which divide it into groups, of which the composing members are alike, though the successive groups are unlike.... To decide which of these agrees most nearly with the observed phenomena of Variation is the first question which we hope, by the Study of Variation, to answer. (Bateson 1894, 15)

He continued that the question of the degree of continuity of evolution had never been decided, but that in the absence of such a decision it had been commonly assumed that the process was a continuous one. But this assumption led to a number of difficulties. There was the problem of how selection could lead to the perpetuation of minute variations. There was the problem of the evolution of structures which were only useful when they were nearly perfect. (He had addressed this problem in 1891 in a discussion of the evolution of irregular flowers adapted to pollination by particular insects.) There was the problem that there are discontinuities between existing species, while the diversities of environments to which they are subject are continuous. All these problems could be solved by supposing that species were formed by the selection of large, discontinuous variations. He therefore set out to show that there was in nature a pool of discontinuous, as well as of continuous, variation on which selection might operate. He concluded that the discontinuity of species resulted from the discontinuity of variation.

This book stimulated Galton in the same year (1894) to elaborate his own views in a paper on "Discontinuity in evolution," which may be taken to represent his mature views on the subject. He first defined the terms "race" and "type": "A race is taken to mean a large body of more or less similar and related individuals, who are separated from analogous bodies by the rarity of transitional forms, and not by any sharp boundary.... The type, or typical centre of a race, ... is to be defined as an ideal form, whose qualities are those of the average of all the members of the race" (Galton 1894, 362–363). The question is how "the typical centre of a race may become changed. At a certain period its position was A; at a second and long subsequent period it was B; by what steps did A change into B? Was it necessarily through the accumulation of a long succession of alterations, individually so small as to be almost imperceptible, though large and conspicuous in the aggregate, or could there ever have been abrupt changes?" (363).

He then countered a "specious and it may be a very misleading argument in favour of the steps being always small" (363–364). It is usually possible to find a series of specimens ranging from A to B in small steps, but it does not follow that the course of evolution followed those steps, since they may not have been typical members of the race when they were alive. Two specimens intermediate between A and B may be fundamentally different in nature, one being a variant of A in the direction of B, and the other a variant of B in the direction of A. Though outwardly alike, their inner difference would be shown by their offspring, which would regress toward the A and B types, respectively. "Therefore, although a museum may contain a full series of intermediate forms between A and B it does not in the least follow that the course of development passed through these forms" (364). He had put forward the same argument in *Natural Inheritance*.

He next put forward his argument about stability of types:

> The causes why the A and B races are such definite entities may be various. In the first place each race has a solidarity due to common ancestors and frequent interbreeding. Secondly, it may be thought by some, though not by myself, to have been pruned into permanent shape by the long-continued action of natural selection. But, in addition to these, I have for some years past maintained that a third cause exists more potent than the other two, and sufficient by itself to mould a race, namely that of definite positions of organic stability. The type A is stable, and so is the type B, but intermediate positions are less stable; therefore I conceive the position of maximum stability to be the essential as well as the most potent agent in forming a typical centre, from which the individuals of the race may diverge and towards which their offspring tend on the whole to regress. (Galton 1894, 364)

The third sentence ("It may be thought by some, though not by myself, to have been pruned into permanent shape by the long-continued action of natural selection") is noteworthy. This was a categorical rejection of Darwinian gradualism.

Finally, he gave several examples of what he meant by positions of organic stability. He began by quoting two instances from Darwin's *Variation of Animals and Plants under Domestication* (1868). The peacock occasionally produces birds of the "japanned" or black-shouldered kind, which differ conspicuously from the common kind and which breed true. No less than six named and several unnamed varieties of the peach have suddenly produced several varieties of nectarine, which also breed true. Galton regarded japanned peacocks and nectarines as sports representing new positions of organic stability. (We now know that nectarines are smooth peaches which lack a dominant gene for hairiness [Bateson 1913], and it seems likely that japanned peacocks likewise are recessive mutants.) He quoted Darwin's generalization, that "though the numerous animals and plants which have given rise to sports are known to have been separated from any common progenitor by a vast number of generations, the varieties they have severally yielded are closely analogous," as evidence that "the competing positions of organic stability are well defined and few in number" (365–366).

As another example, he cited his own work on fingerprints (1892b). He concluded that "they fall into three definite and widely different classes, each of which is a true race in the sense in which that word was defined, transitional forms between them being rare and the typical forms being frequent, while the frequency of deviations from the several typical centres in those respects in which measurement could be applied, correspond approximately with the normal law of frequency. I therefore insisted that the continual appearance of these well-marked and very distinct patterns proved the reality of the alleged positions of organic stability" (366–367). Finally, he cited examples of exceptional human ability. He had himself examined the idiot savant Inaudi, who could perform amazing feats of mental arithmetic: "His parents had no such power; his own remarkable gifts were therefore a sport, and let it be remembered that mental sports of this kind, however large, are none the less heritable.... What has been said about this particular gift of mental arithmetic is equally applicable to every other faculty, such as music and scholarship. Can anybody believe that the modern appearance in a family of a great musician is other than a sport? Is it conceivable that Sebastian Bach derived his musical gifts by atavism, and therefore ultimately from an anthropoid ape? The question is too absurd to answer" (367–368).

He concluded his argument: "These briefly are the views that I have put forward in various publications during recent years, but all along I seemed to have spoken to empty air. I never heard nor have I read any criticism of them, and I believed they had passed unheeded and that my opinion was in a minority of one. It was, therefore, with the utmost pleasure that I read Mr. Bateson's work bearing the happy phrase in its title of 'discontinuous variation,' and rich with many original remarks and not a few trenchant expressions" (369).

Galton's ideas about discontinuity were derived from two sources: his belief in positions of organic stability, and his belief that minor variations were subject to perpetual regression but that sports, which represented a new position of stability, were immune from regression. His theory is subject to several weaknesses. It is not clear how he distinguished between a sport and an individual variation. In practice, he regarded any variation that did not blend in inheritance as a sport, but there is lack of clarity in his usage; he regarded Bach as a sport, but as Pearson pointed out, his pedigree in *Hereditary Genius* shows that he was only the ablest member of a very able musical family. It is unclear why sports should be immune from regression, since they presumably carry the latent heredity of their ancestry. Indeed, Galton was inconsistent in supposing that characters such as eye color in man and coat color in basset hounds, which are presumably as much sports as fingerprint forms, followed the law of ancestral heredity, which is based on regression. In any case, the idea of perpetual regression is inconsistent with the law of ancestral heredity. Galton seems to have conflated two distinct explanations of regression, as the expression of hereditary elements that had been latent since they had been expressed in a more or less distant ancestor, and as the expression of a force leading to reversion to a stable equilibrium.

Speciation and Saltation

The subsequent popularity of saltationist models of evolution, and in particular of de Vries's mutation theory, arose from a lacuna in Darwinian theory. This lacuna was Darwin's failure to distinguish clearly between two types of evolution: phyletic evolution, the gradual adaptive change of a single species under natural selection; and speciation, the splitting of a single species into two species which subsequently diverge. Romanes expressed this distinction thus: "The theory of natural selection has been misnamed; it is not, strictly speaking, a theory of the origin of *species*: it is a theory of the origin—or rather of the cumulative development—of adaptations.... If once this important distinction is clearly perceived, the theory in question is released from all the difficulties which we have been considering. For these difficulties have beset the theory only because it has been made to pose as a theory of the origin of species; whereas, in point of fact, it is nothing of the kind" (1886, 345).

The main difficulty in constructing a theory of speciation is that it requires the prevention of intercrossing between the incipient species because, as Darwin wrote: "Intercrossing plays a very important part in nature in keeping the individuals of the same species, or of the same variety, true and uniform in character" (1859, 103). He also remarked that: "In man's methodical selection, a breeder selects for some definite object, and free intercrossing will wholly stop his work" (102). The homogenizing effect of intercrossing is quite different from the swamping effect of Fleeming Jenkin, despite the confusion of these two effects by Romanes (1886) and others, and is equally applicable to all types of inheritance.

Thus speciation requires the development of some sort of isolating mechanism to prevent intercrossing between incipient species. Moritz Wagner (1868) proposed that geographic isolation was necessary for speciation, but this idea was received without enthusiasm. Darwin wrote in the sixth edition of *The Origin of Species*: "Moritz Wagner … has shown that the service rendered by isolation in preventing crosses between newly-formed varieties is probably greater even than I supposed. But … I can by no means agree with this naturalist, that migration and isolation are necessary elements for the formation of new species" (1872, 196).

Romanes (1886) proposed a theory of physiological selection, in which some variants can only propagate with each other because they are sterile when mated with the parent form. (Polyploid plants are an example, though unknown to Romanes.) Galton proposed a variant on Romanes's theory in which sterility is replaced by sexual preference: "It has long seemed to me that the primary characteristic of a variety resides in the fact that the individuals who compose it do not, as a rule, *care to mate* with those who are outside their pale, but form through their own sexual inclinations a caste by themselves. Consequently that each incipient variety is probably rounded off from among the parent stock by means of *peculiarities of sexual instinct*" (1886d, 395). He suggested that the same argument applied to plants, "if we substitute the selective appetites of the insects which carry the pollen, for the selective sexual instincts of animals" (395).

Romanes's theory depends on the development of a postmating isolating mechanism (inviability or infertility of the offspring of a mating between the two varieties), while Galton's theory depends on a premating isolating mechanism (absence of mating between the two varieties). Both theories are models of what is today called sympatric speciation, in which reproductive isolation can develop between a variant and its parent form without geographical isolation. Since one would only expect instantaneous sympatric speciation if the variant differs substantially from the parent, this idea is predisposed to a saltational model of speciation. De Vries's mutation theory was an influential model of this kind since it purported to describe the mechanism whereby new variants arose that were sterile with their parental type.

De Vries and *The Mutation Theory*

De Vries's theory of intracellular pangenesis, published in 1890, was described in chapter 4. According to this theory, the characters of organisms were determined by hereditary particles called pangenes in the cell nucleus. Each kind of pangene, determining a particular character, exists in many copies. There are two kinds of variability, fluctuating variability (*sensu* de Vries) due to different numbers of the individual kinds of pangenes, and variability due to mutation of pangenes; only the latter is significant in evolution.

De Vries spent the rest of his life trying to verify these ideas experimentally and to extend their evolutionary implications. In the 1890s he performed hybridization experiments between many varieties of plants differing in a single character and obtained results similar to those obtained by Mendel in 1865. He published his results in 1900, acknowledging Mendel's priority but claiming to have rediscovered his laws before he had read his paper. Historians of science are skeptical of his claim to have been an independent rediscoverer of Mendelism and segregation (Campbell 1980, Kottler 1979, Meijer 1985). There is little evidence that he appreciated the significance of the Mendelian ratios before 1900, and it is difficult to see how he derived a theory depending on *paired* hereditary units from his theory of intracellular pangenesis, which he reaffirmed at the beginning of his paper and which assumed *multiple* copies of each kind of pangene. It seems likely that he put a Mendelian interpretation on his results after reading Mendel's paper, but it is still unclear how he reconciled Mendelism with intracellular pangenesis. Perhaps he thought that pangenes of the same kind inherited from the same parent behaved as a unit, but he did not explicitly formulate this auxiliary hypothesis. (This hypothesis was in fact adopted by Weldon [c. 1904] in an unpublished manuscript on evolution. Following Galton's theory of the stirp, he assumed that there were multiple hereditary determinants for each character, and he explained Mendel's results as follows: "This union of alternative or as Mr. Bateson says of allelomorphic determinants in the same germ is on Mendel's view something essentially unstable; and when the plant, produced by such a hybrid germ, proceeds in its turn to form reproductive cells, each such cell is supposed by Mendel to contain only one of these two kinds of determinants; the reproductive cells of each sex, produced by a hybrid plant, are therefore supposed to be divided into two equal sets, one set containing only elements which determine greenness of cotyledons, the other only elements which determine yellowness.")

In *The Mutation Theory* de Vries (1901) developed his theory of evolution by macro-mutations. Fluctuating variability, exemplified in the continuous heritable variation of normally distributed characters such as human height studied by Galton and the biometricians, was due to variation in the number of pangenes of a particular kind. It was subject to selection and provided the breeder with material for his improved races, but it was unimportant in evolution because there could not be selection beyond the species range and because it was subject to regression toward the mean. He concluded that "according to the Mutation theory individual variation has nothing to do with the origin of species. This form of variation ... cannot even by the most rigid and sustained selection lead to a genuine overstepping of the limits of the species and still less to the origin of new and constant characters" (1901, 1:4).

The evolution of new species is due to mutation, a change in the quality rather than the quantity of pangenes, leading to a sudden, discontinuous change: "Species have arisen from one another by a discontinuous, as opposed to a continuous, process. Each new unit, forming a fresh step in this process, sharply and completely separates the new form as an independent species from that from which it sprang. The new species appears all at once; it originates from the parent species without any visible preparation, and without any obvious series of transitional forms" (3).

He distinguished between progressive, retrogressive, and degressive mutations. A *progressive* mutation contributes an entirely new character to the complex of hereditary qualities already present through the creation of a new kind of pangene; it leads to evolutionary progress through an increase in differentiation. A *retrogressive* mutation leads to the disappearance of a character through a change in a particular kind of pangene from the active to the inactive, latent condition; examples are white rather than red flowers or smooth rather than hairy leaves. A *degressive* mutation involves a change in one kind of pangene to an alternative active condition; an example is normal or peloric flower shape.

He thought that there was a fundamental distinction between progressive mutations on the one hand and retrogressive and degressive mutations on the other. In the first case the mutant plant has an extra kind of pangene not present in the non-mutant; in the second, both mutants and non-mutants have the same kinds of pangenes. This makes a marked difference in the behavior of a hybrid cross between the mutant and non-mutant forms. In the second case, the cross follows Mendel's laws. In the first case the hybrid from a cross between a progressive mutant and the wild type tends to be inviable or infertile because of the absence of pairing, but the mutants can breed with each other.

De Vries pointed out that hybrid infertility made progressive mutants likely to make new species instantaneously, and he thought that progressive mutation was the primary cause of speciation. He thought that such mutations occurred mainly during limited mutation periods, during which they were quite frequent so that a breeding population of the new species could arise. The mutation theory accepts Darwin's principle of evolution through natural selection, substituting species selection for individual selection: "According to the Darwinian principle, species-forming variability—mutability— does not take place in definite directions.... The struggle for existence chooses from the mutations at its disposal those which are the best adapted at the moment; in this way alone can their survival be explained" (1901, 1:198–199). It could account for apparently useless specific characters, since a new species could survive if it was not at a selective disadvantage.

De Vries thought that mutability was a periodic phenomenon. Species could stay constant for a long period of time and then enter a period of mutability during which speciation occurred, perhaps as a result of a changing environment. He also thought that he had found a plant in a mutable period in the evening primrose, *Oenothera,* which grew wild near Amsterdam and whose study had suggested the mutation theory to him as early as 1886. *Oenothera* gives rise to new varieties which breed true, in accordance with the sudden origin of new species under the mutation theory, and a large part of his book was devoted to these varieties. The mutation theory, reinforced by his studies of the process in *Oenothera,* attracted much attention because it provided a mechanism for instant sympatric speciation; but it fell into disfavor when the true nature of these mutations became known.

It turned out that, because of its peculiar cytogenetics, all the chromosomes inherited from one parent are, with rare exceptions, transmitted as a unit without recombination (Cleland 1972). Furthermore, recessive lethals have built up in frequency in these chromosome complexes. Consider a variety with genotype AB, having inherited the chromosomal complex A from one parent and B from the other. If two plants of this kind are cross-pollinated, one quarter of the offspring are AA, one quarter BB, and one half AB, since these complexes segregate as a unit. The AA and BB offspring die, so that all the surviving offspring are AB; that is to say, the variety breeds true. A very rare crossover may lead to the creation of a new chromosomal complex A*, but the new variety A*B breeds true. Most of de Vries's mutants were of this type, but the situation is so unusual that it has little evolutionary significance. One of his mutants, the giant variety *gigas*, was due to a doubling of the number of chromosomes to produce a tetraploid; this is a not uncommon method of instant speciation in plants.

Punctuated Equilibria

The idea of discontinuity in evolution is under active debate today. In 1972 two paleontologists, Niles Eldredge and Stephen J. Gould, published a paper on "Punctuated equilibria: An alternative to phyletic gradualism." By phyletic gradualism they meant the traditional view that gradual evolutionary change occurred continuously, but that gaps in the fossil record produced an appearance of discontinuity. Eldredge and Gould suggested instead that the gaps in the fossil record represented real discontinuities. They proposed that relatively rapid evolution occurs during speciation, when a single species splits into two, but that little or no evolution occurs between speciation events. One possible reason for species constancy between speciation events (which they called stasis) is that species respond to environmental change by moving to maintain a constant environment (habitat tracking) rather than by evolutionary change. Speciation, when it occurs, happens in small isolated populations at the edge of the species range, which allows rapid evolutionary change. "If new species arise very rapidly in small, peripherally isolated local populations, then the great expectation of insensibly graded fossil sequences is a chimera. A new species does not evolve in the area of its ancestors; it does not arise from the slow transformation of all its forebears. Many breaks in the fossil record are real. The history of life is more adequately represented by a picture of 'punctuated equilibria' than by the notion of phyletic gradualism. The history of evolution is not one of stately unfolding, but a story of homeostatic equilibria, disturbed only 'rarely' (i.e. rather often in the fullness of time) by rapid and episodic events of speciation" (1972, 84).

This theory has inspired a number of studies, using the relatively rare data for which the fossil record is reasonably complete (Kerr 1995, Ridley 1996). Some of them show a pattern with punctuated equilibria: for example, in Caribbean bryozoans between 8 and 3.5 million years ago, most lineages do not change through time, new species appear suddenly without intermediates, and the ancestral species often persists next to its daughter species. Other studies show a pattern of phyletic gradualism, with evolutionary change occurring continuously. More studies are needed before a general conclusion can be drawn, but it seems likely that punctuated equilibria will prove to be an important feature of evolution. But it is not a saltational theory, at least in the moderate version presented by Eldredge (1995) as opposed to the more extreme ideas of Gould (1980). According to Eldredge, the periods of stasis last of the order of five to ten million years, and the periods of rapid change during speciation of the order of five to fifty thousand years; the latter is a long time on an ecological timescale but a short time on an evolutionary timescale. Indeed Darwin may be claimed as a forerunner of this idea, since he wrote in later editions of *The Origin Species*: "Many species when once formed never undergo any further change but become extinct without leaving modified descendants; and the periods, during which species have undergone modification, though long as measured by years, have probably been short in comparison with the periods during which they retained the same form" (1872, 727).

10

Biometry

> The primary object of Biometry is to afford material that shall be exact enough for the discovery of incipient changes in evolution which are too small to be otherwise apparent.
>
> Galton, *Biometrika*

W. F. R. Weldon and Karl Pearson founded *Biometrika* in 1901, with Galton as consulting editor, as "a journal for the statistical study of biological problems." They called the new science biometry, and outlined its scope in the editorial to the first volume (mainly written by Weldon):

> The starting point of Darwin's theory of evolution is precisely the existence of those differences between individual members of a race or species which morphologists for the most part rightly neglect. The first condition necessary, in order that any process of Natural Selection may begin among a race, or species, is the existence of differences among its members; and the first step in an enquiry into the possible effect of a selective process upon any character of a race must be an estimate of the frequency with which individuals, exhibiting any given degree of abnormality with respect to that character, occur....
>
> As it is with the fundamental phenomenon of variation, so it is with heredity and with selection. The statement that certain characters are selectively eliminated from a race can be demonstrated only by showing statistically that the individuals which exhibit that character die earlier, or produce fewer offspring, than their fellows; while the phenomena of inheritance are only by slow degrees being rendered capable of expression in an intelligible form as numerical statements of the relation between

> parent and offspring, based upon statistical examination of large series of cases, are gradually accumulated.
>
> These, and many other problems, involve the collection of statistical data on a large scale.... The recent development of statistical theory, dealing with biological data on the lines suggested by Mr Francis Galton, has rendered it possible to deal with statistical data of very various kinds in a simple and intelligible way, and the results already achieved permit the hope that simple formulae, capable of still wider application, may soon be found. (*Biometrika* **1,** 1–2)

An accompanying foreword by Francis Galton, quoted in the epigraph to this chapter, defined biometry more succinctly.

Thus biometry was conceived as the application of statistical methods to the study of evolution, particularly the evolution of quantitative characters, and was inspired by Galton's *Natural Inheritance.* The early biometricians rejected Galton's view that selection on continuous characters would be ineffective because of perpetual regression toward the mean. They set out to measure variability, heredity, and selection in natural populations by the statistical methods invented by Galton and developed by Pearson. Their main achievement was the demonstration of selection in natural populations. Their main weakness was their reluctance to accept the validity of Mendelian genetics, which hindered their study of heredity. The resolution of the conflict between the biometricians and the Mendelians led to the construction of a biometrical theory of the evolution of quantitative characters based on Mendelian principles which underpins evolutionary biology today.

It seems fitting to conclude this account of Galton's work on heredity and biometry by briefly considering these developments. They answer the questions about quantitative inheritance that he posed, and they are based, directly or indirectly, on the biometrical foundations laid down in *Natural Inheritance* and on Pearson's work on multiple regression that this work inspired.

The Demonstration of Natural Selection

The work of Weldon in England and of H. C. Bumpus in America at the turn of the nineteenth century was the first direct demonstration of the operation of natural selection, and remained so for many years. An important recent development, based in part on the results of Pearson

(1903b), has clarified the operation of natural selection on several correlated variables.

The Career of W. F. R. Weldon

The careers of William Bateson (1861–1926) and of Raphael Weldon (1860–1906) were closely parallel until about 1890, after which they diverged sharply. They went up to Cambridge at nearly the same time, in 1879 and 1878, respectively, where they came under the influence of Francis Balfour. The fact of evolution, though not the mechanism of natural selection, was quickly accepted after 1859, and the attention of biologists turned to phyletic reconstruction, the tracing of evolutionary lineages. A powerful research tool was provided by Haeckel's law that ontogeny recapitulates phylogeny, meaning that ancestral adult stages are repeated in the embryonic stages of their descendants; thus the study of embryology directly revealed evolutionary history. Balfour was the leading exponent of this methodology in England until his early death in 1882 in an Alpine accident. Both Weldon and Bateson began by doing morphological and embryological research in this tradition, and both became disillusioned with it at about the same time, turning their attention in the early 1890s to the study of variation as the most important component of evolution. In this endeavor, they were influenced by Galton's views in *Natural Inheritance*, but in different ways. Whereas Bateson adopted Galton's belief in saltational evolution (see chapter 9), Weldon retained the gradualist Darwinian view of evolution through natural selection on small individual differences, and he set about trying to detect the action of selection on continuous characters by using Galtonian statistical methods.

Weldon's first biometric work, concerning variability in the common shrimp, was done while he was university lecturer at Cambridge. He acknowledged extensive advice from Galton in analyzing the data (1890a; 1892). In 1891, he moved to a professorship at University College London, where he soon joined forces with Karl Pearson, who described their collaboration: "Weldon and the present writer both lectured from 1 to 2, and the lunch table, between 12 and 1, was the scene of many a friendly battle, the time when problems were suggested, solutions brought, and even worked out on the back of the *menu* or by aids of pellets of bread. Weldon, always luminous, full of suggestions, teeming with vigour and apparent health, gave such an impression to the onlookers of the urgency and importance of his topic that he was rarely, if ever, reprimanded for talking 'shop' " (Pearson 1906).

In 1900 Weldon moved to a professorship at Oxford, but he continued to collaborate with Pearson until his death from pneumonia in 1906. Pearson recalled in particular the biometric teas:

> For some years Francis Galton and his niece [his great-niece Eva Biggs] had come within reach of the biometric holiday workers [Pearson and Weldon] for a few weeks in the summer. We were often some distance from each other as at Bibury, Witney and Oxford. The morning was given to work, then the victoria carried our leader and bicycles the remainder of the party to some inn, in a village if possible with a beautiful church, and there was a biometric tea, at which discussion turned not wholly on work. (Pearson 1930a, 277)

Pearson observed of Galton that "at 'biometric teas' his presence was never over-aweing, indeed it was he who generally started and led the mirth" (1930b, 441). Galton wrote to his sister Bessy from Bibury, where he was staying, in August 1904: "We made an expedition to join our two Professors at tea in a country town. I drove, they and their wives and Eva bicycled. Then we talked 'shop' and other things to our hearts' content and separated after two pleasant hours. We did this every Saturday last year" (Pearson 1930b, 527–528).

The Common Shrimp

Weldon's first biometric work concerned variability in the common shrimp, *Crangon vulgaris* (1890b and 1892). He measured four morphological characters in samples from three widely separate places in England (Plymouth, Southport, and Sheerness). He found that each character in each local race was normally distributed, and he calculated their medians and probable errors. He interpreted these results in the light of the discussion of natural selection in *Natural Inheritance*, in which Galton (1889, 119–124) described in words the conclusion that he had obtained mathematically in "Typical laws of heredity." Suppose that there is noroptimal selection toward an optimal value P, and that a local race is normally distributed with mean $M = P$ and with some probable error Q before selection; after selection it is normally distributed with the same mean but with probable error $Q^* < Q$, but in the next generation before selection it is normal with the original mean and probable error. Galton did not discuss in *Natural Inheritance* the problem of how the mean would track a changing optimal value despite perpetual regression, and he later rejected the idea that such tracking would occur (1894, 364). (See chapter 9: "[A race] may be thought by some, but not by myself, to have been pruned into permanent shape by the long-continued action of natural selection.") Weldon interpreted Galton's remarks on natural selection in *Natural Inheritance* to imply that the mean value would track the local optimal value, and he concluded that his results were in agreement with Galton's theory since the distributions were all normal. He argued that the small differences in mean between the geographical races reflected slightly different optimal values in different places, while the differences in probable error—the latter was higher at Plymouth than at the other two locations—might reflect differences in the intensity of selection, since it was likely that variability before selection was similar for all races.

In a second paper (Weldon 1892) he found the correlations between these characters by Galton's graphical method. He concluded that the correlation between a pair of organs was the same in different races, depending only on the organs under consideration, as Galton had suggested to him at the beginning of the inquiry. He concluded that "a large series of such [correlations] would give an altogether new kind of knowledge of the physiological connexion between the various organs of animals; while a study of those relations which remain constant through large groups of species would give an idea, attainable at present in no other way, of the functional correlations between various organs which have led to the establishment of the great sub-divisions of the animal kingdom" (11). One might argue today that these correlations provide information about the constraints on evolution.

In his two papers on shrimps one has the feeling that, in his enthusiasm for the new statistical methods and encouraged by Galton, Weldon was over-interpreting his results.

The Shore Crab

After moving to London in 1891, Weldon turned his attention from the common shrimp to the shore crab, *Carcinus moenas*. In his next paper (Weldon 1893), he had measured eleven morphometric characters in samples of 1000 adult female crabs from Plymouth Sound and from the Bay of Naples. The results were similar to those on shrimps. Correlations between pairs of characters were similar in the two races. The distributions of individual characters were normal, with one exception, that of frontal breadth in the Naples sample. This distribution was asymmetrical, and Weldon thought that the asymmetry might have arisen from the presence of two races of individuals in the sample. This idea was tested by Pearson (1894), who found that the distribution could be fitted as a mixture of two normal distributions with different means and probable errors, representing two races, the first constituting 41 percent and the second 59 percent of the population (fig. 10.1). Weldon was excited by this result, presumably because he thought that it revealed ongoing evolution for frontal breadth in the Neapolitan population, and possibly incipient speciation. He did not realize that this would require a high degree of reproductive isolation between them, for which there was no evidence.

Weldon concluded this paper with a statement of the fundamental principles of biometry:

> It cannot be too strongly urged that the problem of animal evolution is essentially a statistical problem: that before we can properly estimate the changes at present going on in a race or species we must know accurately (a) the percentage of animals which exhibit a given amount of abnormality

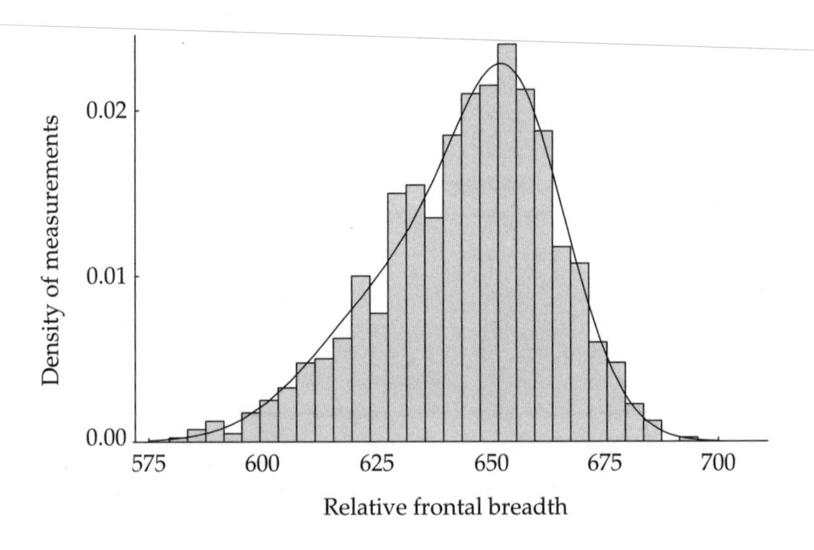

Fig. 10.1. Distribution of relative frontal breadth (1000 x frontal breadth/total length of carapace) in 1000 female crabs from the Bay of Naples (redrawn from Weldon 1893 and Pearson 1894). The curve is the mixture of two normal distributions with means of 630.6 and 654.7, and with probable errors of 12.1 and 8.4, in the proportion of 41:59.

with regard to a particular character; (b) the degree of abnormality of other organs which accompanies a given abnormality of one; (c) the difference between the death rate per cent. in animals of different degrees of abnormality with respect to any organ; (d) the abnormality of offspring in terms of the abnormality of parents and *vice versa*. These are all questions of arithmetic; and when we know the numerical answers to these questions for a number of species we shall know the deviation and the rate of change in these species at the present day—a knowledge which is the only legitimate basis for speculations as to their past history, and future fate. (Weldon 1893, 329)

Weldon had so far addressed only the first two questions raised above. In his next paper he tried to demonstrate selective mortality by comparing the variability of young and adult crabs; selective mortality due to stabilizing selection would be detected from a reduction in variability in adults. He considered two measurements, frontal breadth and right dentary margin, both expressed as thousandths of the carapace length. The results are shown in table 10.1. For frontal breadth, Weldon concluded that Q_F increased at first up to a carapace length of 12.5 mm and then decreased. He explained the initial increase as a confirmation of Darwin's statement, that many variations appear at a late period of development; he provisionally attributed the decrease to stabilizing selection. For the right dentary margin, there was no decrease in adult compared with juvenile values of Q_D, and he concluded that "there is no indication of any destructive agency which acts selectively upon the dentary margin" (1895, 379).

Table 10.1. Half-Interquartile Range for Relative Frontal Breadth (Q_F) and Relative Right Dentary Margin (Q_D) as Function of Carapace Length (C)

C (mm)	Q_F	Q_D
7.5	9.42	8.44
8.5	9.83	8.08
9.5	9.51	9.36
10.5	9.58	8.23
11.5	10.25	8.16
12.5	10.79	8.05
13.5	10.09	8.68
40+ (Adult)	9.96	9.28

Source: Weldon 1895

Weldon supposed that the reduction in Q_F from 10.79 at $C = 12.5$ mm to 9.96 in adults was due to noroptimal selection, such that crabs with deviation d from the mean value at $C = 12.5$ had fitness

$$w(d) = \exp\left(-hd^2\right) \qquad [1]$$

and selective death rate $1 - w(d)$ before reaching adulthood, but he made a numerical error in calculating h. (He gave the correct formula for h, except that he omitted a pair of brackets, but he mistakenly calculated h as 0.015 instead of 0.00034.) In any case, the evidence of selection seems very weak. In particular, no allowance was made for the fact that the shape of the carapace changed during growth in such a way that the mean value of the relative frontal breadth, M_F, declined from about 850 in the smallest crabs to 605 in adults. The ratio 100 Q_F/M_F, similar to the coefficient of variation, shows no evidence of a decrease in adult crabs; it is 1.11 in the smallest crabs, 1.38 when $C = 12.5$, and 1.65 in adult crabs. The small decrease in Q_F with increasing size in table 10.1 can be attributed to the change in shape.

Weldon continued to work on crabs and presented strong evidence of directional selection on frontal breadth in his presidential address to the Zoological Section of the British Association in 1898. He first withdrew his previous hypothesis of stabilizing selection since it "neglected several important facts which I now know, and was open to other objections" (1898, 897). He began afresh by showing that there had been a marked decrease in frontal breadth in crabs of all sizes in Plymouth Sound between 1893 and 1898 (see fig. 10.2), and he suggested that this was due to selection. A large breakwater had been built across the entrance to Plymouth Sound, as a result of which increasing amounts of silt had been deposited in the sound from the rivers draining into it and from the expanding towns around it. This physical change had been accompanied by the disappearance of animals that used to live in the sound, but which were at that time found only outside the area affected by the breakwater. Weldon suggested that crabs with a narrow frontal breadth could filter muddy water more efficiently and were therefore more likely to survive under these conditions. To test this hypothesis he did a number of experiments of keeping crabs in water containing fine mud in suspension. At the end of an experiment, the dead were separated from the living, both were measured, and relative frontal breadth was expressed as a deviation from the value predicted for its length. In every case in which the experiment was done with china clay as fine as that brought

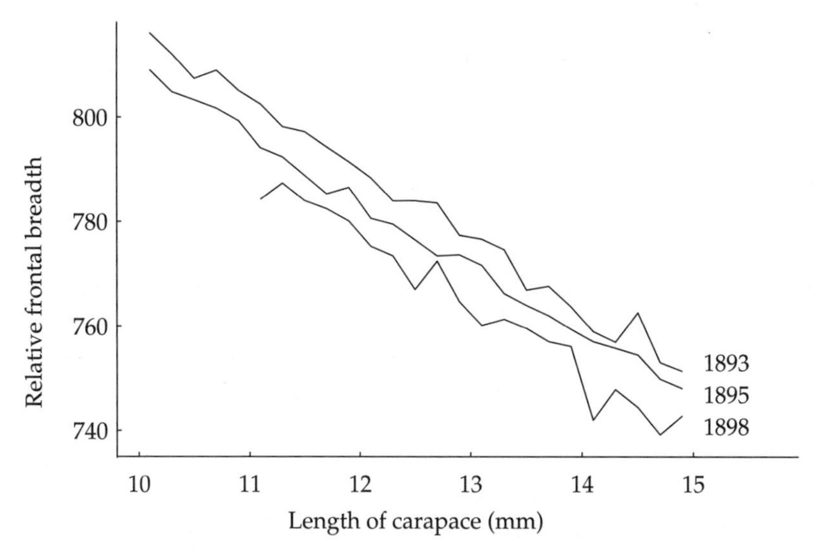

Fig. 10.2. Secular decrease in relative frontal breadth in crabs from Plymouth Sound (data from Weldon 1898)

down by the rivers, the dead crabs were on average broader than the survivors, but when the experiment was repeated with coarser clay, the death rate was smaller and was not selective. Weldon gave numerical results only for one experiment with fine mud, in which there were 248 crabs, of which 154 died. The difference between the mean frontal breadth deviations of the dead crabs and the survivors was five thousandths (which is statistically significant).

Weldon concluded that "we have here a case of Natural Selection acting with great rapidity, because of the rapidity with which the conditions of life are changing" (900), but this conclusion was criticized:

> Weldon's observations may be completely explained by variations in the amount or rate of growth. The difference in different years would be at once explained if the amount of change in frontal breadth was constant for each molt, while the amount of growth was variable. The fact is that in 1893 crabs of a given frontal breadth were larger than in 1895 and 1898; and I have shown that the summer of 1893 was exceptionally fine and warm. Either the warmth alone, or warmth and food together, very probably made the crabs grow more in that year for the same number of molts. On this view the broad-fronted crabs died in the experiments with clay and mud because they were younger and weaker. (Cunningham 1899, quoted in Kellogg 1907, 161)

This criticism by the British Lamarckian J. T. Cunningham (Bowler 1983) would today be regarded as tendentious, but it illustrates the difficulty of demonstrating natural selection conclusively.

Stabilizing Selection in Snails

Weldon returned to the question of stabilizing selection in 1901 in a paper on the snail *Clausilia laminata*. Its shell is essentially a tube coiled round an axis, which is laid down from its narrow end by the growing animal. Thus the upper whorls of an adult shell form a permanent record of the condition of the young shell. Weldon collected 100 adult and 100 young snails from a beech wood in Germany and ground them down to reveal a flat longitudinal section through the center of the shell. For each snail he measured the distance from its tip to the point where two successive whorls met and the angular distance of that point from a fixed reference point, and he repeated these measurements for all meeting points.

Comparison of the mean and the standard deviation of these distances at six corresponding points present in both age groups showed that there was no difference in the mean between young and adult shells, but that the standard deviation was consistently higher in young shells. Weldon concluded that there had been stabilizing selection on the shell, with selection against young shells that deviated from the mean value in either direction:

Table 10.2. Standard Deviation of Radial Length at Different Angular Distances in 100 Adult and 100 Young Snails

Angular distance in right angles	Standard deviation of radial length		F ratio
	Adult snails	Young snails	
–8 to –10	0.121	0.141	1.36
–6 to –8	0.134	0.153	1.30
–4 to –6	0.134	0.155	1.34
–2 to –4	0.146	0.169	1.34
0 to –2	0.156	0.170	1.19
0 to 2	0.170	0.174	1.05

Source: Weldon 1901

> At every point ... the variability of young shells is greater than that of adults.... The chances against an apparent excess of variability in a sample of young shells, so large as that recorded in the table, are very great, unless we admit that there is a real difference in variability between the newly-formed whorls of growing shells and the corresponding whorls of adults; and the necessary consequence of such an admission is that the variability of these newly-formed whorls is reduced after their formation by a process which destroys individuals with abnormal shells more rapidly than others, so that a process of 'periodic selection' [stabilizing selection] occurs. (Weldon 1901, 121)

This was an elegant attempt to demonstrate stabilizing selection, but examination of his evidence for the reduction in variability suggests that he exaggerated its statistical significance. Table 10.2 shows the estimated standard deviations of radial length (y) at different angular distances (x), together with the F ratios that I calculated. (F is the ratio of the variance of young to that of adult snails.) The upper 10 percent point of the F distribution with 99 degrees of freedom in both numerator and denominator is 1.33, so that three of these six values are just significant and one is almost significant at the 10 percent level. If these were independent tests, this would provide strong evidence of a decrease in the variance in adult snails, but they are not independent since all the F ratios are based on repeated measurements on the same set of snails. Since measurements on the same snail at different angular distances are likely to be highly correlated, the F ratios are also likely to be highly correlated. Thus the evidence of stabilizing selection from these data is weak. Furthermore, Weldon (1903) failed to confirm these findings in the closely related species *Clausilia itala* from Italy.

Bumpus's Sparrows

The American ornithologist Hermon C. Bumpus had a particular interest in evolution, and he recognized that he had been given a rare opportunity to study the effect of natural selection: "A possible instance of the operation of natural selection, through the process of the elimination of the unfit, was brought to our notice on February 1 of the present year (1898), when, after an uncommonly severe storm of snow, rain, and sleet, a number of English sparrows were brought to the Anatomical Laboratory of Brown University. Seventy-two of these birds revived; sixty-four perished; and it is the purpose of this lecture to show that the birds which perished, perished not through accident, but because they were physically disqualified, and that the birds which survived, survived because they possessed certain physical characters" (1898, 209).

Bumpus measured nine characters on each bird (see table 10.3) and compared the mean values of the survivors and the dead. He concluded that they differed in several characters: "I think these tests prove that there are fundamental differences between the birds which survived and those which perished. While the former are shorter and weigh less (*i.e.,* are of smaller body), they have longer wing bones, longer legs, longer sternums, and greater brain capacity. These characters are in accordance with our ideas of physical fitness; their defective development is evidently a mark of inferiority, and we are justified in concluding that the birds so handicapped failed to pass one of Nature's rigorous tests and perished" (213–214).

Bumpus then considered a second kind of selection, remarking that "the fact that the birds which perished had in the *average* longer and larger bodies, and shorter head, wing, and leg bones, does not tell all the story of selective elimination" (214). He observed that the most extreme

Table 10.3. Statistical Analysis of Bumpus's Data on Sparrows

	Males		Females	
Character	t test	F ratio	t test	F ratio
Length	–4.48***	1.05	–0.99	1.36
Wing span	–0.15	1.34	–0.39	1.86
Weight	–0.82	1.48	–1.74	2.37*
Head length	+0.99	0.89	–0.20	1.37
Humerus length	+1.79	1.45	+0.36	2.51*
Femur length	+1.45	0.89	+0.69	1.97
Tibio-tarsus length	+1.04	0.82	+1.07	2.59*
Head width	+0.70	1.08	–0.32	1.93
Sternum length	+1.42	0.82	–0.12	2.30*
First principal component	+0.24	1.15	–0.18	2.60*
Second principal component	–6.79***	1.14	–2.29*	1.04
Degrees of freedom	85	35, 50	47	27, 20

* Significant at 5% level
*** Significant at 0.1% level

Note: t test = (mean of survivors – mean of non-survivors)/standard error; F ratio = variance of non-survivors/variance of survivors.

birds, in either direction, tended to be found among the dead; for example, both the longest bird and the shortest bird were found in this group. To quantify this idea, he calculated the mean deviation in the two groups, and found that it was larger among the dead for eight out of the nine characters. He concluded: "The process of selective elimination is most severe with extremely variable individuals, no matter in what direction the variations may occur. It is quite as dangerous to be conspicuously above a certain standard of organic excellence as it is to be conspicuously below the standard. It is the *type* that nature favors" (219). Bumpus's work is usually cited as an example of this kind of selection—stabilizing selection—which he regarded as selection in favor of the type.

Bumpus did not do any tests of significance (they had not been invented in 1898), but he gave the complete data, so that a statistical analysis can be performed (see table 10.3). The t test is a test for directional selection. The only significant result among the nine characters is a highly significant reduction in the length of male survivors. The F ratio is a test for stabilizing selection. There is no evidence of this type of selection in males, but there are several significant values in females. It is difficult to find an overall significance level because of the high correlations between all the characters. A common statistical approach is to do a principal components analysis, in which the components are uncorrelated with each other. The results for the first two components are shown near the bottom of the table. All the loadings of the first component are positive, so that it represents overall size, and accounts for about 60 percent of the variance. There is no evidence of directional selection on this component, but there is evidence of stabilizing selection in females (significant at the 5 percent level) but not in males. The loadings of the second component differ slightly in the two sexes, but they have in common positive values for length and weight and negative values for femur length and tibio-tarsus length; it may be interpreted as an index of size in relation to leg length, and accounts for about 10 per cent of the variance. There is very strong evidence of directional selection for this component in males, and significant evidence of selection in the same direction in females; there is no evidence of stabilizing selection. In conclusion, there is strong evidence of directional selection, particularly in males, against birds with long bodies and short legs; there is some evidence of stabilizing selection for overall size, but it is weakened by the fact that it can be demonstrated only in one sex. A similar conclusion was reached by Grant (1972) and by Lande and Arnold (1983), though the latter authors obtained and interpreted the second principal component in a slightly different way.

Statistical analysis of their data shows that both Weldon and Bumpus had suggestive evidence of directional selection but that their evidence for stabilizing selection was much weaker. This may be because it is much more difficult to demonstrate a significant difference between two variances than between two means. For example, the mean length of the surviving male sparrows was 159 mm compared with 162 mm for the dead male sparrows. The difference of 3 mm, which is about 2 percent of the mean value, is highly significant. On the other hand, the upper 5 percent point of the corresponding F distribution is 1.67, so that, to be significant, the variance of the non-survivors would have to be 67 percent higher than that of the survivors.

Multivariate Selection

So far we have considered the effect of selection on a single character on the assumption that it could be considered in isolation. In fact, as Darwin and Galton knew well (see chapter 5), different characters are usually quite highly correlated with each other, and this must be taken into account in understanding the selective forces operating at a phenotypic level. If a change occurs in the mean value of a character following a selective incident, how can one tell whether it is due to direct selection on that character, or whether it is an indirect effect due to selection on another correlated character? Some understanding of this question developed from work on animal breeding, but the application of multivariate concepts to evolutionary problems was first made in important articles by Lande (1979) and Lande and Arnold (1983), based in part on the results of Pearson (1903b).

Consider k correlated characters with phenotypic variance-covariance matrix **V**. It is convenient to suppose that the characters have each been standardized to have unit variance, so that **V** has unity down the leading diagonal and has the phenotypic correlations off the diagonal. The *selection differential* for the ith character, s_i, is the observed change in this character as a result of selection. This results both from direct selection on this character and from indirect effects due to selection on other correlated characters. The force of directional selection acting directly on the ith character, independent of selection on other characters, can be estimated by the *selection gradient*, β_i. It is shown in the appendix to this chapter that the selection gradients can be calculated from the equation

$$\boldsymbol{\beta} = \mathbf{V}^{-1}\mathbf{s}, \qquad [2]$$

where $\boldsymbol{\beta}$ and **s** are the vectors of selection gradients and selection differentials respectively.

Lande and Arnold (1983) illustrated this methodology on a sample of 94 pentatomid bugs which had been knocked down into Lake Michigan by a storm and were then washed ashore; 55 of them were dead and 39 were alive. They made four measurements (head width, thorax width, scutellum length, and wing length) on each individual. After logarithmic transformation of each measurement and standardization by division by the standard deviation, the variance-covariance matrix, which is the same as the correlation matrix, was

$$\mathbf{V} = \begin{bmatrix} 1 & .72 & .50 & .60 \\ .72 & 1 & .59 & .71 \\ .50 & .59 & 1 & .62 \\ .60 & .71 & .62 & 1 \end{bmatrix}. \qquad [3]$$

The selection differentials found by Lande and Arnold (1983) and the selection gradients calculated from eq.[2] are shown in table 10.4. (The latter differ slightly from those found by Lande and Arnold, perhaps because of rounding errors. Calculation of the inverse of **V** is facilitated by use of a computer package.)

These data reveal some striking differences between selection differentials and gradients. In particular, there was no significant selection differential on thorax width although the selection gradient was highly

Table 10.4. Standardized Selection Differentials (s) and Gradients (β) for a Hemipteran Bug

Character	*s*	β
Head width	–0.11	+0.02
Thorax width	–0.06	+0.53**
Scutellum length	–0.28*	–0.15
Wing length	–0.43**	–0.72**

Source: Lande and Arnold 1983

* Significant at the 5% level
** Significant at the 1% level

significant; thus there was substantial direct selection to increase thorax width, but this was masked in the selection differential by the indirect effect of selection to reduce wing length, which is highly correlated with thorax width ($r = 0.71$). The selection gradients reveal that selection favored bugs with proportionally small wings in relation to thorax length. Lande and Arnold (1983) suggest that in turbulence and/or precipitation such bugs may have been better fliers and have spent less time in the lake during the storm.

These authors also analyzed Bumpus' data in this way, and found the standardized directional selection gradients separately for each sex. The only significant values were for weight in both sexes ($\beta = -0.27^{***}$ for males and -0.52^{**} for females) and for total length in males ($\beta = -0.52^{***}$).

Another striking application is provided by the work of Peter and Rosemary Grant and their collaborators on Darwin's finches. Over the past quarter of a century they have ringed almost the entire population of the medium ground finch *Geospiza fortis* on the island of Daphne Major in the Galapagos and measured six morphological variables (weight, wing span, tarsus length, bill length, bill depth, and bill width) on each bird (Grant 1986 and 1991, Weiner 1994). After a logarithmic transformation, the measurements were standardized by dividing by their standard deviations. The phenotypic variance-covariance matrix calculated by Boag (1983) was:

$$\mathbf{V} = \begin{bmatrix} 1 & .64 & .60 & .69 & .76 & .75 \\ .64 & 1 & .56 & .60 & .64 & .60 \\ .60 & .56 & 1 & .52 & .54 & .52 \\ .69 & .60 & .52 & 1 & .81 & .78 \\ .76 & .64 & .54 & .81 & 1 & .88 \\ .75 & .60 & .52 & .78 & .88 & 1 \end{bmatrix}. \qquad [4]$$

Directional selection on these characters was observed during two periods of severe drought leading to shortage of food (seeds) in 1977 and 1985. Breeding ceased and a large percentage of the population died, 85 percent in the first drought and 68 percent in the second. Grant and Grant (1995) have summarized their data on selection and the evolutionary response to it, which are re-analyzed in table 10.5. The first and third columns show the selection differentials, s, standardized after a logarithmic transformation. The standardized selection gradients, β, in the

Table 10.5. Directional Selection in 1977 and 1985 in the Medium Ground Finch

	1977		1985	
	s	β	*s*	β
Weight	+0.74	+0.59	–0.11	+0.03
Wing length	+0.72	+0.46	–0.08	+0.01
Tarsus length	+0.43	–0.14	–0.09	–0.04
Bill length	+0.54	–0.05	–0.03	+0.34
Bill depth	+0.63	+0.31	–0.16	–0.23
Bill width	+0.53	–0.35	–0.17	–0.23

Source: After Grant and Grant 1995

second and fourth columns, showing the targets of selection, were obtained from eq.[2].

The targets of selection in 1977 were large body size (weight and wing length) and a deep, narrow bill. The latter enabled birds to crack large, hard seeds, which were all the food available. Grant (1991) remarks that during normal wet seasons, many plants produce an abundance of small seeds, while a few other plants produce a much smaller number of large seeds; as the finches deplete the supply of small seeds during a drought, they turn increasingly to the large seeds. The selective advantage of large body size independently of bill depth is less clear; possibly larger birds survived better because they could dominate smaller finches in social interactions at restricted sources of food or moisture (Grant 1991). The target of selection in 1985 was quite different. There was no direct selection on body size, but there was strong selection for birds with long, slender bills (neither deep nor wide). There were almost no large seeds at the beginning of the drought, so that during the drought there was selection for finches that were better at dealing with the few small seeds which were the only food available (Grant and Grant 1993).

Quantitative Genetics

The early biometrical work on selection described above was phenomenological and was not based on any underlying biological theory. Under the influence of Karl Pearson, the biometricians believed that "the problems of evolution were in the first place statistical, and in the second place statistical, and only in the third place biological" (Pearson to Galton in 1897, see Pearson 1930a, 128). After the rediscovery of Mendelism in 1900, the next stage in the development of biometry was the reconciliation between biometry and Mendelism, and the construction of a statistical theory of quantitative genetics based on Mendelian principles. This allowed the question of the response to selection in the next and subsequent generations to be placed on a sound basis.

The Multiple Factor Hypothesis

Mendel (1866) suggested a multifactorial theory for the inheritance of flower color in the bean *Phaseolus*. He crossed *Phaseolus nanus*, with white flowers, to *Ph. multiflorus*, with purple-red flowers. The hybrid offspring resembled the second parent, but the flower color was less intense. Among thirty one plants in the F_2 generation, one had white flowers while the remaining thirty plants developed flower colors which were of various grades from purple-red to pale violet. He continued:

> Even these enigmatical results, however, might probably be explained by the law governing *Pisum* if we might assume that the colour of the flowers and seeds of *Ph. multiflorus* is a combination of two or more entirely independent colours, which individually act like any other constant character in the plant. If the flower-colour A were a combination of the individual characters $A_1 + A_2 + \cdots$ which produce the total impression of a purple coloration, then by fertilisation with the differentiating character, white colour, a, there would be produced the hybrid unions $A_1a + A_2a + \cdots$. According to the above assumption, each of these hybrid colour unions would be independent, and would consequently develop quite independently from the others. It is then easily seen that from the combinations of the separate developmental series a complete colour-series must result. If, for instance, $A = A_1 + A_2$, then the hybrids A_1a and A_2a form the developmental series—
>
> $A_1 + 2A_1a + a$
> $A_2 + 2A_2a + a$
>
> The members of this series can enter into nine different combinations, and each of these denotes another colour—
>
> | | | |
> |---|---|---|
> | 1 A_1A_2 | 2 A_1aA_2 | 1 A_2a |
> | 2 A_1A_2a | 4 A_1aA_2a | 2 A_2aa |
> | 1 A_1a | 2 A_1aa | 1 aa |

> Should the colour development really happen in this way, we could offer an explanation of the case above described, viz. that the white flowers and seed-coat colour only appeared once among thirty-one plants of the first [F_2] generation. This colouring appears only once in the series, and could therefore also only be developed once in the average in each sixteen, and with three colour characters only once even in sixty-four plants. (Mendel 1866, 367–368)

Mendel was therefore suggesting that the occurrence of only one white-flowered plant out of thirty-one in the F_2 generation, rather than one in four expected if white flower-color was due to a single recessive factor, could be explained if flower-color was determined by two or three independent factors, with white flowers only occurring if the recessive factor was present in double dose for each factor (the predicted frequency being one in sixteen with two factors and one in sixty-four with three factors). This model could also explain the range of colors in the remaining plants in the F_2 generation, if there was some degree of partial dominance. (Mendel used a different notation from that of the modern geneticist. In particular, he wrote the homozygote for A_1 and A_2 as A_1A_2 rather than $A_1A_1A_2A_2$ and he did not distinguish between the two recessive white factors as, say, a_1 and a_2. Olby [1979] has argued that these notational differences show that "Mendel was no Mendelian" in the sense that he did not have the concept that hereditary elements occurred in pairs, only one of which entered each germ cell. Hartl and Orel [1992] have vigorously, and in my view successfully, refuted this iconoclastic suggestion.)

Confirmation of this model was provided after the rediscovery of Mendelism by the work of H. Nilsson-Ehle in Sweden and of E. M. East and his associates in the United States. Nilsson-Ehle (1909) did a Mendelian analysis of a cross between a red-glumed and a white-glumed variety of wheat. All the offspring in the F_1 generation had red glumes, though they were less intense in color than the red parent. The F_2 plants were more variable in color, though none of them were white. Among 78 selfed F_2 plants, 8 gave ratios of 3 red:1 white (as if they were heterozygotes for a single red gene), 15 gave 15:1 ratios (as if they were double heterozygotes), 5 gave approximate 63:1 ratios (as if they were triple heterozygotes), while 50 gave only red offspring (as if they were homozygotes). These are close to the Mendelian expectation for three segregating loci, with white a triple heterozygote (Wright 1968).

East and his associates provided convincing support for the multiple factor hypothesis by calculating the means and variances of quantitative characters in crosses between inbred strains of various plant species, in particular maize and tobacco. For example, East (1916) reported the results of an experiment on two varieties of an ornamental species of tobacco. The mean flower lengths in these two varieties were 93 and 40 mm, respectively, with rather small variances. The mean flower length in the F_1 was intermediate with a small variance; the mean flower length in the F_2 generation was again intermediate, but with a greatly increased variance. Since tobacco is self-fertilizing the two parents may be assumed to be homozygous at all loci; the F_1 plants are heterozygous at any loci at which the two parents differ. Thus the variability in the parents and in F_1 must be entirely of environmental origin, but the great increase in variability in F_2 can be attributed to the segregation of Mendelian genes.

If only a single gene were involved, the F_2 distribution should consist of three nonoverlapping distributions, representing the two homozygotes (the two parental varieties) and the heterozygote (F_1). This was clearly not the case since the F_2 distribution was approximately normal, like the F_1 and the two parental distributions. The multiple factor hypothesis postulates that the parents differed in several genes controlling flower length. It can be calculated from the difference between the parental means and the increase in the variance in the F_2 generation that the difference in flower length between the two varieties is largely of genetic origin and is controlled by about ten loci, each of which has a rather small effect, together with a small degree of environmental variability. It can easily be seen how an approximately normal distribution is produced under this model in F_2 when a small amount of environmental variability is superimposed. In reality, of course, the situation will be complicated by factors such as dominance, unequal gene effects, linkage, and so on.

The Hardy-Weinberg Law

Consider a Mendelian locus with two alleles, A and a, with arbitrary dominance. Suppose that the relative frequencies of the three genotypes, AA, Aa, and aa, among adults of both sexes in a particular generation are x, y, and z. Write $p = x + y/2$ for the frequency of the allele A and $q = 1 - p$ for the frequency of a; these are the frequencies of the two types of gametes produced by these adults. Under random mating, the chance of a union between two A gametes to form an AA offspring in the next generation is p^2, the chance of a union between an *A* and an *a* gamete to form an *Aa* offspring is $2pq$, and the chance of a union between two *a* gametes to form an *aa* offspring is q^2; these frequencies remain constant in all future generations in the absence of selection, apart from random genetic drift due to finite population size. This is the Hardy-Weinberg law. It is important in population genetics theory today because it means that the effect of selection can be studied by tracking gene frequencies among zygotes from one generation to the next, which is simpler than tracking genotype frequencies. It was even more important in the early history of Mendelian genetics because it showed that gene frequencies and genotype frequencies remained constant from one generation to the next in the absence of selection.

This law was first stated by Yule (1902), in the special case of equal gene frequencies, in the paper reconciling the ancestral law with Mendelism, which was discussed in chapter 8. He supposed that an *AA* race was crossed to an *aa* race, and that the resulting hybrids were then allowed to mate at random as one race during succeeding generations. In the first generation all individuals were *Aa*, and in the next generation the genotype frequencies would be 1/4, 1/2, and 1/4, or 1:2:1, from Mendel's results. He remarked that these genotype frequencies would stay the same in subsequent generations: "If all these are again intercrossed at random the composition remains the same. 'Dominant' and 'recessive' gametes are equally frequent, and consequently conjugation of a 'dominant' gamete will take place with a 'recessive' as frequently as with another 'dominant' gamete" (225). This is the random union of gametes argument used above. Pearson (1904a) obtained the same result in this special case (see chapter 7), but neither of them thought at this time of extending the argument to arbitrary gene frequencies.

In an interesting paper in 1903, Castle began by making some ill-judged criticisms of the work of the biometricians on the law of ancestral heredity, which elicited a robust rebuttal from Pearson (1904b). He then made some valuable calculations on the problem that he mistakenly thought that Yule (1902) had addressed (see chapter 8). He supposed that a pure *AA* race was mated to an *aa* race, and that in subsequent generations there was random mating among the remainder of the population after eliminating all the recessive *aa* individuals. In the first generation the whole population was *Aa*. In the following generations he found that the proportions of the three genotypes before selection were in the ratios 1:2:1, 4:4:1, 9:6:1, 16:8:1, and so on. These are the Hardy-Weinberg ratios with $p = 1/2,\ 2/3,\ 3/4,\ 4/5$, and so on. He also found that if selection ceased at some stage there would be no further change in these frequencies; for example, if the breeder eliminates recessives twice only, ceasing selection after the third generation, the population genotypes in the fourth and all subsequent generations are in the ratios 9:6:1. Thus Castle had obtained the Hardy-Weinberg ratios for some particular cases other than $p = 1/2$, he had made the first study of the effect of selection under a Mendelian model, and he had shown that there was no further change, and in particular no regression to the mean, when selection ceased. But these results attracted little attention, perhaps because there was no indication of their generality, perhaps because they were overshadowed by the weakness of the first part of the paper.

In 1908 Bateson's collaborator, R. C. Punnett, gave a talk to the Royal Society of Medicine on "Mendelism in relation to disease" (see chapter 8), and was asked by Yule in the discussion to resolve this paradox: "Assuming that brown ... eye-colour was dominant over blue, if matings of persons of different eye-colours were random (and that was very nearly true), it was to be expected that in the population there would be three persons with brown eyes to one with blue; but that was not so. There were more blues than browns. The same applied to the examples of brachydactyly. The author said that brachydactyly was dominant. In the course of time one would then expect, in the absence of counteracting factors, to get three brachydactylous persons to one normal, but that was not so" (Punnett 1908, 165). Thus Yule, who had previously obtained the Hardy-Weinberg law in the special case of equal gene frequencies, thought that this was the only stable equilibrium. (This was a common misunderstanding, as witnessed by Galton's unpublished essay discussed at the end of chapter 4.) Punnett could only reply: "Mr. Yule wondered why the nation was not slowly becoming brown-eyed and brachydactylous, since those characters were both dominant. So it might be for all he knew, but this made no difference to the mode of transmission of eye-colour or brachydactyly" (167).

On his return to Cambridge, Punnett presented this problem to the mathematician G. H. Hardy, with whom he had become friendly, in part because they used to play cricket together (Punnett 1950). Hardy immediately derived the general form of the law with arbitrary gene frequencies, and he concluded that "there is not the slightest foundation for the idea that a dominant character should show a tendency to spread over a whole population, or that a recessive should tend to die out" (1908, 49). It seems that either Punnett or Hardy had misunderstood Yule to mean that the dominant character would completely eliminate its recessive counterpart. Hardy's letter to *Science* was published in July 1908. The German obstetrician Wilhelm Weinberg (see Stern 1962) had already obtained this result earlier in the year in connection with an attempt to apply Mendelism to the inheritance of twinning in man, and in 1909 he extended the result to multiple alleles. Weinberg's studies on the frequency of twins and higher multiple births are the best studies ever published on this subject (Bulmer 1970), but his independent discovery of the Hardy-Weinberg law was not recognized until later.

Mendelian Theory of Quantitative Genetics

We are now in a position to consider how the theory of quantitative genetics can be built on the assumption of Mendelian heredity. The theory developed from a difficult but groundbreaking paper by Fisher (1918), though it had been foreshadowed by Yule (1906) and by Weinberg (1910). Modern accounts are given by Bulmer (1980), Falconer and Mackay (1996), and Lynch and Walsh (1998). I give a brief overview, assuming random mating in an outbred population and ignoring complications due to epistasis.

The phenotypic value of an individual is determined partly by that individual's genotype and partly by environmental factors. We may therefore write

$$y = g + e, \qquad [5]$$

where y is the phenotypic value, g is the genotypic value, the average value of all individuals with the same genotype, and e is an environmental deviation. In an inbred population divided into a number of pure lines it is possible to observe the genotypic value directly (chapter 7), but this is not possible in an outbred population. But we may assume that g and e are independent, or at least uncorrelated, so that the phenotypic variance can be decomposed into genetic and environmental components:

$$\mathrm{Var}(y) = \mathrm{Var}(g) + \mathrm{Var}(e). \qquad [6]$$

A major objective is to estimate these components of variance from observations on correlations between relatives, but before doing this we must consider a further decomposition of the genetic variance into additive and dominance components:

$$\text{Var}(g) = \text{Var}(a) + \text{Var}(d). \qquad [7]$$

I illustrate this decomposition from an example on the dwarfing gene in the mouse, known as 'pygmy' (symbol pg), taken from Falconer and Mackay (1996), shown in table 10.6. The first row shows the three genotypes, the second row their average weights, and the third row their frequencies, assuming equal gene frequencies. The mean weight is $\mu = 11$, and the genetic variance is $\text{Var}(g) = 9$. The additive value in the fifth row is the best linear approximation to the deviation of the genotypic value from the mean. It can be calculated from the linear regression of the genotypic value g on the number of + alleles n. The slope of this line is 4, so that the regression is

$$E(g|n) = 11 + 4(n-1). \qquad [8]$$

Hence

$$a = 4(n-1). \qquad [9]$$

The dominance deviation in the last row is the deviation of g from its linear predictor, so that

$$g = \mu + a + d. \qquad [10]$$

Since a and d are uncorrelated, the genetic variance has been decomposed into two additive components:

$$\text{Var}(a) = 8, \text{Var}(d) = 1, \text{Var}(g) = 9. \qquad [11]$$

This decomposition can be made in the same way for other gene frequencies, though the numerical results of course differ.

Table 10.6. Decomposition of Genotypic Value for the Dwarfing Gene in the Mouse, Assuming Equal Gene Frequencies

Genotype	++	+pg	pg pg
g = mean weight (gm)	14	12	6
Frequency	0.25	0.5	0.25
$g - \mu$ = deviation from mean	+3	+1	–5
a = additive value	+4	0	–4
d = dominance deviation	–1	+1	–1

Table 10.7. Identity by Descent between Relatives

Relationship	P_0	P_1	P_2	ρ
Identical twins	0	0	1	1
Parent-child	0	1	0	0.5
Sibs	0.25	0.5	0.25	0.5
Grandparent–grandchild, half-sibs, uncle–nephew	0.5	0.5	0	0.25
Great-grandparent–great-grandchild, cousins	0.75	0.25	0	0.125

For a quantitative character determined by several loci, this exercise could in principle be carried out for each locus and the variances summed over all loci to give the total components of variance for the character. The problem is to estimate these components from the correlations between relatives which provide the main source of information about them. The concept of identity by descent can be used to do this. Two genes are said to be identical by descent if one of them has been derived by direct replication from the other or if both are copies of the same gene in a common ancestor. Identical genes must, barring mutation, be alike in state, that is to say they must be the same alleles. It is this fact that causes the resemblance between relatives, who share some pairs of identical genes.

Table 10.7 shows the probabilities P_i that a particular pair of relatives share i pairs of identical genes (i = 0, 1 or 2), together with the coefficient of relatedness ρ, defined as $\rho = P_1 + P_2/2$. A mother transmits one gene at any locus to each of her children, so that mother and child are certain to have exactly one pair of identical genes at every locus. A pair of sibs is equally likely to have identical or nonidentical maternal genes, and likewise they are equally likely to have identical or nonidentical paternal genes; since these events are independent, sibs have 0, 1, or 2 pairs of identical genes with probabilities 1/4, 1/2, and 1/4. The other results in the table are obtained in a similar way.

It is shown in texts on quantitative genetics that the covariance between a pair of relatives due to identity by descent is

$$\text{Genetic covariance} = \rho\text{Var}(a) + P_2\text{Var}(d). \qquad [12]$$

Table 10.8. Correlations between Relatives for Abdominal Bristle Number in Drosophila melanogaster

Relationship	Correlation
Parent-child	0.25 ± 0.03
Sibs	0.26 ± 0.03
Half-sibs	0.12 ± 0.03

Source: Clayton et al. 1957

The correlation is obtained by division by the phenotypic variance. As an example, table 10.8 shows correlations between relatives for abdominal bristle number in *Drosophila melanogaster*. The data suggest that there is little or no dominance variance, and that about half the variance is of (additive) genetic and about half of environmental origin. The heritability of a character, denoted by h^2, is defined as the ratio of the additive genetic to the phenotypic variance:

$$h^2 = \mathrm{Var}(a)/\mathrm{Var}(y). \quad [13]$$

The best estimate of this quantity for abdominal bristle number is $h^2 = 0.5$.

The Response to Selection

The correlation between parent and offspring is $h^2/2$. Hence the regression coefficient of offspring on the mid-parental value is h^2. If we select the parents so that the selection differential, the difference in the mean value after and before selection, is s, the predicted response to selection in the next generation is h^2s:

$$R = h^2 s. \quad [14]$$

This is the fundamental result in selection theory, known as the breeder's equation. For example, Clayton, Morris, and Robertson (1957) selected for high bristle number in *Drosophila melanogaster*. The average bristle number in the base population was 35.3, the selected parents had a mean value of 40.6, and their offspring had a mean value of 37.9 bristles. Thus the selection differential was 40.6 – 35.3 = 5.3 bristles, and the response to selection was 37.9 – 35.3 = 2.6 bristles, in good agreement with the predicted response of 0.5 x 5.3 = 2.65 bristles. If the heritability is estimated from the regression of offspring on mid-parent, the breeder's equation is tautological, provided that this regression is linear, and does not depend on any genetic assumptions. A genetic model is essential to predict the response to selection over several generations. In particular, the response should be fixed when selection is relaxed under a Mendelian model with no epistatic interactions between loci, but may be subject to reversion under epistasis (chapter 8, Bulmer 1980) or counterbalancing natural selection.

The heritability is the regression coefficient of additive genetic value on phenotypic value, so that eq.[14] can be interpreted as meaning that the additive value is, on average, transmitted without dilution from one generation to the next. For this reason the additive value is often known as the breeding value. This fact can be used to predict the response to selection in the multivariate case. It is shown in the appendix that the vector of predicted responses **R** is

$$\mathbf{R} = \mathbf{G}\mathbf{V}^{-1}\mathbf{s}, \qquad [15]$$

where **G** is the genetic variance-covariance matrix whose (i, j)th element is

$$G_{ij} = \mathrm{Cov}(a_i, a_j), \qquad [16]$$

the covariance between the breeding values of the ith and jth characters.

This theory has been applied in table 10.9 to compare the observed and predicted responses to selection for the data on Darwin's finches described above. The genetic variance-covariance matrix was estimated by Boag (1983) as

$$\mathbf{G} = \begin{bmatrix} .85 & .77 & .66 & .72 & .73 & .82 \\ .77 & .89 & .52 & .73 & .74 & .70 \\ .66 & .52 & .65 & .47 & .55 & .47 \\ .72 & .73 & .47 & .67 & .67 & .69 \\ .73 & .74 & .55 & .67 & .82 & .80 \\ .82 & .70 & .47 & .69 & .80 & .90 \end{bmatrix}. \qquad [17]$$

The diagonal elements are the heritabilities, calculated from the regression of offspring on mid-parent. The off-diagonal elements G_{ij} (with $i \neq j$) can be calculated either from the regression of the ith measurement in the offspring on the jth mid-parental measurement or vice versa; the average of these two regressions was taken. (I have recalculated the

Table 10.9. Response to Selection in 1977 and 1985 in the Medium Ground Finch

	1977		1985	
	R Observed	*R* Predicted	*R* Observed	*R* Predicted
Weight	+0.70	+0.67	–0.38	–0.11
Wing length	+0.43	+0.74	–0.48	–0.08
Tarsus length	+0.30	+0.52	–0.11	–0.08
Bill length	+0.46	+0.63	+0.10	–0.08
Bill depth	+0.62	+0.64	–0.31	–0.15
Bill width	+0.42	+0.64	–0.20	–0.15

Source: After Grant and Grant 1995

values in Boag 1983, table 6, to convert them from genetic correlations to genetic covariances.) The phenotypic variance-covariance matrix was given in eq.[4], the selection differentials in table 10.5, and the predicted response was calculated from eq.[15]. The observed response to selection was calculated from tables 2 and 4 of Grant and Grant (1995) and from Boag (1983, table 6) as

$$R \text{ observed} = \frac{\log_{10}(m'/m)}{\text{SD}}, \qquad [18]$$

where m and m' are the mean values before selection and in the next generation respectively, in the original units of measurement, and SD is the standard deviation after a logarithmic transformation.

The observed response to selection after the drought of 1977 was predicted well by eq.[15] for weight and bill depth, but was less than predicted for the other four characters. The reason for this discrepancy is unclear. Grant and Grant (1995) obtained a better fit by using only the 36 offspring hatched in 1978 whose parents were included in the selection analysis rather than the unrestricted sample of 135 offspring born in 1978 used here. There are considerable discrepancies between the observed and predicted responses to selection after the drought of 1985, and in particular there was a much larger decrease in body size (weight and wing length) in offspring born in 1987 than predicted. Grant and Grant (1995) attribute this decrease to unfavorable conditions for growth in 1987. A general problem with natural experiments is that conditions cannot be controlled as they can be in the laboratory.

Coda

Finally, it may be of interest to review Galton's statistical theory of quantitative inheritance in the light of the Mendelian theory. The first question that Galton set out to answer was: Why are many quantitative characters normally distributed, and how do they remain normal with the same mean and variance from one generation to the next? He thought that human stature was normally distributed from the central limit theorem because "[it] is not a simple element, but a sum of the accumulated lengths or thicknesses of more than a hundred bodily parts" (1889, 83–84). This is not entirely satisfactory since the lengths of the different bones are correlated. A more convincing explanation is that stature is determined by a large number of genes which are approximately additive in their effect and which are statistically independent through segregation and assortment. He thought that the statistical properties of the distribution remained constant from one generation to the next because the reduction in the variability due to regression toward the mean was balanced by the increase due to variability between members of the same family. But he provided no explanation why these two processes should balance. Under Mendelian heredity, all the statistical properties of the phenotypic distribution of a random mating population remain constant from one generation to the next because, by the Hardy-Weinberg law, all the genotype frequencies remain the same.

Galton's second question was: What was the average share contributed to the personal features of the offspring by each ancestor severally? His answer to this question was discussed in chapter 8, on the law of ancestral heredity, where we saw that this question was motivated by his belief in an out-dated model of inheritance with latent and patent elements, but that the law could be interpreted in a model-free sense as the regression of an individual's phenotypic value on ancestral values. Under Mendelism the correlation between an individual and an ancestor i generations ago is

$$r_i = (\tfrac{1}{2})^i h^2, \qquad [19]$$

which are the same as Galton should have calculated from his model with the heritability h^2 substituted by p, the chance that a hereditary element is patent (eq.[9] in chapter 8). The regression coefficients can be calculated from eqs.[10] and [42] in chapter 8, illustrated in table 8.2. The important point is that these correlations are of the form that guarantees that there is no further regression back to the mean if selection ceases.

The intuitive reason under Mendelism is that selection has acted on the offspring generation by changing gene frequencies, which do not change further in the absence of selection. Two important provisos are that there is no epistasis due to interaction between the effects of different loci (see table 9.1) and that there is little or no non-linearity due to dominance, either because it is absent or because there is a large number of loci. Galton himself believed that there would be perpetual regression toward the mean after selection ceases because of his ignorance of multiple regression theory (see chapter 9).

Galton's third question was: How could the nearness of kinship between different types of relative be measured? His answer, shown in table 7.6, was the regression or correlation between their phenotypic values, assumed to be the same for all traits; it was estimated from observation for first-degree relatives (though the value for sibs is exaggerated) and calculated by erroneous logic for other relatives. The Mendelian answer is the coefficient of relatedness shown in the last column of table 10.7. It is the correlation between the additive genetic values of the relatives. It is not the same as the correlation between their phenotypic values, except in the absence of dominance and environmental variability, but it has been found useful in population genetics, particularly in the theory of kin selection. Galton's use of the (presumed) phenotypic correlation reflects his disregard of environmental variability.

Galton's main contribution to quantitative inheritance was his discovery of regression to the mean, but he was confused about the reason for this phenomenon. Johannsen (1903) discovered the correct explanation for self-fertilizing populations by showing that there was no regression toward the mean within a pure line and that regression to the mean in a mixed population of several pure lines was due to the masking of an individual's genotypic value by non-heritable environmental variability (chapter 7). Yule (1902, 1906) extended this explanation of regression toward the mean to outbred populations under Mendelian inheritance, and pointed out that dominance would also contribute to regression. Fisher expressed this explanation clearly in a letter to Leonard Darwin in 1932, except that he exaggerated the importance of dominance relative to environmental variability:

> Regression, as the word was used by Galton and the Biometers, i. e. regression to the mean, must have at least three contributory causes:-
>
> (a) If the relation of the child to only one parent is considered, regression is due to the contribution of the other parent, for the reason that tall

men will on the average have not so tall wives.... To avoid this obvious cause of regression Galton was led to use the 'mid-parent'.

(b) Non-inherited fluctuations due to environment will cause a group of parents selected for height above the average to have more than their share of those whose stature has been enhanced, and less than their share of those whose stature has been stunted by environmental circumstances. Their children, therefore, if reared on the average in an average environment, will be shorter than their parents for this reason. As far as I can judge, this makes a very unimportant contribution to the regression observed.

(c) The main regression from the 'mid-parent' in man seems to be due to dominance, which may be regarded as similar in its effects to environmental fluctuations, seeing that it, like them, disguises to some extent the genetic nature, so that we select a little amiss, and do not find the whole of what we saw in the parents reproduced in the children. (Bennett 1983, 148)

This explanation was not open to Galton because he disregarded environmental variability and because he was unaware of Mendelism when he was investigating quantitative inheritance. He might have constructed a similar explanation with patent and latent elements replacing dominant and recessive genes (see the section on "Galton's theory as it should have been" in chapter 8). There are passages in which he came close to doing this, but full understanding of the reasons for regression to the mean eluded him; his most perceptive attempt was his discussion of predicting stature from the length of a thigh-bone in his popular account of correlation (see chapter 6).

Appendix: Multivariate Selection Theory

Selection Differentials and Selection Gradients

The problem is most simply studied by using the idea of the covariance between fitness and phenotypic values, following Robertson (1966) and Price (1970 and 1972); I consider only directional selection, resulting in changes in mean values. Consider first a single variable y under some form of selection such that $W(y)$ is the fitness of an individual with phenotype y, and write $w(y) = W(y)/E(W)$ for the relative fitness, so that $E(w) = 1$. The mean value of y before selection is

$$\mu = \int y f(y) dy = \mathrm{E}(y), \qquad [20]$$

and the mean value after selection is

$$\mu^* = \frac{\int y w(y) f(y) dy}{\int w(y) f(y) dy} = \frac{\mathrm{E}(wy)}{\mathrm{E}(w)} = \mathrm{E}(wy). \qquad [21]$$

The covariance between w and y is

$$\mathrm{Cov}(w, y) = \mathrm{E}(wy) - \mathrm{E}(w)\mathrm{E}(y) = \mu^* - \mu = s, \qquad [22]$$

where s is the selection differential, the change in the mean value as a result of selection. Thus the selection differential is the covariance between the variable and its relative fitness.

This argument is easily extended to k variables to show that the vector of selection differentials $\mathbf{s}$ is equal to the vector of covariances:

$$\mathbf{s} \equiv \begin{bmatrix} s_1 \\ s_2 \\ \cdot \\ \cdot \\ s_k \end{bmatrix} = \begin{bmatrix} \mathrm{Cov}\,(w, y_1) \\ \mathrm{Cov}\,(w, y_2) \\ \cdot \\ \cdot \\ \mathrm{Cov}\,(w, y_k) \end{bmatrix} \equiv \mathrm{Cov}\,(w, \mathbf{y}). \qquad [23]$$

Consider the regression of w on $\mathbf{y}$, the best predictor of an individual's relative fitness from his or her phenotypic values. If this regression is linear, so that

$$\mathrm{E}(w|\mathbf{y}) = \alpha + \beta_1 y_1 + \beta_2 y_2 + \cdots \beta_k y_k, \qquad [24]$$

the regression coefficients β_i are given from eq.[25] in chapter 6 by

$$\boldsymbol{\beta} = \mathbf{V}^{-1}\mathrm{Cov}\,(w, \mathbf{y}) = \mathbf{V}^{-1}\mathbf{s}. \qquad [2\ \mathrm{bis}]$$

Even if the true regression is not linear, this gives the best linear approximation to it. Thus the regression coefficient β_i provides a measure of the force of directional selection on the ith character; it is called the selection gradient for the ith character because it is the average gradient of the relative fitness in this direction.

Premultiplying eq.[2] by $\mathbf{V}$,

$$\mathbf{s} = \mathbf{V}\boldsymbol{\beta}. \qquad [25]$$

For example, the first selection differential is

$$s_1 = V_{11}\beta_1 + V_{12}\beta_2 + V_{13}\beta_3 + \cdots V_{1k}\beta_k. \qquad [26]$$

The first term on the right hand side is the direct effect of selection on the first character in changing its mean value, while the remaining terms are the indirect effects on the first character of selection on other characters.

The Response to Selection

Write $\mathbf{a} = (a_1, \ldots, a_k)$ for the breeding values corresponding to the different variables in the generation under selection, and consider the regression of a_1 on $\mathbf{y}$, which is linear under multivariate normality:

$$\mathrm{E}(a_1|\,\mathbf{y}) = \alpha_1 + \beta_{11}y_1 + \beta_{12}y_2 + \cdots + \beta_{1k}y_k. \qquad [27]$$

The regression coefficients $\boldsymbol{\beta}_1 = (\beta_{11}, \ldots, \beta_{1k})$ are given from eq.[25] in chapter 6 by

$$\boldsymbol{\beta}_1 = \mathbf{V}^{-1}\mathrm{Cov}(a_1, \mathbf{y}) = \mathbf{V}^{-1}\mathrm{Cov}\,(a_1, \mathbf{a}). \qquad [28]$$

Since breeding values are transmitted without dilution from one generation to the next, the response to selection for the first variable is

$$R_1 = \beta_{11} + \beta_{12} + \cdots + \beta_{1k}s_k = \boldsymbol{\beta}_1^{\mathrm{T}}\mathbf{s}, \qquad [29]$$

where $\boldsymbol{\beta}_1^{\mathrm{T}}$, the transpose of $\boldsymbol{\beta}_1$, is the row vector of regression coefficients. Repeating this argument for a_2, a_3, and so on, and stacking answers on top of one another, leads to eq.[15].

References

The Galton and Pearson papers held in the Manuscripts and Rare Books Room of the D. M. S. Watson Library at University College London contain a wealth of unpublished material. The website www.galton.org reproduces many of Galton's books and papers, together with related material. An entry like "Galton, F. [1853] 1889" denotes a work first published in 1853 for which I have used a later edition, believed to be equivalent, published in 1889; it is referenced in the text as "Galton 1853."

Airy, G. B. 1861. *On the Algebraic and Numerical Theory of Errors of Observations and the Combination of Observations.* Macmillan, London.

Annan, N. 1955. The Intellectual Aristocracy. In J. H. Plumb (ed.), *Studies in Social History: A Tribute to G. M. Trevelyan,* 241–287. Longmans, Green, London.

Barlow, N. 1958. *The Autobiography of Charles Darwin, 1809–1882.* Collins, London.

Bateson, W. 1894. *Materials for the Study of Variation Treated with Especial Regard to Discontinuity in the Origin of Species.* Macmillan, London.

Bateson, W. 1900a. Letter in Galton papers, Box 198, quoted by Gillham 2001, 305.

Bateson, W. 1900b. Problems of heredity as a subject for horticultural investigation. *Journal of the Royal Horticultural Society* **25,** 54–61.

Bateson, W. 1902. *Mendel's Principles of Heredity: A Defence.* Cambridge University Press, Cambridge.

Bateson, W. 1913. *Mendel's Principles of Heredity.* 3rd impression with additions. Cambridge University Press, Cambridge.

Beavan, C. 2002. *Fingerprints: Murder and the Race to Uncover the Science of Identity.* Fourth Estate, London.

Bell, G. 1982. *The Masterpiece of Nature.* Croom Helm, London.

Bell, G. 1988. *Sex and Death in Protozoa.* Cambridge University Press, Cambridge.

Bennett, J. H. (ed.). 1983. *Natural Selection, Heredity, and Eugenics: Including Selected Correspondence of R. A. Fisher with Leonard Darwin and Others.* Clarendon Press, Oxford.

Blacker, C. P. 1952. *Eugenics: Galton and After.* Duckworth, London.

Boag, P. T. 1983. The heritability of external morphology in Darwin's ground finches (*Geospiza*) on Isla Daphne Major, Galapagos. *Evolution* **37,** 877–894.

Boole, G. 1854. *An Investigation of the Laws of Thought.* Macmillan, London.

Bouchard, T. J., and P. Propping (eds.). 1993. *Twins as a Tool of Behavioral Genetics.* Wiley, Chichester, England. Dahlem Workshop LS 53.

Bowler, P. J. 1983. *The Eclipse of Darwinism.* Johns Hopkins University Press, Baltimore.

Brody, N. 1992. *Intelligence.* Academic Press, San Diego.

Bryan, G. H. 1904. Average number of kinsfolk in each degree. *Nature* **71,** 9 and 101.

Buckle, H. T. 1857. *History of Civilization in England.* Vol. 1. J. W. Parker, London.

Bulmer, M. G. 1970. *The Biology of Twinning in Man.* Clarendon Press, Oxford.

Bulmer, M. G. 1980. *The Mathematical Theory of Quantitative Genetics.* Clarendon Press, Oxford.

Bulmer, M. 1994. *Theoretical Evolutionary Ecology.* Sinauer, Sunderland, MA.

Bulmer, M. 1998. Galton's law of ancestral heredity. *Heredity* **81,** 579–585.

Bulmer, M. 1999. The development of Francis Galton's ideas on the mechanism of heredity. *Journal of the History of Biology* **32,** 263–292.

Bumpus, H. C. 1898. The elimination of the unfit as illustrated by the introduced sparrow, *Passer domesticus. Biological Lectures, Marine Biological Laboratory Wood's Hole,* 209–226.

Burbridge, D. 1994. Galton's 100: An exploration of Francis Galton's imagery studies. *British Journal for the History of Science* **27,** 443–463.

Burbridge, D. 2001. Francis Galton on twins, heredity and social class. *British Journal for the History of Science* **34,** 323–340.

Burchfield, J. D. 1990. *Lord Kelvin and the Age of the Earth.* University of Chicago Press, Chicago.

Burkhardt, R. W. 1979. Closing the door on Lord Morton's mare: The rise and fall of telegony. *Studies in the History of Biology* **3,** 1–21.

Burleigh, M. 2000. *The Third Reich: A New History.* Macmillan, London.

Burns, M., and M. N. Fraser. 1966. *Genetics of the Dog.* Oliver and Boyd, Edinburgh.

Campbell, M. 1980. Did de Vries discover the law of segregation independently? *Annals of Science* **37,** 639–655.

Carlson, E. A. 2001. *The Unfit: A History of a Bad Idea.* Cold Spring Harbor Laboratory Press.

Castle, W. E. 1903. The laws of heredity of Galton and Mendel, and some laws governing race improvement by selection. *Proceedings of the American Academy of Arts and Sciences* **39,** 223–242.

Clayton, G. A., J. A. Morris, and A. Robertson. 1957. An experimental check on quantitative genetical theory. I. Short-term responses to selection. *Journal of Genetics* **55,** 131–151.

Cleland, R. E. 1972. *Oenothera: Cytogenetics and Evolution.* Academic Press, New York.

Clutton-Brock, J. 1999. *A Natural History of Domesticated Mammals.* Cambridge University Press, Cambridge.

Cookson, G., and C. A. Hempstead. 2000. *A Victorian Scientist and Engineer: Fleeming Jenkin and the Birth of Electrical Engineering.* Ashgate, Aldershot, UK.

Cowan, R. S. 1977. Nature and nurture: The interplay of biology and politics in the work of Francis Galton. *Studies in the History of Biology* **1,** 133–208.

Cross, W. E. 1996. *The Bell Curve* and transracial adoption studies. In J. L. Kincheloe, S. R. Steinberg, and A. D. Gresson (eds.), *Measured Lies: The Bell Curve Examined,* 331–342. St. Martin's Press, New York.

Crow, J. F., and N. E. Morton. 1955. Measurement of gene frequency drift in small populations. *Evolution* **9,** 202–214.

Cunningham, J. T. 1899. *Natural Science* **14,** 38–45.

Darbishire, A. D. 1909. An experimental estimation of the theory of ancestral contributions in heredity. *Proceedings of the Royal Society* B **81,** 61–79.

Darwin, C. 1859. *On the Origin of Species by Means of Natural Selection.* Murray, London. Reprinted 1964, Harvard University Press, Cambridge, MA.

Darwin, C. 1868. *The Variation of Animals and Plants under Domestication.* 2 vols. Murray, London.

Darwin, C. 1871a. *The Descent of Man, and Selection in Relation to Sex.* 2 vols. Murray, London. Reprinted 1981, Princeton University Press, Princeton, NJ.

Darwin, C. 1871b. Pangenesis. *Nature* **3,** 502–503.

Darwin, C. 1872. *The Origin of Species by Means of Natural Selection.* 6th edn. In M. Peckham (ed.), 1959, *The Origin of Species by Charles Darwin: A Variorum Text,* University of Pennsylvania Press, Philadelphia.

Darwin, C. 1875 *The Variation of Animals and Plants under Domestication.* 2nd edn. 2 vols. Murray, London. Reprinted 1998, Johns Hopkins University Press, Baltimore from the American edition (Appleton, New York, 1883).

Darwin, C. 1876. *The Effects of Cross and Self Fertilisation in the Vegetable Kingdom.* Murray, London.

Darwin, F. 1887. *The Life and Letters of Charles Darwin.* 3 vols. Murray, London.

Darwin, G. H. 1911. Sir Francis Galton, 1822–1911. *Proceedings of the Royal Society* B **84,** x–xvii.

Davis, A. S. 1871. The "North British Review" and the origin of species. *Nature* **5,** 161.

De Candolle, A. 1873. *Histoire des sciences et des savants depuis deux siècles.* H. Georg, Geneva.

Desmond, A. 1989. *The Politics of Evolution.* University of Chicago Press, Chicago.

De Vries, H. [1890] 1910. *Intracellular Pangenesis.* Translation by C. S. Gager. Open Court, Chicago.

De Vries, H. 1900. Das Spaltungsgesetz der Bastarde. *Berichte der deutschen botanischen Gesellschaft* **18,** 83–90.

De Vries, H. [1901] 1910. *The Mutation Theory.* 2 vols. Translation by J. B. Farmer and A. D. Darbishire. Kegan Paul, Trench, Trubner, London.

Dickson, J. H. 1890. Letter in Galton papers, Box 236/4.

Dunn, L. C. 1973. Xenia and the origin of genetics. *Proceedings of the American Philosophical Society* **117,** 105–111.

East, E. M. 1916. Studies on size inheritance in *Nicotiana. Genetics* **1,** 164–176.

Eldredge, N. 1995. *Reinventing Darwin: The Great Debate at the High Table of Evolutionary Theory.* Wiley, New York.

Eldredge, N., and S. J. Gould. 1972. Punctuated equilibria: An alternative to phyletic gradualism. In T. J. M. Schopf (ed.), *Models in Paleobiology,* 82–115. Freeman, Cooper, San Francisco.

Falconer, D. S., and T. F. C. Mackay. 1996. *Introduction to Quantitative Genetics.* 4th edn. Longman, Harlow, Essex, UK.

Fancher, R. E. 1985. *The Intelligence Men: Makers of the IQ Controversy.* Norton, New York.

Fancher, R. E. 1998. Biography and psychodynamic theory: Some lessons from the life of Francis Galton. *History of Psychology* **1,** 99–115.

Fancher, R. E. 2001. Eugenics and other "secular religions." In C. D. Green, M. Shore, and T. Tea (eds.), *The Transformation of Psychology,* 3–20. APA Books, Washington, DC.

Farrar, F. W. 1870. Review of *Hereditary Genius. Fraser's Magazine* **2,** 251–265.

Fisher, R. A. 1918. The correlation between relatives on the supposition of Mendelian inheritance. *Transactions of the Royal Society of Edinburgh* **52,** 399–433.

Fisher, R. A. 1930. *The Genetical Theory of Natural Selection.* Cambridge University Press, Cambridge. Reprinted 1999, Oxford University Press, Oxford.

Fisher, R. A. [1935] 1966. *The Design of Experiments.* 8th edn. Oliver and Boyd, Edinburgh. Reprinted 1990, Oxford University Press, Oxford.

Fisher, R. A. [1956] 1973. *Statistical Methods and Scientific Inference.* 3rd edn. Oliver and Boyd, Edinburgh. Reprinted 1990, Oxford University Press, Oxford.

Ford, E. B. [1980] 1998. Some recollections pertaining to the evolutionary synthesis. In E. Mayr and W. B. Provine (eds.), *The Evolutionary Synthesis,* 2nd edn, 334–342. Harvard University Press, Cambridge, MA.

Forrest, D. W. 1974. *Francis Galton: The Life and Work of a Victorian Genius.* Elek, London.

Froggatt, P., and N. C. Nevin. 1971a. Galton's "Law of Ancestral Heredity": Its influence on the early development of human genetics. *History of Science* **10,** 1–27.

Froggatt, P., and N. C. Nevin. 1971b. The "Law of Ancestral Heredity" and the Mendelian-Ancestrian controversy in England, 1889–1906. *Journal of Medical Genetics* **8,** 1–36.

Fukuyama, F. 2002. *Our Posthuman Future: Consequences of the Biotechnology Revolution.* Profile Books, London.

Galton, D. 2001. *In Our Own Image: Eugenics and the Genetic Modification of People.* Little, Brown, London.

Galton, F. 1852. Recent expedition into the interior of South-Western Africa. *J. R. Geog. Soc.* **22,** 140–163.

Galton, F. [1853] 1889. *Tropical South Africa.* 2nd edn. Ward, Lock, London (Minerva).

Galton, F. 1855. *The Art of Travel; Or, Shifts and Contrivances Available in Wild Countries.* Murray, London.

Galton, F. [1861] 1889. Visit to North Spain at the time of the eclipse. Reprinted in Galton [1853] 1889, 217–240.

Galton, F. 1863. A development of the theory of cyclones. *Proceedings of the Royal Society* **12,** 385–386.

Galton, F. 1865a. Hereditary talent and character. *Macmillan's Magazine* **12,** 157–166, 318–327.

Galton, F. 1865b. The first steps towards the domestication of animals. *Transactions of the Ethnological Society of London* **3,** 122–138.

Galton, F. 1869. *Hereditary Genius.* Macmillan, London. Reprinted 1979, Friedmann, London.

Galton, F. 1871a. Experiments in pangenesis, by breeding from rabbits of a pure variety, into whose circulation blood taken from other varieties had previously been largely transfused. *Proceedings of the Royal Society* **19,** 393–410.

Galton, F. 1871b. Pangenesis. *Nature* **4,** 5–6.

Galton, F. 1871c. Gregariousness in cattle and in men. *Macmillan's Magazine* **23,** 353–357.

Galton, F. 1872a. Statistical inquiries into the efficacy of prayer. *Fortnightly Review* **12,** 125–135.

Galton, F. 1872b. On blood-relationship. *Proceedings of the Royal Society* **20,** 394–402.

Galton, F. 1872c. Blood-relationship. *Nature* **6,** 173–176.

Galton, F. 1873. On the causes which operate to create scientific men. *Fortnightly Review* **13,** 345–351.

Galton, F. 1874a. *English Men of Science: Their Nature and Nurture.* Macmillan, London. Reprinted 1970, Cass, London.

Galton, F. 1874b. On a proposed statistical scale. *Nature* **9,** 342–343.

Galton, F. 1875a. The history of twins, as a criterion of the relative powers of nature and nurture. *Journal of the Anthropological Institute* **5,** 391–406.

Galton, F. 1875b. A theory of heredity. *Journal of the Anthropological Institute* **5,** 329–348.

Galton, F. 1877. Typical laws of heredity. *Nature* **15,** 492–495, 512–514, 532–533. Also in *Proceedings of the Royal Institution* **8,** 282–301.

Galton, F. 1878. Composite portraits made by combining those of many different persons into a single figure. *Journal of the Anthropological Institute* **8,** 132–148.

Galton, F. 1879. The geometric mean, in vital and social statistics. *Proceedings of the Royal Society* **29,** 365–367.

Galton, F. 1880. Mental imagery. *Fortnightly Review* **28,** 312–324.

Galton, F. [1883] 1907. *Inquiries into Human Faculty and Its Development.* 2nd edn. Dent, London (Everyman).

Galton, F. 1884. *Record of Family Faculties.* Macmillan, London.

Galton, F. 1885a. The measure of fidget. *Nature* **32,** 174–175.

Galton, F. 1885b. Presidential Address, Section H, Anthropology. *British Association Report* **55,** 1206–1214.

Galton, F. 1885c. Regression towards mediocrity in hereditary stature. *Journal of the Anthropological Institute* **15,** 246–263.

Galton, F. 1886a. Family likeness in stature. *Proceedings of the Royal Society* **40,** 42–73.

Galton, F. 1886b. President's Address. *Journal of the Anthropological Institute* **15,** 489–499.

Galton, F. 1886c. Family likeness in eye-colour. *Proceedings of the Royal Society* **40,** 402–416.

Galton, F. 1886d. The origin of varieties. *Nature* **34,** 395–396.

Galton, F. 1887a. Pedigree moth-breeding, as a means of verifying certain important constants in the general theory of heredity. *Transactions of the Entomological Society of London,* Part 1, 19–34.

Galton, F. 1887b. Good and bad temper in English families. *Fortnightly Review* **42,** 21–30.

Galton, F. 1887c. Presidential address. *Journal of the Anthropological Institute* **16,** 387–402.

Galton, F. 1888. Co-relations and their measurement, chiefly from anthropometric data. *Proceedings of the Royal Society* **45,** 135–145.

Galton, F. 1889. *Natural Inheritance.* Macmillan, London.

Galton, F. 1890a. Dice for statistical experiments. *Nature* **42,** 13–14. Reprinted in Stigler 1999.

Galton, F. [1890b] 1989. Kinship and correlation. *Statistical Science* **4,** 81–86.

Galton, F. 1890c. Sexual generation and cross fertilisation. Ms in Galton papers, Box 340A.

Galton, F. 1891. The patterns in thumb and finger marks. *Philosophical Transactions of the Royal Society* B **182,** 1–23.

Galton, F. 1892a. *Hereditary Genius.* 2nd edn. Macmillan, London. Prefatory chapter reprinted 1979, Friedmann, London.

Galton, F. 1892b. *Finger Prints.* Macmillan, London.

Galton, F. 1894. Discontinuity in evolution. *Mind* **3,** 362–372.

Galton, F. 1896. The service of sex. Ms in Galton papers, Box 340E.

Galton, F. 1897a. The average contribution of each several ancestor to the total heritage of the offspring. *Proceedings of the Royal Society* **61,** 401–413.

Galton, F. 1897b. Rate of racial change that accompanies different degrees of severity in selection. *Nature* **55,** 605–606.

Galton, F. 1897c. Hereditary colour in horses. *Nature* **56,** 598–599.

Galton, F. 1897d. Retrograde selection. *The Gardeners' Chronicle,* 15 May.

Galton, F. 1901. The possible improvement of the human breed under the existing conditions of law and sentiment. *Nature,* **64,** 659–665.

Galton, F. 1904a. Distribution of successes and of natural ability among the kinsfolk of Fellows of the Royal Society. *Nature* **70,** 354–356.

Galton, F. 1904b. Average number of kinsfolk in each degree. *Nature* **70,** 529, 626; **71,** 30.

Galton, F. [1904c] 1909. Eugenics. Its definition, scope and aims. In *Essays in Eugenics,* 35–43, Eugenics Education Society, London.

Galton, F. 1905. Number of strokes of the brush in a picture. *Nature* **72,** 198.

Galton, F. c. 1905. On the large advantage of biparental over uniparental generation. Ms in Galton papers, Box 146.

Galton, F. 1908. *Memories of My Life.* Methuen, London.

Galton, F., and E. Schuster. 1906. *Noteworthy Families.* John Murray, London.

Galton, L. 1830–97. *Annual Record.* Ms in Galton papers, Boxes 53–55.

Gayon, J. 1998. *Darwinism's Struggle for Survival: Heredity and the Hypothesis of Natural Selection.* Cambridge University Press, Cambridge.

Gigerenzer, G., Z. Swijtink, T. Porter, L. Daston, J. Beatty, and L. Krüger. 1989. *The Empire of Chance.* Cambridge University Press, Cambridge.

Gillham, N. W. 2001. *The Triumph of the Pedigree and the Birth of Eugenics: A Life of Sir Francis Galton.* Oxford University Press, New York.

Gould, S. J. 1980. Is a new and general theory of evolution emerging? *Paleobiology* **6,** 119–130.

Gould, S. J. 1991. Fleeming Jenkin revisited. In *Bully for Brontosaurus,* 340–353. Norton, New York.

Gould, S. J. 1996. *The Mismeasure of Man.* 2nd edn. Penguin, London.

Gould, S. J., and R. C. Lewontin. 1979. The spandrels of San Marco and the panglossian paradigm: A critique of the adaptationist program. *Proceedings of the Royal Society* B **205,** 581–598.

Gökyigit, E. A. 1994. The reception of Francis Galton's *Hereditary Genius* in the Victorian periodical press. *Journal of the History of Biology* **27,** 215–240.

Grant, B. R., and P. R. Grant 1993. Evolution of Darwin's finches caused by a rare climatic event. *Proceedings of the Royal Society* B **251,** 111–117.

Grant, M. 1916. *The Passing of the Great Race.* Scribner, New York.

Grant, P. R. 1972. Centripetal selection and the house sparrow. *Systematic Zool.* **21,** 23–30.

Grant, P. R. 1986. *Ecology and Evolution of Darwin's Finches.* Princeton University Press, Princeton, NJ.

Grant, P. R. 1991. Natural selection and Darwin's finches. *Scientific American* October 1991, 60–65.

Grant, P. R., and B. R. Grant. 1995. Predicting microevolutionary responses to directional selection on heritable variation. *Evolution* **49,** 241–251.

Guttorp, P. 1995. Three papers on the history of branching processes. *International Statistical Review* **63,** 233–245.

Hacking, I. 1990. *The Taming of Chance.* Cambridge University Press, Cambridge.

Haldane, J. B. S. 1954. The measurement of natural selection. *Proceedings of the IXth International Congress of Genetics, Caryologia* (Suppl.), 480–487.

Hamilton, W. D. 1971. Geometry for the selfish herd. *Journal of theoretical Biology* **31,** 295–311.

Hardin, G. 1959. *Nature and Man's Fate.* Holt, Rinehart and Winston, New York.

Hardy, G. H. 1908. Mendelian proportions in a mixed population. *Science* **28,** 49–50.

Harris, J. R. 1998. *The Nurture Assumption.* Bloomsbury, London.

Hartl, D. L., and V. Orel. 1992. What did Gregor Mendel think he discovered? *Genetics* **131,** 245–253.

Hauser, G. 1993. Galton and the study of fingerprints. In M. Keynes (ed.), *Sir Francis Galton, FRS: The Legacy of His Ideas,* 144–157. Macmillan, London.

Herrnstein, R. J., and C. Murray. 1994. *The Bell Curve: Intelligence and Class Structure in American Life.* Free Press, New York.

Heyde, C. C., and E. Seneta. 1977. *Statistical Theory Anticipated.* Springer-Verlag, New York.

Hilts, V. L. 1975. A guide to Francis Galton's *English Men of Science. Transactions of the American Philosophical Society* **65,** 1–85.

Hitler, A. *Mein Kampf.* [1925] 1992. Translation by R. Manheim. Pimlico, London.

Huxley, T. H. 1860. The origin of species. In *Darwiniana,* 22–79. Macmillan, London.

Jeal, T. 1973. *Livingstone.* Heinemann, London. Republished 1993 by Pimlico, London.

Jenkin, F. 1867. The origin of species. *North British Review* **46,** 277–318.

Johannsen, W. 1903. *Über Erblichkeit in Populationen und in reinen Linien.* Gustav Fischer, Jena.

Johannsen, W. 1911. The genotype concept of heredity. *American Naturalist* **45,** 129–159.

Jordan, J. 2001. *Josephine Butler.* Murray, London.

Kellogg, V. L. 1907. *Darwinism Today.* Henry Holt, New York.

Kendall, D. G. 1966. Branching processes since 1873. *Journal of the London Mathematical Society* **41,** 385–406.

Kendall, D. G. 1975. The genealogy of genealogy: Branching processes before (and after) 1873. *Bulletin of the London Mathematical Society* **7,** 225–253.

Kerr, R. A. 1995. Did Darwin get it all right? *Science* **267,** 1421–1422.

Kershaw, I. 1998. *Hitler 1889–1936: Hubris.* Penguin, London.

Kershaw, I. 2000. *Hitler 1936–1945: Nemesis.* Penguin, London.

Kevles, D. J. 1995. *In the Name of Eugenics.* Harvard University Press, Cambridge, MA.

Keynes, J. M. 1937. Some economic consequences of a declining population. *Eugenics Review* **29,** 13–17.

Keynes, M. 1993. Sir Francis Galton—A man with a universal scientific curiosity. In M. Keynes (ed.) *Sir Francis Galton,FRS: The Legacy of His Ideas,* 1–32. Macmillan, London.

King-Hele, D. 1998. *Erasmus Darwin: A Life of Unequalled Achievement.* De la Mare, London.

Kleinwächter, L. 1871. *Die Lehre von den Zwillingen.* Haerpfer, Prague.

Kottler, M. J. 1979. Hugo de Vries and the rediscovery of Mendel's laws. *Annals of Science* **36,** 517–538.

Krebs, J. R. and N. B. Davies. 1993. *An Introduction to Behavioural Ecology.* 3rd edn. Blackwell, Oxford.

Krüger, L., L. J. Daston, and M. Heidelberger (eds.). 1987. *The Probabilistic Revolution.* 2 vols. MIT Press, Cambridge, MA.

Kühl, S. 1994. *The Nazi Connection: Eugenics, American Racism, and German National Socialism.* Oxford University Press, New York.

Lande, R. 1979. Quantitative genetic analysis of multivariate evolution, applied to brain-body size allometry. *Evolution* **33,** 402–416.

Lande, R., and S. J. Arnold. 1983. The measurement of selection on correlated characters. *Evolution* **37,** 1210–1226.

Laplace, P. S. 1814. *Essai philosophique sur les probabilités.* Courcier, Paris.

Lewes, G. H. 1859. *The Physiology of Common Life.* Blackwood, Edinburgh.

Li, C. C. 1955. *Population Genetics.* University of Chicago Press, Chicago.

Lombardo, P. 2002a Eugenic sterilization laws. www.eugenicsarchive.org; accessed 12 July 2002.

Lombardo, P. 2002b Eugenic laws against race mixing. www.eugenicsarchive.org; accessed 12 July 2002.

Lombardo, P. 2002c Eugenic laws restricting immigration. www.eugenicsarchive.org; accessed 12 July 2002.

Lovell, M. S. 1998. *A Rage to Live: A Biography of Richard and Isabel Burton.* Little, Brown, London.

Lucas, P. 1847. *Traité philosophique et physiologique de l'hérédité naturelle dans les états de santé et de la maladie du système nerveux.* J. B. Baillière, Paris.

Lyell, C. 1830. *Principles of Geology.* Vol. 1. Murray, London. Reprinted 1990, University of Chicago Press, Chicago.

Lyell, C. 1832. *Principles of Geology.* Vol. 2. Murray, London. Reprinted 1991, University of Chicago Press, Chicago.

Lynch, M., and B. Walsh. 1998. *Genetics and Analysis of Quantitative Traits.* Sinauer, Sunderland, MA.

MacKenzie, D. 1981. *Statistics in Britain, 1865–1930: The Social Construction of Scientific Knowledge.* Edinburgh University Press, Edinburgh, UK.

Mascie-Taylor, C. G. N. 1993. Galton and the use of twin studies. In M. Keynes (ed.), *Sir Francis Galton, FRS: The Legacy of His Ideas,* 119–143. Macmillan, London.

Maynard Smith, J. 1978. *The Evolution of Sex.* Cambridge University Press, Cambridge.

Maynard Smith, J., R. Burian, S. Kauffman, P. Alberch, J. Campbell, B. Goodwin, R. Lande, D. Raup, and L. Wolpert. 1985. Developmental constraints and evolution. *Quarterly Review of Biology* **60,** 265–287.

Mayr, E. 1982. *The Growth of Biological Thought.* Harvard University Press, Cambridge, MA.

Mazumdar, P. M. H. 1992. *Eugenics, Human Genetics and Human Failings: The Eugenics Society, Its Sources and Its Critics in Britain.* Routledge, London.

McAlister, D. 1879. The law of the geometric mean. *Proceedings of the Royal Society* **29,** 367–376.

Meijer, O. G. 1985. Hugo de Vries no Mendelian? *Annals of Science* **42,** 189–232.

Mendel, G. [1866] 1913. Experiments in plant-hybridisation. Translation in Bateson 1913, 335–379.

Merivale, H. 1870. Galton on hereditary genius. *Edinburgh Review* **132,** 100–125.

Merriman, M. 1877. *Elements of the Method of Least Squares.* Macmillan, London.

Mill, J. S. 1843. *A System of Logic.* In J. M. Robson (ed.), 1973, *Collected Works of J. S. Mill,* vols 7 and 8, Toronto.

Mill, J. S. 1846. *A System of Logic.* 2nd edn. In J. M. Robson (ed.), 1973, *Collected Works of J. S. Mill,* vols 7 and 8, Toronto.

Mill, J. S. [1873] 1989. *Autobiography.* Penguin Books, London.

Moorehead, A. 1960. *The White Nile.* Hamish Hamilton, London.

Morris, S. W. 1994. Fleeming Jenkin and *The Origin of Species*: A reassessment. *British Journal for the History of Science* **27,** 313–343.

Muller, H. J. 1964. The relation of recombination to mutational advance. *Mutation Research* **1,** 2–9.

Nesselroade, J. R., S. M. Stigler, and P. B. Baltes. 1980. Regression towards the mean and the study of change. *Psychological Bulletin* **88,** 622–637.

Nilsson-Ehle, H. 1909. Kreuzunguntersuchungen an Hafer und Weizen. *Lunds Universitets Årsskrift* n.s., series 2, vol. 5, no. 2, 1–122.

Olby, R. 1979. Mendel, no Mendelian? *History of Science* **17,** 53–72.

Olby, R. 1985. *Origins of Mendelism.* 2nd edn. University of Chicago Press, Chicago.

Olby, R. 1987. William Bateson's introduction of Mendelism to England: A reassessment. *British Journal for the History of Science* **20,** 399–420.

Pakenham, T. 1991. *The Scramble for Africa 1876–1912.* Weidenfeld and Nicolson, London.

Pearson, K. 1894. Contributions to the mathematical theory of evolution. *Philosophical Transactions of the Royal Society* **185,** 71–110.

Pearson, K. 1896. Mathematical contributions to the mathematical theory of evolution.—III. Regression, heredity, and panmixia. *Philosophical Transactions of the Royal Society* **187,** 253–318.
Pearson, K. 1898. Mathematical contributions to the mathematical theory of evolution. On the law of ancestral heredity. *Proceedings of the Royal Society* **62,** 386–412.
Pearson, K. 1899. Theory of genetic or reproductive selection. *Philosophical Transactions of the Royal Society* A **192,** 259–278.
Pearson, K. 1900a. Mathematical contributions to the theory of evolution.—On the law of reversion. *Proceedings of the Royal Society* **66,** 140–164.
Pearson, K. 1900b. *The Grammar of Science.* 2nd edn. Walter Scott, London.
Pearson, K. 1903a. The law of ancestral heredity. *Biometrika* **2,** 211–236.
Pearson, K. 1903b. Mathematical contributions to the theory of evolution. XI. On the influence of natural selection on the variability and correlation of organs. *Philosophical Transactions of the Royal Society* **200,** 1–66.
Pearson, K. 1904a. Mathematical contributions to the theory of evolution. XII. On a generalised theory of alternative inheritance, with special reference to Mendel's laws. *Philosophical Transactions of the Royal Society* A **203,** 53–86.
Pearson, K. 1904b. A Mendelian's view of the law of ancestral inheritance. *Biometrika* **3,** 109–112.
Pearson, K. 1906. Walter Frank Raphael Weldon, 1860–1906. A memoir. *Biometrika* **5,** 1–52.
Pearson, K. 1908. On a mathematical theory of determinantal inheritance from suggestions and notes of the late W. F. R. Weldon. *Biometrika* **6,** 80–93.
Pearson, K. 1909. On the ancestral gametic correlations of a Mendelian population mating at random. *Proceedings of the Royal Society Lond.* B **81,** 225–229.
Pearson, K. 1910. Further remarks on the law of ancestral heredity. *Biometrika* **8,** 239–243.
Pearson, K. 1914. *The Life, Letters and Labours of Francis Galton.* Vol. 1, *Birth 1822 to Marriage 1853.* Cambridge University Press, Cambridge.
Pearson, K. 1924. *The Life, Letters and Labours of Francis Galton.* Vol. 2, *Researches of Middle Life.* Cambridge University Press, Cambridge.
Pearson, K. 1930a. *The Life, Letters and Labours of Francis Galton.* Vol. 3A, *Correlation, Personal Identification and Eugenics.* Cambridge University Press, Cambridge.

Pearson, K. 1930b. *The Life, Letters and Labours of Francis Galton.* Vol. 3B, *Characterisation, Especially by Letters; Index.* Cambridge University Press, Cambridge.

Pearson, K. 1930c. On a new theory of progressive evolution. *Annals of Eugenics* **4,** 1–40.

Pearson, K., and A. Lee. 1899. On the inheritance of fertility in mankind. *Philosophical Transactions of the Royal Society* A **192,** 279–290.

Pearson, K., and A. Lee. 1903. Laws of inheritance in man. *Biometrika* **2,** 357–362.

Peel, R. A. (ed.). 1998. *Essays in the History of Eugenics.* Galton Institute, London.

Plomin, R., J. C. DeFries, G. E. McClearn, and M. Rutter. 1997. *Behavioral Genetics.* 3rd edn. Freeman, New York.

Porter, T. M. 1986. *The Rise of Statistical Thinking.* Princeton University Press, Princeton, NJ.

Price, G. R. 1970. Selection and covariance. *Nature* **227,** 520–521.

Price, G. R. 1972. Extension of covariance selection mathematics. *Annals of Human Genetics* **35,** 485–490.

Proctor, R. N. 1988. *Racial Hygiene: Medicine under the Nazis.* Harvard University Press, Cambridge, MA.

Provine, W. B. 1971. *The Origins of Theoretical Population Genetics.* University of Chicago Press, Chicago.

Punnett, R. C. 1908. Mendelism in relation to disease. *Proceedings of the Royal Society of Medicine* **1,** 135–168 (Epidemiological Section).

Punnett, R. C. 1950. Early days of genetics. *Heredity* **4,** 1–10.

Quetelet, A. 1849. *Letters Addressed to H. R. H. the Grand Duke of Saxe Coburg and Gotha, on the Theory of Probabilities as Applied to the Moral and Political Sciences.* Translation by O. G. Downes. Layton, London.

Raverat, G. 1952. *Period Piece: A Cambridge Childhood*. Faber and Faber, London.

Rende, R. D., R. Plomin, and S. G. Vandenberg. 1990. Who discovered the twin method? *Behavior Genetics* **20,** 277–285.

Ridley, Mark. 1996. *Evolution.* 2nd edn. Blackwell, Oxford.

Ridley, Matt. 1999. *Genome.* Fourth Estate, London.

Robertson, A. 1966. A mathematical model of the culling process in dairy cattle. *Animal Production* **8,** 93–108.

Robinson, R. 1990. *Genetics for Dog Breeders.* 2nd edn. Pergamon, Oxford.

Romanes, G. J. 1886. Physiological selection: An additional suggestion on the origin of species. *Journal of the Linnean Society of London* **19,** 337–411.

Romanes, G. J. 1893. *An Examination of Weismannism.* Longmans, London.
Ruse, M. 1999. *Mystery of Mysteries: Is Evolution a Social Construction?* Harvard University Press, Cambridge, MA.
Rushton, J. P. 1997. Race, intelligence, and the brain: The errors and omissions of the 'revised' edition of S. J. Gould's *The Mismeasure of Man* (1996). *Personality and Individual Differences* **23,** 169–180.
Schofield, R. E. 1963. *The Lunar Society of Birmingham.* Clarendon Press, Oxford.
Seal, H. L. 1967. The historical development of the Gauss linear model. *Biometrika* **54,** 1–24.
Smith, B. M. D. 1967. The Galtons of Birmingham: Quaker gun merchants and bankers, 1702–1831. *Business History* **9,** 132–150.
Smith, C. 1998. *The Science of Energy: A Cultural History of Energy Physics in Victorian Britain.* University of Chicago Press, Chicago.
Spaeth, J. 1860. Studien über Zwillingen. *Zeitschrift der Wiener Gesellschaft Ärzte* **16,** 225–231, 241–244.
Spaeth, J. 1862. Studies regarding twins (English translation). *Edinburgh Medical Journal* **7,** 841–849.
Spottiswoode, W. 1861. On typical mountain ranges: An application of the calculus of probabilities to physical geography. *Journal of the Royal Geographical Society* **31,** 149–154.
Stafford, R. A. 1989. *Scientist of Empire: Sir Roderick Murchison, Scientific Exploration and Victorian Imperialism.* Cambridge University Press, Cambridge.
Stern, C. 1962. Wilhelm Weinberg, 1862–1937. *Genetics* **47,** 1–5.
Stigler, S. M. 1986. *The History of Statistics: The Measurement of Uncertainty before 1900.* Harvard University Press, Cambridge, MA.
Stigler, S. M. 1999. *Statistics on the Table: The History of Statistical Concepts and Methods.* Harvard University Press, Cambridge, MA.
Stock, G. 2002. *Redesigning Humans: Choosing Our Children's Genes.* Profile Books, London.
Stove, D. C. 1995. *Darwinian Fairytales.* Avebury, Aldershot, UK.
Sulloway, F. J. 1996 *Born to Rebel.* Little, Brown, London.
Sweeney, G. 2001. "Fighting for the good cause": Reflections on Francis Galton's legacy to American hereditarian psychology. *Transactions of the American Philosophical Society* **91,** part 2.
Swinburne, R. G. 1965. Galton's law—formulation and development. *Annals of Science* **21,** 15–31.
Uglow, J. 2002. *The Lunar Men: The Friends Who Made the Future, 1730–1810.* Faber and Faber, London.

Venn, J. 1866. *The Logic of Chance.* Macmillan, London.

Vorzimmer, P. 1963. Charles Darwin and blending inheritance. *Isis* **54,** 371–390.

Wagner, M. 1868. *Die Darwin'sche Theorie und das Migrationgesetz der Organismen.* Duncker and Humblot, Leipzig. Translation by J. L. Laird, 1873, *The Darwinian Theory and the Law of the Migration of Organisms.* Stanford, London.

Watson, H. W., and F. Galton. 1874. On the probability of the extinction of families. *Journal of the Anthropological Institute* **4,** 138–144.

Watson, H. W. 1891. Observations on the law of facility of errors. *Proceedings of the Philosophical Society of Birmingham* **7,** 289–318.

Weinberg, R. A., S. Scarr, and I. D. Waldman. 1992. The Minnesota transracial adoption study: A follow-up of IQ test performance at adolescence. *Intelligence* **16,** 117–135.

Weinberg, W. 1908. Über den Nachweis der Vererbung beim Menschen. *Jahreshefte des Vereins für Vaterländische Naturkunde in Württemburg* **64,** 369–382.

Weinberg, W. 1909. Über Vererbungsgesetze beim Menschen. *Zeitschrift für Induktive Abstammungs- und Vererbungslehre* **1,** 377–392, 440–460; **2,** 276–330.

Weinberg, W. 1910. Weitere Beiträge zur Theorie der Vererbung. *Archiv für Rassen- und Gesellschafts-Biologie* **7,** 35–49, 169–173. Translation in W. G. Hill (ed.), 1984, *Benchmark Papers in Quantitative Genetics,* Part 1, 42–57. Van Nostrand Reinhold, New York.

Weindling, P. 1989. *Health, Race and German Politics between National Unification and Nazism 1870–1945.* Cambridge University Press, Cambridge.

Weiner, J. 1994. *The Beak of the Finch.* Jonathan Cape, London.

Weismann, A. [1885] 1889. The continuity of the germ-plasm as the foundation of a theory of heredity. In *Essays upon Heredity,* 161–248. Translation by E. B. Poulton, S. Schönland, and A. E. Shipley. Clarendon Press, Oxford.

Weismann, A. [1887] 1889. On the number of polar bodies and their significance in heredity. In *Essays upon Heredity,* 333–384. Translation by E. B. Poulton, S. Schönland, and A. E. Shipley. Clarendon Press, Oxford.

Weismann, A. [1892] 1893. *The Germ-Plasm: A Theory of Heredity.* Translation by W. N. Parker and H. Ronnfeldt. Walter Scott, London.

Weismann, A. 1904. *The Evolution Theory.* Translation by J. A. and M. R. Thomson. 2 vols. Edward Arnold, London.

Weiss, S. F. 1990. The race hygiene movement in Germany 1904–1945. In M. B. Adams, (ed.), *The Wellborn Science: Eugenics in Germany, France, Brazil, and Russia,* 8–68. Oxford University Press, New York.

Weldon, W. F. R. 1890a. Letter in Galton papers, Box 340A.

Weldon, W. F. R. 1890b. The variations occurring in certain Decapod Crustacea.—I. *Crangon vulgaris. Proceedings of the Royal Society* **47,** 445–453.

Weldon, W. F. R. 1892. Certain correlated variations in *Crangon vulgaris. Proceedings of the Royal Society* **51,** 2–21.

Weldon, W. F. R. 1893. On certain correlated variations in *Carcinus moenas. Proceedings of the Royal Society* **54,** 318–329.

Weldon, W. F. R. 1895. An attempt to measure the death-rate of *Carcinus moenas* with respect to a particular dimension. *Proceedings of the Royal Society* **57,** 360–379.

Weldon, W. F. R. 1898. Presidential address to the zoological section of the British Association. *Transactions of the British Association,* 887–902.

Weldon, W. F. R. 1899. Letter to Pearson dated April 12 1899 in Pearson papers.

Weldon, W. F. R. 1901. A first study of natural selection in *Clausilia laminata* (Montagu). *Biometrika* **1,** 109–124.

Weldon, W. F. R. 1902. Mendel's laws of alternative inheritance in peas. *Biometrika* **1,** 228–254.

Weldon, W. F. R. 1903. Note on a race of *Clausilia itala* (von Martens). *Biometrika* **3,** 299–307.

Weldon, W. F. R. c. 1904. Ms in Pearson papers, Box 264.

Weldon, W. F. R., and K. Pearson. 1903. Inheritance in *Phaseolus vulgaris. Biometrika* **2,** 499–503.

Willis, M. B. 1989. *Genetics of the Dog.* Witherby, London.

Wright. S. 1968. *Evolution and the Genetics of Populations. I. Genetic and Biometric Foundations.* University of Chicago Press, Chicago.

Yule, G. U. 1902. Mendel's laws and their probable relations to intra-racial heredity. *New Phytologist* **1,** 193–207, 222–238.

Yule, G. U. 1906. On the theory of inheritance of quantitative compound characters on the basis of Mendel's laws—A preliminary note. *Report of the 3rd International Conference on Genetics,* 140–142.

Zirkle, C. 1946. The early history of the idea of the inheritance of acquired characters and of pangenesis. *Transactions of the American Philosophical Society* **35,** 95–141.

Index

Acquired characters, inheritance of
accepted by Darwin, 105–107, 112–113, 276
rejected: by Galton, 32, 103, 105–107, 114, 119–123; by Weismann, 132–133
Activation theory of development, 134–135
Albert, Lake, 25
Alternative inheritance, 139
and ancestral law, 243–244, 249, 254–256
Ancestral law, 34, 104–105, 127–128, 214, 209, 231, 238–274
Galton's derivation: in 1885, 241–244; in 1897, 244–247; as it should have been, 247–250
and Mendelism, 257–272
Pearson's interpretation, 104–105, 250–257, 261–266
Weldon's interpretation, 259–261
Yule's interpretation, 266–272
Andersson, Charles (companion in S. W. Africa), 11, 14, 16
Anticyclone, FG's discovery of, 31
Arnaud, Bey, 9
Arnold, S. J., 311–314
Art of Travel (1855), 17, 38
Atavism, *see* Reversion
Athenaeum Club, 21, 36

Balfour, Francis, 301
Basset hounds, 240–241, 244, 251–252, 255, 292
Bateson, William
and discontinuous evolution, 288–289, 291, 301
and Mendelism, 140, 257–259, 262, 271, 294, 320
Bienaymé, I. J., 160
Biggs, Eva (great-niece), 10, 20, 302
Biometric teas, 302
Biometrika, 87, 221, 259, 299–300
Biometry, 299–331
Biparental inheritance, 46, 103–105, 128, 136–137, 160–167, 251, 257
Birmingham General Hospital, 5, 39
Blacker, C. P., 84, 86
Blending inheritance, 138–141
and ancestral law, 244, 254
and problem of swamping, 141–146
Boole, George, 200–201
Boulton, Matthew, 4, 7
Boulton, Montague (companion in Near East), 7, 10
British Association, 27–35
FG's contributions: composite photography, 34–35; finger-prints, 35; inventions, 28, 30; list of, 29; mental images and number-forms, 32–33; meteorology, 30–31
Brock, Sir Laurence, 87
Buck v. *Bell*, 89

Bugs, selection on, 313–314
Bumpus's sparrows, natural selection in, 309–312, 314
Burton, Richard, 21, 23–26, 38
Butler, Josephine (sister-in-law), 39–40
Butler family, 19, 36, 39–40, 58

Castle, W. E., and the Hardy-Weinberg law, 319–320
Ceylon (ox), 13–14, 151
Churchill, Winston, 86
Common shrimp, selection in, 302–303
Correlation, 191–196
Correlations between relatives, 232, 324

D'Alembert, example of hereditary ability, 47
Damaraland, 11–17, 150–152
Darwin, Charles
 attitudes to slavery and eugenics *vs* FG's, 39, 98
 belief in inheritance of acquired characters, 106–107, 112–113
 early interaction with FG, 5
 FG impressed by *Origin*, 32, 42
 gradualism, 275–276
 impressed by: *Hereditary Genius*, 57; *Tropical South Africa*, 16–17
 and problem of swamping, 141, 144
 proposes FG for Fellowship of the Royal Society, 21
 sweet pea experiments, 213–214
 theory of pangenesis, 102, 108–113
Darwin, Erasmus, 3–5, 154
Darwin, George, 21, 39– 40
Darwin, Leonard, 84–86, 328
Darwin, Robert, 5
Darwin's finches, selection in, 314–315, 325–327
Davenport, C. B., 87–88, 91, 100
De Candolle, Alphonse, 59–61, 156, 201
Determinantal inheritance, Weldon's theory of, 260–261
De Vries, Hugo
 activation theory of development, 134–135
 intracellular pangenesis, 133–136
 mutation theory, 294–297
 rediscoverer of Mendelism, 257, 294
Dickson, Hamilton, 189–191, 196–198
Discontinuity in evolution, 275–298
Dissection theory of development, 134–135
Domestication of animals, 16, 32, 147–150

East, E. M., 317–318
Edgeworth, F. Y., 202–203, 207
English Men of Science: Their Nature and Nurture (1874), 56, 60–63
Etosha salt-pan, 14
Eugenics, 79–101
 in America, 87–92
 in Britain, 84–87
 Galtonian, 79–84
 in Germany, 92–98
 rationale, 98–101
Evolution of sex, 160–167
Exclusive inheritance, *see* Alternative inheritance
Extinction of surnames, 156–160

Farrar, W. F., review of *Hereditary Genius*, 58–59
Fertility of heiresses, 153–156
Fingerprints, 35, 40, 181, 291
Finger Prints (1892), 35, 181, 291
Fisher, R. A.
 blending inheritance, 138
 branching process, 160
 eugenics, 84–87
 quantitative genetics, 269, 321, 328–329
 statistics, 203–206
Fluctuating variability
 sensu de Vries, 135, 221, 295
 sensu Johannsen, 221–223
Forrest, D. W. (*The Life and Work of a Victorian Genius*), 1, 10, 22–23, 27, 34, 37

Galton, Adèle (sister), 4, 10, 40
Galton, Douglas, 11
Galton family, 1–6

Galton, Francis, inventions, 28–30
  Beauty map, 30
  composite photography, 34–35
  dice for statistical experiments, 184
  finger-printing, 35
  Galton's sun-signal, 28
  quincunx, 182–184
  spectacles for divers, 30
Galton, Francis, life, 1–41
  character and political views, 36–41
  childhood, 1–6
  education: at Birmingham General hospital, 5; at King Edward's School Birmingham, 4–5; at King's College London, 5; at Trinity College Cambridge, 5–6
  marriage to Louisa Butler, 18–21
  travels: in Eastern Europe, 6–7; in the Near East, 7–11; in South West Africa, 11–18; vacation tours and alpinism, 18–21
Galton, Francis, papers
  The average contribution of each several ancestor to the total heritage of the offspring (1897), 240, 244–247, 251, 253
  Average number of kinsfolk in each degree (1904), 77
  Composite portraits (1878), 34
  Co-relations and their measurement, chiefly from anthropometric data (1888), 191–196
  A development of the theory of cyclones (1863), 31
  Dice for statistical experiments (1890), 184
  Discontinuity in evolution (1894), 288–291, 303
  Distribution of successes and of natural ability among the kinsfolk of Fellows of the Royal Society (1904), 75
  Eugenics. Its definition, scope and aims (1904), 84, 100
  Experiments in pangenesis (1871), 116–118
  Family likeness in eye-colour (1886), 224, 243–244
  Family likeness in stature (1886), 184–191
  The first steps towards the domestication of animals (1865), 147–150
  The geometric mean, in vital and social statistics (1879), 180, 196
  Good and bad temper in English families (1887), 224
  Gregariousness in cattle and in men (1871), 150–153
  Hereditary colour in horses (1897), 241
  Hereditary talent and character (1865): eugenics, 79–81; inheritance of ability, 44–46; inter-racial differences, 67; mechanism of heredity, 103–108, 132
  The history of twins, as a criterion of the relative powers of nature and nurture (1875), 64–67
  Kinship and correlation (1890), 192–196
  The measure of fidget (1885), 30
  Mental imagery (1880), 32–33
  Number of strokes of the brush in a picture (1905), 37
  On a proposed statistical scale (1874), 178, 180
  On blood-relationship (1872), 119–127, 136–138
  On the causes which operate to create scientific men (1873), 59–60
  On the large advantage of biparental over uniparental generation (c. 1905, unpublished), 166–167
  The origin of varieties (1886), 293
  Pangenesis (1871), 118–119
  The patterns in thumb and finger marks (1891), 181
  Pedigree moth-breeding, as a means of verifying certain important constants in the general theory of heredity (1887), 284–285
  Presidential address, Anthropological Institute (1887), 145–146
  Presidential Address, Section H,

Galton, Francis, papers (*cont.*)
Anthropology (British Association) (1885), 224, 226, 228–231
President's Address, Anthropological Institute (1886), 229, 233
Rate of racial change that accompanies different degrees of severity in selection (1897), 282–283
Recent expedition into the interior of South-Western Africa (1852), 16
Regression towards mediocrity in hereditary stature (1885), 184–191, 214–215, 224, 239, 242, 281
Retrograde selection (1897), 284–285
The service of sex (1896, unpublished), 121, 131, 164–165
Sexual generation and cross fertilisation (1890, unpublished), 163–164
Statistical inquiries into the efficacy of prayer (1872), 51
A theory of heredity (1875), 119–123, 125–127, 136, 161–163
Typical laws of heredity (1877), 182–184, 211–218
Visit to North Spain at the time of the eclipse (1861), 19
Galton, Louisa (née Butler), 18–21, 28, 57
Galton, Violetta (née Darwin, mother), 3–5
Galton-Watson process, 156–160
Genealogy, 2
Geometric mean, 187, 196
Germinal selection, *see* Weismann
Germ-plasm, continuity of, *see* Weismann
Gillham, N. W. (*Sir Francis Galton*), 1, 27
Goldschmidt, Richard, 96
Göring, Hermann, 94
Grant, J. A., 25–26
Grant, Madison, 90
Grant, P. R., 311, 314–315, 326
Gregariousness, evolution of, 150–153
Grundy, Burton's description of FG, 38

Haeckel's law, 301
Haldane, J. B. S., 160, 217
Hardy, G. H., 264, 320–321
Hardy-Weinberg law, 166, 262, 318–321, 327
Heiresses, fertility of, 153–156
Hereditary Genius (1869), 42–78
in all professions, 50–54
in English judges, 48–50
eugenics, 81–82
fertility of heiresses, 153–156
normal distribution, 173–175
pangenesis discussed, 114–116
quantitative theory of heredity, 210–211
reception of, 57–60
stability of type, 277–279, 281–282
transmission of ability through male and female lines, 54–57
Herschel, John, 108, 199
Herschel, William, 35
Himmler, Heinrich, 97
Hitler, Adolf, 94–96, 98
Hooker, J., 22, 213
Humor, FG's sense of, 41
Huxley, Aldous, 100
Huxley, T. H., 22, 37
on discontinuous evolution 144, 276, 279

Identity by descent, 323
Inheritance
of abdominal bristle number in *Drosophila*, 324
of coat color in basset hounds, 240–241
of coat color in horses, 255–256, 260
of eye color in man, 87, 224, 243–244, 249, 254–256, 271
of flower color in beans, 316–317
of flower length in tobacco, 318
of glume color in wheat, 317
of human height, 224–231
Inquiries into Human Faculty and Its Development (1883)
domestication of animals, 16, 148
eugenics, 79, 82–83, 282
mental images, 33

Jenkin, Fleeming
stability of type, 278
swamping, 138, 141–146, 293
Johannsen, W.,
ancestral heredity, 267
experiments with beans and pure line theory, 218–224, 267, 328
genotype concept of heredity, 122, 133
Jonker, Hottentot chief, 13

Kilimanjaro, Mt., 23–24
King Edward's School, Birmingham, 4–5
King's College London, 5–7
Kinship
Galton's theory, 231–232, 337
Mendelian theory, 323–324, 328
Kölreuter, Joseph, 104

Lande, R., 311–314
Laplace, P. S., 169, 198–202
Latent elements
and ancestral law, 239, 247, 250, 260
in Darwin's theory of pangenesis, 111, 122, 131, 139
in Galton's theory of heredity, 66, 102–103, 115–131, 136–137, 237
and regression to mean, 286–287
Laughlin, Harry, 88–92
Law of ancestral heredity, *see* Ancestral law
Lethbridge, Millicent (niece), 40–41
Livingstone, David, 11, 26
Locke, John, 69
Lunar Society, 4
Lyell, C., 22, 275, 278

McAlister, Sir Donald, 180, 196
Markham, Clements, 22, 38
Memories of My Life (1908)
African exploration, 23, 25, 38
British Association, 27–28, 30–31, 33
correspondence with Darwin, 57
early life, 3–6, 10–11, 36, 39
eugenics, 83, 99
heredity, 213
Mendel, 141
normal distribution, 173, 181, 189, 196, 201
travels, 11–12, 15–16, 20
Watson, Rev. Henry, 157
Mendel and Mendelism
ancestral law, 257–272
De Vries as rediscoverer, 135–136, 257, 294–297
eugenics, 87, 89
explains Darwin's experiments, 112
Galton's reaction, 140–141, 166–167
instantaneous regression, 288
quantitative genetics, 71, 316–331
see also Bateson, Fisher
Mental images, 32–33
Merivale, Herman, review of *Hereditary Genius*, 57–58, 68
Mill, John Stuart
contrasted with Galton, 69–70
on probability, 199–200
Muller's ratchet, 162–163
Murchison, Sir Roderick, 22–23, 25–26
Mytton, Jack, 10, 68

Nangoro, King of Ovampos, 14–15
Natural Inheritance (1889)
human height, 224–226, 228–229, 235
kinship, 231
law of ancestral heredity, 241, 242, 250
mechanism of heredity, 127–131, 136, 139
natural selection, 302
normal distribution, 175–184
organic stability, 279
regression and the bivariate normal distribution, 187, 191, 198
sweet peas, 213–214
Nature and nurture, 44, 60–67, 71–73
Naudin, Charles, and his theory of segregation, 110–111
Ngami, Lake, 11, 15–16
Nilsson-Ehle, H., 273
Normal distribution, 173–184
bivariate, 184–196
Noteworthy Families (1906), 75–76

Number-forms, 33
Number of kinsfolk, 48–49, 74–78

Okavango River, 16
Omanbonde, Lake, 13
Ovampoland, 12, 14–15, 18

Pangenesis
Darwin's theory, 108–114
De Vries's intracellular pangenesis, 133–136
Galton's experiments on rabbits, 116–119
Galton's modification of Darwin's theory, 119–123
Galton's reaction to Darwin's theory, 114–119
Galton's theory of heredity based on, 210–211
Parkyns, Mansfield, meets Galton in Khartoum, 9–10
Patent elements, *see* Latent elements
Pearson, Karl
ancestral law, 104–105, 250–257, 261–266
Life, Letters and Labours of Francis Galton, 1 and throughout
statistical work, 202–204, 206–208, 234–235
see also Weldon
Penrose, Lionel, 84
Perpetual regression, 71, 276, 281–288, 292, 300, 303, 328
fallacy of, 285–288
experiments to verify, 284–285
Petrie, Prof. Flinders, 20
Petrus, Mrs., Hottentot beauty, 13
Photography, composite, 34–35
Ploetz, Alfred, 92–93, 101
Priestley, Joseph, 4
Probability, meaning of, 168, 196–202
Punctuated equilibria, 297–298
Punnett, R. C., 271–272, 320

Quagga, Lord Morton's, 113
Quakers, 1, 3, 36
Quantitative genetics, 315–327
Mendelian theory, 321–324
Quetelet, Adolphe, 168–173, 180–181, 212
Quincunx, 182–184

Rabbits, transfusion experiments, 116–119
Record of Family Faculties (1884), 224–227, 233–235
Regression, 184–191, 206–208
and ancestral law, 238–274
in beans, 218–224
fraternal, 233–237
in human height, 229–231
in sweet peas, 212–215
see also Perpetual regression
Reversion,
to ancestral values, 107–108, 110–112, 115–116, 128, 131
to mean, 184, 213, 215, 218
Pearson's law, 254–257
to stable type, 279, 292
see also Regression
Ripon Falls, 25
Romanes, G. J.
on Galton's theory of the stirp, 132
physiological selection, 292–294
Royal Geographical Society, 22–27
exploration in Central Africa: Burton-Speke, Speke-Grant, and Livingstone expeditions, 23–27
FG awarded gold medal in 1854, 16
FG belonged to progressive wing, 22–23
Royal Society
FG elected Fellow, 21
grandfathers of FG both Fellows, 4
papers at, 137–138, 193
questionnaires to Fellows 61, 75

Segregation,
Galton's flirtation with, 136–138
Mendelian, 110, 259–261, 268
Naudin's hypothesis, 110–111,
Weismann's hypothesis, 136–137
Selection
in: bugs, 313–314; crabs, 303–308;

Darwin's finches, 314–315, 325–327; *Drosophila*, 324–325; moths, 284; shrimp, 302–303; snails, 308–309; sparrows, 309–312, 314
differential, 312–315, 324–327, 329–331
directional, 306, 311–315, 329–331
gradient, 312–315, 329–331
multivariate, 312–315, 329–331
noroptimal, 217, 302–303, 306
response, 324–327, 331
stabilizing, 305–306, 308–309, 311–312
Sex, *see* Evolution
Shaw, Norton, 22
Shore crab, selection in, 303–308
Similarities between relatives
under stirp theory, 123–127
Snails, stabilizing selection in, 308–309
Speke, J. H., 23–26
Spencer, Herbert, 21, 37
Spottiswoode, William, 173
Stability of type, 181, 277–285, 290–292
Stanley, H. M., 26–27
Stirp, defined, 102
Surnames, *see* Extinction
Swamping, problem of, 138, 141–146, 293
Sweet pea experiments, 212–215

Tanganyika, Lake, 23–24, 26–27, 38
Telegony, 113
Transportation hypothesis, 102, 121–123, 130, 131, 134, 139
Trinity College Cambridge, 5, 9, 19, 157
Tropical South Africa (1853), 11–16
Twins, 44, 60, 63–67, 71–72, 126–127, 137, 321, 323

Venn, John, 200–201
Victoria, Lake, 23–25

Wallace, A. R., 22, 57
Watson, Rev. Henry, 156–160, 196–198, 201
Watt, James, 4
Weber-Fechner law, 180, 196
Wedgwood, Josiah, 4
Weinberg, W., 321
see also Hardy-Weinberg law
Weismann, August
continuity of the germ-plasm, 105, 132–133
dissection theory of development, 134
germinal selection, 121
segregation, 136–137
Weldon, W. F. R.
the ancestral law and Mendelism, 221–223, 238, 244, 258–261, 294–295
career, 301–302
on Galton's essays on the evolution of sex, 137, 164–165, 246
studies of natural selection: in common shrimp, 302–303; in shore crab, 303–308; in snails, 308–309
Withering, William, 4
Wright, Sewall, 138, 317

Xenia, 113, 134

Yule, G. U., 207
and ancestral law, 258, 266–272
and Hardy-Weinberg law, 319–321